T0213391

AN INTRODUCTION TO OPTICAL STELLAR INTERFEROMETRY

During the last two decades, optical stellar interferometry has become an important tool in astronomical investigations requiring spatial resolution well beyond that of traditional telescopes. This is the first book to be written on the subject. The authors provide an extended introduction discussing basic physical and atmospheric optics, which establishes the framework necessary to present the ideas and practice of interferometry as applied to the astronomical scene. They follow with an overview of historical, operational and planned interferometric observatories, and a selection of important astrophysical discoveries made with them. Finally, they present some as-yet untested ideas for instruments both on the ground and in space which may allow us to image details of planetary systems beyond our own.

This book will be used by advanced students in physics, optics, and astronomy who are interested in the ideas and implementations of astronomical interferometry.

ANTOINE LABEYRIE is Professor at the Collège de France. During his distinguished career he has made many fundamental contributions to high-resolution optical astronomy.

STEPHEN G. LIPSON is Chair of Electro-Optics and Professor of Physics at Technion–Israel Institute of Technology, Haifa. He is co-author of *Optical Physics, 3rd Edition* (Cambridge University Press, 1995).

PETER NISENSON (1941–2004) studied physics and optics before becoming a professional astronomer at the Harvard Smithsonian Center for Astrophysics. His achievements include developing image detectors that can measure individual photon events.

AN INTRODUCTION TO OPTICAL STELLAR INTERFEROMETRY

A. LABEYRIE, S. G. LIPSON, AND P. NISENSON

CAMBRIDGE
UNIVERSITY PRESS

CAMBRIDGE
UNIVERSITY PRESS

University Printing House, Cambridge CB2 8BS, United Kingdom

Published in the United States of America by Cambridge University Press, New York

Cambridge University Press is part of the University of Cambridge.

It furthers the University's mission by disseminating knowledge in the pursuit of
education, learning and research at the highest international levels of excellence.

www.cambridge.org
Information on this title: www.cambridge.org/9781107656468

© A. Labeyrie, S. G. Lipson, and P. Nisenson 2006

First published 2006
First paperback edition 2013

A catalogue record for this publication is available from the British Library

ISBN 978-0-521-82872-7 Hardback
ISBN 978-1-107-65646-8 Paperback

Cambridge University Press has no responsibility for the persistence or accuracy of
URLs for external or third-party internet websites referred to in this publication,
and does not guarantee that any content on such websites is, or will remain, accurate
or appropriate.

Contents

Illustrations

Preface

Although the optical telescope is the most venerated instrument in astronomy, it developed relatively little between the time of Galileo and Newton and the beginning of the twentieth century. In contrast to the microscope, which enjoyed considerable conceptual development during the same period from the application of physical optics, telescopes suffered from atmospheric disturbances, and therefore physical optics was considered irrelevant to their design. The realization that wave interference could be employed to overcome the atmospheric resolution limit was first recorded by Fizeau and put into practice by Michelson around 1900, but his experience then lay dormant until the 1950s. Since then, first in radio astronomy and later in optical and infrared astronomy, interferometric methods have improved in leaps and bounds. Today, many optical interferometric observatories around the world are adding daily to our knowledge about the cosmos.

The aim of this book is to build on a basic knowledge of physical optics to describe the ideas behind the various interferometric techniques, the way in which they are being put into practice in the visible and the infrared regions of the spectrum, and how they can be projected into the future. Some techniques consist of optical additions to existing large telescopes; others require complete observatories which have been built specially for interferometry. Today all these are being used to make accurate measurements of stellar angular positions, to discern features on stellar surfaces and to study the structure of clusters and galaxies. Tomorrow, maybe they will be able to image planetary systems other than our own. To this end, many new ideas are being generated and tested with the eventual aim of looking at an extrasolar Earth-like planet, either from the ground or from a space platform.

The book contains some introductory chapters on basic optics, which establish an unsophisticated physical and mathematical framework which is used to discuss the various ideas and instruments presented in the later chapters. It is hoped that, despite the inevitable use of mathematics, the physical principles of the astronomical interferometric techniques in the following chapters will be clear. In the final

Antoine Labeyrie and Stephen Lipson

chapters, some astrophysical results achieved by interferometry are discussed, and some untested future ideas are presented. The level of detail is hopefully sufficient for senior undergraduate and graduate students who are interested in understanding the ideas and implementations of astronomical interferometry. We have attempted to give fair credit to all those whose work has substantially advanced the field, without overloading the book with references to every detail.

Peter Nisenson first conceived of this book in 2002, and asked us to join him in writing it. Sadly, he never lived to see its publication, but he was active in determining its layout and he wrote fairly complete drafts of two chapters. As a result of this, we decided to continue the work as a memorial to his life-long dedication to astronomy, although his further contributions are sorely missing.

Many people have helped us in collecting and understanding the material presented, and have spent time showing us round their interferometric observatories. SGL wishes in particular to thank Dr Erez Ribak, from whom he has learnt such a lot through innumerable discussions on optics and astronomical interferometry. He is also grateful to Mark Colavita, Amir Giveon, David Snyder Hale, Chris Haniff, Pierre Kern, Nachman Lupu and Nils Turner for their time, help and comments. AL wishes to thank the late Prof. André Lallemand and Pierre Charvin for their early support. Emile Blum, James Lequeux, Françoise Praderie and Arthur Vaughan gave crucial encouragement and Deane Peterson also encouraged, in the critical early stages, part of the work described in the book.

In addition, we should like to thank Laurent Koechlin, John Davis, Chris Haniff, Chris Dainty, Andrew Booth and Noam Soker, who have read and made useful comments on parts of the manuscript. Itzik Klein carried out the experiments described in section 4.6 and Carni Lipson drew some of the figures. We are also grateful to the many authors and journals for permission to reproduce figures and data, as indicated in the figure captions. SGL wishes to acknowledge the support of the Norman and Helen Asher Space Science Institute at Technion, and the hospitality of the Kavli Institute for Theoretical Physics, UCSB, where part of the manuscript was researched and written.

We should also like to thank our wives and families for their understanding during the periods when we have been necessarily absorbed in research and writing.

Antoine Labeyrie
Stephen Lipson
Plateau de Calern, August 2005.

Peter Nisenson, 1941–2004

This book was Peter Nisenson's idea. Peter received his BS degree from Bard University in New York, and continued with post-graduate work in Physics and Optics at Boston University. He was then employed as an optical scientist by the Itek Corporation in Lexington, MA, where he worked for 14 years. Both of us first met him there in 1973. At that time he was working on a programmable optical memory device (the PROM) which used a photoconducting crystal as a recording medium. When he first heard about speckle interferometry, he realized that this device could carry out the required Fourier transform on-line and therefore provide directly the image power spectrum. The three of us then met for the first time; an observing plan was proposed, supported by Itek, and carried out at Kitt Peak in December 1973. Although this particular project was not successful, it was probably the turning point in Pete's life, at which he decided to become a professional astronomer. In the same period, under the leadership of John Hardy with whom he had a life-long friendship, his group at Itek became heavily involved in adaptive optics, an involvement which led to his making some important measurements of atmospheric optical properties.

Pete was for five years a Research Associate at the Center for Earth and Planetary Physics at Harvard before joining the Harvard Smithsonian (CfA) in 1982, where he remained for 22 years. He worked on various exciting and innovative projects including the development of programs for high-resolution image reconstruction of solar and other astrophysical data, using speckle interferometry. Together with Costas Papaliolios and Steven Ebstein he developed the "Precision Analog Photon Address" (PAPA) image detector, which gave the digital addresses of individual photon events, one of a new generation of image detectors for astronomy, and which he used extensively for speckle interferometry.

He highlighted the use of interferometry during observations in Chile of the Supernova SN1987a, which received television prominence, as well as several

Peter Nisenson

publications. He then became involved in the extrasolar planet search, and contributed greatly to the creation and successful use of the "Advanced Fiber Optic Echelle" (AFOE) spectrograph, a technique that has been successfully used in recent years to discover several planets outside of our solar system. In addition he developed original concepts for imaging extrasolar planets using high-dynamical-range apodization techniques. He possessed a commanding knowledge of optics and his ability to envisage alternative ways to achieve a goal was invaluable to many projects.

During his period at CfA he became involved with the IOTA interferometer. In 2002 he originated the idea of writing a textbook about optical interferometric astronomy, with the feeling that this was becoming a mature technique and was already beginning to provide important astrophysical data. This book is the result, and is dedicated to his memory.

As a young man, Peter chose the study of math and physics over becoming a professional cello player but continued a life-long love of music. He was an avid golf and tennis player and an active member of the Harvard College Observatory Tennis Club for years. Somewhat of a terror on the tennis court, he nevertheless delighted in encouraging others amongst his colleagues to join in the game, and the HCO Tennis Clinic has been named in his memory. Peter had been in poor health for the last year of his life, but throughout this time he stayed as active as his condition allowed him to be. He was survived by his wife Sarah (Sally), his son Kyle and his daughter Elizabeth.

1

Introduction

1.1 Historical introduction

The Earth orbits a star, the Sun, at a distance of 140 million km, and the distance to the next closest star, α-Centauri, is more than $4 \cdot 10^{13}$ km. The Sun is one star in our galaxy, the Milky Way. The Milky Way has 10^{11} stars and the distance from the Sun to its center is $2.5 \cdot 10^{17}$ km; it is one galaxy in a large group of galaxies, called the Local Group and the distance to the next nearest group, called the Virgo Cluster, is about $5 \cdot 10^{20}$ km. The Universe is made up of a vast number of clusters and superclusters, stretching off into the void for enormous distances. How can we learn anything about what's out there, and how can we understand its nature?

We can't expect to learn anything about distant galaxies, black holes or quasars, or even the nearest stars by traveling to them. We can maybe explore our own solar system but, for the foreseeable future, we will learn about the Universe by using telescopes, on the ground and in space.

The principal methods of astronomy are spectroscopy and imaging. Spectroscopy measures the colors of light detected from distant objects. The strengths and wavelengths of spectral features tell us how an object is moving and what is its composition. Imaging tells us what an object looks like. Because distant stars are so faint, the critical characteristic of a telescope used for spectroscopy is its light-gathering power and this is determined principally by its size, or "collecting area." For imaging, the critical characteristic is its resolution. In general, we don't know the distance to the objects we are looking at; we can only measure the angle they subtend at the location of the observer. So we use the term "angular" rather than spatial resolution to characterize the imaging capability of a telescope. In principle, the larger the telescope aperture, the better is its inherent resolution. However, in practice, telescopes operating on the ground, observing through the Earth's turbulent atmosphere, are limited by atmospheric turbulence.

The inherent or *diffraction-limited* angular resolution limit of a telescope is determined by the ratio λ/D between the wavelength, λ, of the light used for the observation and the diameter, D, of the telescope aperture. A 10-m telescope, like the Keck telescopes on Mauna Kea in Hawaii, has an inherent angular resolution limit of about 10 milli-arcseconds in visible light ($\lambda = 500$ nm). 10 milli-arcseconds is the angle subtended at Earth by a soccer ball on the surface of the Moon. The Earth's atmosphere degrades the resolution so much that the typical resolution is only about 1 arcsecond, which is no better than the inherent resolution of a 10-cm telescope. Even at the site with the best seeing in the world, in the remote mountains on the Antarctic continent, the best seeing is 0.15 arcsecond. Techniques such as adaptive optics and speckle interferometry have been developed for measuring the effects of the atmosphere and correcting them, which improve ground-based resolution to the inherent limit. By putting a telescope in space, the atmospheric disturbance problems can indeed be avoided, so that the inherent diffraction-limited angular resolution of about λ/D can be obtained. However, one recalls the costly design disaster in the early 1990s whereby the 2.4-m Hubble Telescope optics had to be corrected by the addition of the COSTAR system before this could be realized.

The closest star to us, α-Centauri, is actually a triple-star system, and its largest component is a very close analog to our Sun. Its angular diameter is only 7 milli-arcseconds, less than the resolution limit of the largest telescope. The second closest star like our Sun, τ-Ceti, is three times farther away and has an angular diameter only 2.5 milli-arcseconds. Clearly, if we want to study any of these in detail, we need much better angular resolution. In the near future there will not be any telescopes much larger than the Kecks; studies for a 100-m telescope are currently being carried out, but whether anything approaching this size is technically or financially feasible is an open question. Interferometry is the proven approach to obtaining higher resolution without having to build enormous telescopes. In an interferometer, we coherently combine the light from two or more telescopes or apertures. The angular resolution is then given by λ/B, where B is the baseline, or largest edge-to-edge separation between the telescopes. So we can take several small telescopes and separate them by large distances in order to achieve resolution comparable to that of a large telescope with diameter B. Of course, producing images from such an array is more difficult than with a conventional telescope, but it can and has been done.

Stellar interferometry was first suggested by H. Fizeau in 1868. In a report to the French Académie des Sciences on the judgement of the Bordin prize, offered in 1867 for an essay on methods to determine the direction of the vibrations of the aether in polarized light, Fizeau remarked that interference fringes produced from a source of finite dimensions must necessarily be smeared by an amount depending on the size of the source. He suggested that observation of the smearing, or lack

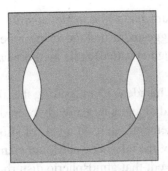

Fig. 1.1. Mask used by Stéphan on the Marseilles telescope. This mask provides a pair of identical apertures with the largest separation possible.

of clarity, of the fringes created by a star through a large telescope whose aperture was masked by a pair of well-separated apertures, could be used to put an upper limit on its angular dimensions. The challenge was taken up by M. Stéphan within a few years. Stéphan was the director of the Marseilles observatory, whose 80-cm reflector was at that time the largest in the world. He masked the aperture in a manner which gave two identical apertures with the largest possible separation (figure 1.1) and indeed observed fringes crossing the now enlarged image of a star. In fact all the stars which he observed eventually showed fringes, from which he deduced an upper limit to their diameter of 0.16 arcsec, which is indeed true (Stéphan 1874). But in doing so, he improved the practical atmospherically limited resolution by almost one order of magnitude.

More than a decade later, A. A. Michelson (1890) came up with the same idea of using a pair of small openings in a mask covering the aperture of a telescope to produce fringes, and thereby to measure the profile of a star. Michelson nowhere refers to the earlier works of Fizeau or Stéphan, and Lawson (1999) has discussed the question of whether the French work was indeed known to Michelson. It would appear that although Michelson had visited Paris in 1881 for an extended study period and may have met Fizeau, there is no evidence of their having discussed this question. Michelson's paper in 1890 not only reinvented the idea of stellar interferometry, but describes in detail how it should be carried out by measuring the fringe visibility, a concept there defined for the first time, as a function of aperture separation. He discusses the expected results not only for uniform disk-like and binary stars, but also for limb-darkened stars, using a model previously developed to describe the intensity profile of the Sun. The technique he used for these calculations was to superpose the interference fringes created from each point on the extended disk of the star. The concept of the coherence function, usually used for such calculations today, arose decades later from the work of F. Zernike in 1938. From the instrumental point of view, Michelson pointed out that for useful measurements of stellar diameters to be made, apertures separated by up to 10 m would be required, and suggested a method using a beam-splitter by which this could be carried out, although eventually this method was not used till the modern era of stellar interferometry.

Michelson (1891) tested his ideas by measuring the diameters of the four major satellites of Jupiter, whose diameters had already been determined by other methods, and got excellent agreement. He used a pair of slits with variable spacing to mask the 12-inch aperture of the telescope at Mount Hamilton. He confirmed what Stéphan had already observed, that atmospheric disturbances cause the fringes to shift around, but that they can be followed by an observer's eye and their visibility is not much degraded by the atmosphere. He describes the effect quite graphically in his book *Studies in Optics*, quoted at the beginning of chapter 5. This work was followed up by K. Schwartzschild (1896) and J. A. Anderson (1920) who used the same technique to measure the separation of many binary pairs.

The experiments of Stéphan and Michelson showed one way to achieve diffraction-limited resolution from a telescope. In doing so, they lost the true image-creating capability of the telescope, and this is a loss which is today still proving irksome. Michelson's experiments were intended as a preliminary trial for a much more ambitious project, which would improve on the diffraction limit considerably. This project took another 25 years before bearing fruit.

The instrument which evolved is now known as the Michelson stellar interferometer, a name which distinguishes it from the probably more famous Michelson interferometer used for the Michelson–Morley experiment and the optical determination of the standard meter which earned him the Nobel Prize in 1907. A sketch of the optics of the stellar interferometer and a photograph of it mounted on the 100-inch telescope at the Mount Wilson Observatory in California are shown in figure 1.2. The actual apertures of the interferometers were two 6-inch mirrors mounted on a rigid beam 20-ft long attached to the telescope normal to the line of sight, and whose separation could be changed at will. Using a periscope type of construction, the light from these apertures is brought to within the telescope aperture, and the beams intersect in the image plane, forming an image of the star, diffraction-limited by the 6-inch apertures, crossed by Young's fringes. A great advantage of this arrangement over that used by Stéphan and in Michelson's experiments on Jupiter's satellites, was that the angle of intersection of the beams, and therefore the fringe-spacing, was independent of the separation of the apertures. Although finding white-light fringes in such an enormous system might seem to be an impossible task, since the paths must be equalized to about one micron, in fact the entrance mirrors could be positioned geometrically to better than one millimeter and then a path-compensator next to the observing position was used for final equalization.

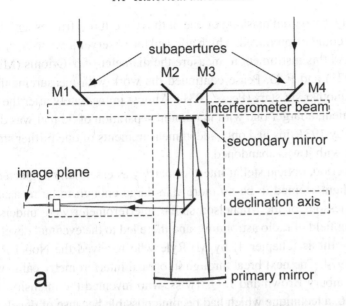

Fig. 1.2. Michelson's 20-foot beam stellar interferometer. (a) Optical diagram; (b) a photograph of the instrument, as it is today in the Mount Wilson Museum (reproduced by permission of the Huntington Library).

This compensator consisted of a glass plate in one beam and, in the other, a pair of glass wedges which could slide laterally on one another to form a parallel-sided plate with variable thickness. A second pair of fairly closely spaced apertures was also observed through the same path-correction optics, allowing comparison fringes with high visibility to be created from the same source and to be observed simultaneously. By partial masking of one of these apertures, fringes with less than unit visibility could be produced, which allowed the observer's eye to be calibrated. Michelson used this instrument to measure the diameter of α-Orionis (Michelson and Pease 1921) and F. G. Pease continued the work with measurements of the diameters of another six stars (Pease 1931). Almost immediately after the success of this instrument, a larger one with maximum separation of 50 feet was designed and built, but by 1931 this had only added measurements of one further star to the list, and work with it was abandoned.

After this period, optical stellar interferometry was essentially neglected for 30 years, since there seemed to be no more stars large enough to be measured this way. But the lessons of the Michelson stellar interferometer were understood in the blossoming field of radio astronomy, and there led to the eventual development of aperture synthesis (chapter 4) by M. Ryle who received the Nobel Prize for this work in 1974. The next breakthrough in optical interferometry came with the work of R. Hanbury Brown and J. Q. Twiss, who invented the intensity interferometer in 1956, a technique which had become possible because of developments in electronics and photodetectors during the Second World War (Hanbury Brown and Twiss 1956). This was implemented using two independent telescopes instead of Michelson's two entrance mirrors and employed separations up to 166 m at the Narrabri Observatory near Sydney, Australia (Hanbury Brown 1974). The theory and results of this method are the subject of chapter 7. But this technique too had its limitations, and was found to be practical with the existing equipment only for stars brighter than magnitude 2.5. When all available interesting stars in the southern hemisphere had been measured, the technique was abandoned around 1974, mainly because of growing successes in a resurgence of optical Michelson stellar interferometry using separated telescopes, which showed greater promise.

One of the strengths of intensity interferometry is that it is oblivious to atmospheric disturbances. This is because the technique works by correlating fluctuations in light intensity at frequencies up to 200 MHz. Since atmospheric fluctuations are limited to frequencies less than 1 kHz, these can easily be filtered out. But, in 1970 A. Labeyrie suggested a revolutionary technique of observation which actually took advantage of the randomness of the atmospheric fluctuations in order to get diffraction-limited images with a conventional large-aperture telescope. This technique, called "speckle interferometry" (Labeyrie 1970), started a revival in interest in high-resolution optical astronomical imaging, described in chapter 6. It was shortly followed by the first successful coherent combination of two telescopes separated by a baseline of 13.8 m (Labeyrie 1975). Once this proof of principle was demonstrated, several groups started construction of large-scale interferometers having several subapertures, and the rest of the story is described in the succeeding chapters.

1.2 About this book

The plan of this book can be summarized as follows. The following three chapters are devoted to basic principles of optics and interferometry. Chapter 2 introduces the relevant ideas in a very qualitative manner; chapter 3 is devoted to a quantitative discussion of interference, diffraction and coherence, and chapter 4 to aperture synthesis. Chapter 5 discusses atmospheric statistics and turbulence, with a section on adaptive optics which has recently begun to play a role in stellar interferometry. Chapter 6 describes passive techniques used to achieve maximum resolution from single-aperture telescopes, including speckle interferometry and aperture masking. Chapter 7 is devoted to intensity interferometry. Chapter 8 discusses the techniques employed in modern amplitude (Michelson stellar) interferometry with descriptions of the observatories around the world that have been built for this purpose. Chapter 9 describes the "hypertelescope", a way in which many fixed apertures in a very sparse array can in principle be combined to give the equivalent of a very large steerable telescope. Chapter 10 is devoted to three types of instrument devoted to extrasolar planet detection, based on apodization, coronagraphy and interferometric nulling. Chapter 11 aims to present a selection of significant scientific results which have been obtained by them. Finally, chapter 12 discusses future ground- and space-based interferometry systems, also aimed mainly at detecting and imaging planets around distant stars.

Much of the material in the book can be found in recently published reviews. Those we have found particularly relevant and helpful are by Roddier (1988), Lawson (2000), Baldwin and Haniff (2002), Saha (2002) and Monnier (2003). However, the subject is developing continuously, and many of the details in the book will already be outdated before it is published. A useful way of keeping up to date with developments is throught the Optical Long-Baseline Interferometry News (OLBIN) website, http://olbin.jpl.nasa.gov/ edited by P. R. Lawson.

References

Anderson, J. A. (1920). *Astrophys. J.*, **51**, 263.
Baldwin, J. E. and C. A. Haniff (2002). *Phil. Trans. R. Soc. Lond. A*, **360**, 969.
Hanbury Brown, R. and R. Q. Twiss (1956). *Nature*, **178**, 1046.
Hanbury Brown, R. (1974). *The Intensity Interferometer*, London: Taylor and Francis.
Labeyrie, A. (1970). *Astron. Astrophys.*, **6**, 85.
Labeyrie, A. (1975). *Astrophys. J.*, **196**, L71.
Lawson, P. R. (2000) ed. *Principles of Long Baseline Stellar Interferometry*, NASA-JPL.
Michelson, A. A. (1890). *Phil. Mag.*, **30**, 1.
Michelson, A. A. (1891). *Nature*, **45**, 160.
Michelson, A. A. and F. G. Pease (1921). *Astrophys J.*, **53**, 249.
Monnier, J. D. (2003). *Rep. Prog. Phys.*, **66**, 789.

Pease, F. G. (1931). *Ergebnisse der Exakten Naturwissenschaften*, **10**, 84.
Roddier, F. (1988). *Phys. Rep.*, **170**, 97.
Saha, S. K. (2002). *Rev. Mod. Phys.*, **74**, 551.
Schwartzschild, K. (1896). *Astr. Nachrichten*, **139**, 3335.
Stéphan, M. (1874). *Comptes Rendus*, **78**, 1008.

2

Basic concepts: a qualitative introduction

2.1 A qualitative introduction to the basic concepts and ideas

Interferometric astronomy is founded on the basic principles of interference of light waves, which were first conceived in the seventeenth century by Christiaan Huygens, based on experimental evidence by F. M. Grimaldi and Robert Hooke. Interference itself was studied quantitatively in the nineteenth century, beginning with Thomas Young and Augustin Fresnel, and quickly blossomed into the major subject of physical optics. The first application of interference to astronomy was proposed by Hyppolyte Fizeau in the middle of that century.

The purpose of this chapter is to cover the basic ideas of optical interference relevant to astronomy in a qualitative manner. It is followed by two chapters describing the concepts in a more mathematical way. Some of the tools (particularly Fourier analysis), which are essential to detailed understanding of the subject, but may well be quite familiar to many readers, are described in Appendix A.

2.1.1 Young's experiment (1801-3)

This book is about the application of interference to optical astronomy. The possibility of interference is the major distinction between particles and waves as mechanisms for transporting energy and momentum, and the phenomenon of "destructive interference," in which two disturbances cancel one another out under specific conditions, is peculiar to waves and cannot occur with particles. Although it might seem that energy is somehow being destroyed under these conditions, we always find that the energy which appears to have been lost when two waves interfere destructively appears somewhere else in the system, so that there is, almost miraculously, never any problem with its conservation.

Interference was observed with water waves, for example, long before it was formally described by Thomas Young in England in 1801 (Magie, 1935). Moreover,

9

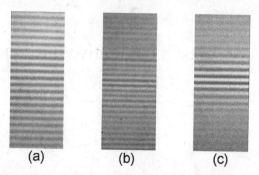

(a)   (b)   (c)

Fig. 2.1. Young's fringes between light passing through two pinholes separated vertically: (a) from a monochromatic source; (b) from a polychromatic line source; (c) from a broad-band source.

the well-known interference phenomenon of beats between sound waves must have been known and used much earlier. But of course there is a world of difference between observing and using a phenomenon and understanding its origins. The idea that light could be described as a wave motion had been proposed over a hundred years earlier by Robert Hooke and by Christiaan Huygens, who had invented the idea of a *wavelength*. However, Huygens had for some reason not alighted on the idea of interference, and his ideas had been more-or-less eclipsed by the weight of Newton's authority which claimed that light was a particle phenomenon.[†] In 1803, Young described an experiment in which a sunbeam, selected by passing sunlight through a small pinhole, fell on a narrow strip of card (1/30 inch wide) and the light beams spreading out from the two edges overlapped on a screen placed after the strip. There he saw a set of bands, "interference fringes," of which the center one was white and the outer ones colored. Cutting out the light from either of the two .beams caused the fringes to disappear, so that in the positions of the dark bands the intensity had been *reduced* by adding the second beam while it had been increased in the bright bands. Indeed, Young also remarked on the fact that when one beam was only partially obscured the fringes remained, but with lesser contrast. In a later series of lectures, he idealized this experiment to a pair of pinholes or narrow parallel slits, which is the way in which it is generally presented today (figure 2.1). He used the interference phenomenon, albeit in the form of Newton's rings, to determine for the first time the approximate value of the wavelengths of light of different colors. Young's experiments, carried out more than 100 years after Newton, were analyzed in mathematical detail by a French contemporary, Augustin Fresnel. Because Newton was so revered in England, the further development of the subject

[†] Newton had explained his observation of what are now known as Newton's rings in his *Opticks*, Book II, proposition XII, on the basis of the corpuscular theory, by imbuing his particles with internal vibrations. He thereby came very close to rediscovering Huygens' wave theory. The particles incident on a surface were considered to have "Fits of easy Reflexion" and "Fits of easy Transmission", the interval between them being what we now recognize as half of the wavelength.

Fig. 2.2. Template for preparing your own double slit. Photocopy this diagram onto a viewgraph transparency at 30% of full size, to give a slit spacing of about 1 mm.

of physical optics, essentially based on Fresnel's work, was continued mainly on the European continent by scientists such as Arago, Fraunhofer, Fizeau and Foucault.

All the physical optics that we need to know in order to understand astronomical interferometry can be gleaned from a careful and quantitative examination of the interference fringes that Young observed, and that is what interferometric astronomers do.

2.1.2 Using Young's slits to measure the size of a light source

The principle of stellar interferometry using a pair of separated telescopes can be illustrated by a simple experiment which you, the reader, are encouraged to carry out for yourself. The only apparatus that you need is that of Young: a pair of parallel slits with a separation of about 1 mm, and some monochromatic light sources. You can prepare the slits by cutting them out of thick paper or foil, or by photocopying figure 2.2 onto an overhead transparency with a linear size reduction to about 30% and high contrast. The light sources should be street lamps at various distances[†]. These are your "stars."

Now cover one eye, and with the slit pair as close as possible in front of the other look at a typical urban night scene including sodium or mercury discharge street lamps. If the slits are horizontal, as in figure 2.2, you will see each source stretched vertically into a set of Young's fringes. The fine periodic fringes are modulated by a rather coarse envelope, resulting from the widths of the individual slits, which can be ignored in the present discussion; it is the fine fringes that we are interested in. You will notice that the farthest lamps, which appear smallest, give the clearest fringes. Looking at closer sources, which appear larger and brighter, the fringes

---

[†] Young would have had to put salt into a flame to create a monochromatic source. Today we get such sources free of charge, because the most common high-intensity street lights provide approximately monochromatic light at 589 nm (Na) and 546 nm (Hg). But even white-light sources give enough interference fringes to demonstrate the effects.

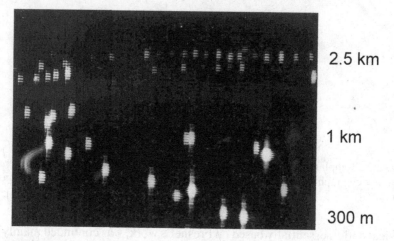

Fig. 2.3. A typical observation of an urban night scene photographed through a pair of slits separated vertically by about 1 mm. Approximate distances to the street lights are shown on the right.

become more blurred and lose their contrast. You will not see any fringes around the closest lamps, only maybe the envelope (figure 2.3).

Your first reaction to this might be that it is obvious. The fringes around all the lamps have the same angular spacing, which is determined by the slit separation and the wavelength. If the source angular size is small compared with this fringe spacing, the fringes are clear and have good contrast. But if the source size appears comparable with the fringe spacing, when each point on the source creates its own set of fringes, the result is blurring by the size of the source and a consequent reduction of contrast. And if the source size appears larger than the fringe spacing, one does not see the fringes at all. This, in principle, is all there is to stellar interferometry using two receivers, which are simulated by the two slits. The experiment is made quantitative by measuring the contrast of the fringes as a function of the spacing between the slits. The slits can also be rotated in their plane so as to measure dimensions in different orientations.

Let's do a simple calculation to see how it works. First remember that what we see on the retina of the eye is angular dimension; that is why distant objects seem smaller than close ones. The wavelength λ of a sodium street lamp is about $0.6 \, \mu$m. The interference fringes for a slit separation d have angular positions θ_m given by the well-known formula (see any book on elementary physical optics) $m\lambda = d \sin \theta_m$ from which the fringe spacing is, assuming small angles, λ/d. Therefore we expect the fringes to be smeared out when the angular size of the source becomes comparable with this. For a 1-mm slit separation, this is $6 \cdot 10^{-4}$ radians. Now a typical dimension for a street lamp is 0.2 m, and if it is situated at distance L meters its angular size is $0.2/L$. The fringes are therefore visible clearly only

Fig. 2.4. Waves on a still pond, photographed at (a) $t = 0$, (b) $t = 2$ and (c) $t = 4$ sec. The radius r of a selected wavefront, measured from the source point, is shown on each of the pictures.

if $0.2/L \ll 6 \cdot 10^{-4}$ or $L \gg 330$ m. To get better resolution of the size of the source, we should use a larger slit spacing; using the eye, d cannot be greater than the diameter of the dilated pupil, a few mm, so that resolution much better than the $6 \cdot 10^{-4}$ rad is difficult to achieve. A star with one of the largest angular diameters is Betelgeuse, of order 0.05 arcsec, or $2.5 \cdot 10^{-7}$ radians. To be useful in astronomy, the method therefore has to be capable of measurements in this range; the double slit must have spacing much in excess of $\lambda/(2.5 \cdot 10^{-7}) \approx 2$ m. That is what makes astronomical interferometry in the visible so difficult and has presented many technical problems, some of which have only recently been overcome.

In the rest of this chapter and the two which follow, we shall introduce the concepts which allow us to make the above discussion quantitative, and to turn what might be considered a rough method of estimating stellar diameters into a means of obtaining stellar images with resolution much greater than that of any single-aperture telescope.

2.2 Some basic wave concepts

In this section we shall introduce several of the concepts and quantities needed in the book. Of course, many readers will already be quite familiar with these.

We consider an extended medium or field (in one, two or three dimensions) which is in equilibrium. This could be, for example, a straight stretched wire, the smooth horizontal surface of a pond, air at uniform pressure, or an electric field defined by some boundary configuration. A disturbance is somehow created at a point, for example by throwing a stone into the pond. This disturbance propagates outwards from its point of origin in a way which is determined by the dynamics of how the medium or field responds to changes. In the case of the pond, the disturbance propagates as a growing group of circles around the source (figure 2.4). Although the circles appear to move outwards, the water surface is disturbed locally but there is no outward flow, as you can see by watching a bit of debris floating on the surface, ·

which just bobs up and down as the wave passes it. Just the same, the wave motion transfers energy and momentum from the source.

The waves on the pond can be used to illustrate several other wave concepts. The stone hitting the water surface excites a group of waves, within which the surface profile is approximately sinusoidal. The line following a particular maximum of the wave (a growing circle in this case) is called a *wavefront*. The wavefront has a velocity called the *wave* or *phase velocity*; this can be deduced by comparing figure 2.4(a) and (b) and measuring the distance that the wavefront has traveled in the time between the photographs. It also has a *group velocity* which is the distance traveled in unit time by the envelope of the group of waves excited by the stone. The group and wave velocities may not be equal; in the case of water surface waves they are not, but for electromagnetic waves in free space the two velocities are equal.

Within the group of waves one can determine the *wavelength*, λ, which is the distance between adjacent wavefronts[†]. The *amplitude*, A, is the distance by which the surface at height h deviates from the equilibrium (flat) surface at h_0. For a simple sinusoidal wave the up and down amplitudes have the same value. Notice that the amplitude of the wave gets smaller as the wave propagates outwards; it is not a constant of the motion. If we concentrate on the movement of the bit of debris on the surface at a certain point, we can determine the wave's *frequency*, f, which is the number of oscillations per second, and its *phase*, ϕ. To measure the phase, we need to define a zero of time (which is arbitrary, but must be the same for all related measurements) and then to express the sinusoidal motion of the surface at that point as

$$h - h_0 = A \cos(2\pi f t + \phi). \tag{2.1}$$

Then a more general definition of a wavefront is a line or surface along which the argument of the cosine is constant (for example zero, where the wave has its maximum value A). Now clearly at a given moment (e.g. one of the snapshots in figure 2.4) the wave repeats itself after distance λ in the propagation direction (x). This corresponds to a change of 2π in phase, so that a more general description of the wave is

$$h(x, t) - h_0 = A \cos[2\pi(f t - x/\lambda) + \phi]. \tag{2.2}$$

We can relate the values of x and t on a given wavefront, for which the phase of (2.2) is a constant, by $f t - x/\lambda = $ const., leading to phase or wave velocity $v = f\lambda$.

[†] Actually, in figure 2.4 you can see that the wave group is quite complicated, with a longer wavelength at its leading edge and a shorter one at the tail. But this is a property of water waves and is not generally true.

2.2.1 Plane waves

In order to extend the above discussion to dimensionality higher than one, it is convenient to define two variants on the frequency and wavelength which will simplify the analysis. The first is the product $\omega \equiv 2\pi f$, which is called the *circular frequency* and has units of radians/sec. The second is the *wavevector* $\mathbf{k}$, which is a vector of length $2\pi/\lambda$ in the direction of propagation of the wave (normal to the wavefronts). It has dimensions of inverse length[‡]. We then write a generalization of (2.2) in the form:

$$h(\mathbf{r}, t) - h_0 = A \cos[\omega t - \mathbf{k} \cdot \mathbf{r} + \phi], \qquad (2.3)$$

where $\mathbf{r}$ is position in real space. You can see that the wavefronts now correspond to $\mathbf{k} \cdot \mathbf{r} = \text{const}$. This defines a plane normal to the wavevector $\mathbf{k}$. The wave (2.3) is called a *plane wave* and is the basic building block of physical optics.

Finally, we define the *polarization* of the wave as the direction of the disturbance with respect to the direction of propagation of the wave; i.e. we let h and A become vectors. If the disturbance is normal to the propagation direction, the wave is *transverse* and since there are two independent normals, there are two polarizations. Waves on a string are the easiest example of transverse waves to appreciate; there is no motion along the string, only normal to it, and two normal directions are possible. Electromagnetic waves, the subject of this book, are also transverse waves with two orthogonal polarizations, although the reason for this is more obscure. In a compressible medium, such as a gas, motion along the wave propagation direction is possible, causing periodic increases and decreases in density and pressure compared to the uniform gas. Then we have a *longitudinal wave*. The polarization is along the propagation direction, but is not usually referred to as such; one simply says that longitudinal waves have no polarization properties.

2.2.2 Huygens' principle: propagation of limited or distorted waves, and gravitational lensing

As early as 1658 Huygens proposed a construction which would allow the propagation of a wave to be followed through apertures, round obstacles and from one transparent medium into another. His idea, which was substantially justified (i.e. up to a few details which did not alter the main idea) by Kirchhof two hundred years

---

[‡] It should be emphasized that λ, despite its having a length and direction, is not a vector, since if one looks at a projection of the wave on a direction inclined at angle α to the propagation axis, it has wavelength $\lambda/\cos\alpha$ which is not right for a vector. A vector $\mathbf{V}$ would have projection $V\cos\alpha$ and clearly $2\pi/\lambda$ behaves correctly in this respect.

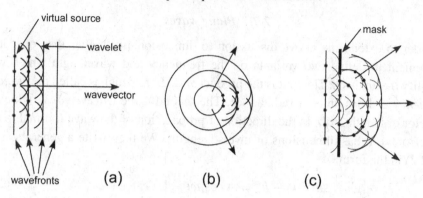

Fig. 2.5. Huygens' principle applied to (a) propagation of a plane wave, (b) propagation of a spherical wave, (c) diffraction after passage through an aperture mask.

later, was very simple; and it is still very useful for complicated wave propagation problems. We consider each point on a wavefront to be a new source of waves, like the stone in the pond in figure 2.4. After propagation for a wavelength in the forward direction, the set of new wavefronts of these waves defines an envelope, which is the next wavefront. Huygens called these *wavelets*, although this term has recently been adopted for something rather different. Huygens ignored the backwards propagation of his wavelets, with the justification that they cannot be relevant to a forwards propagating wave. The construction confirms what is pretty obvious for propagation of a plane wave or for a circular wave of the sort shown in the figure, but when there is an obstacle – for example an aperture – it shows quite convincingly why diffraction occurs (figure 2.5), although the well-known intensity variations do not appear. This theme will be developed in section 3.1.

A beautiful example of how Huygens' principle can give us an insight into the behavior of wave propagation in an inhomogeneous medium is the phenomenon of "gravitational lensing." Without going into the relativistic explanation, let us just quote the result that a light wave traversing the space in the region of a massive body of mass M behaves as if the space had a refractive index which is larger than 1 by an amount $2GM/c^2r$ proportional to the gravitational potential, where r is the distance from the body and G and c are fundamental constants. This causes the region around the body to behave like a rather strange lens, whose thickness varies logarithmically with the radius, rather than quadratically (like r^2) as in a normal lens (Schneider et al. 1992). The result is that the plane front of a wave passing such a body becomes distorted and acquires a dimple in that region. The way in which the dimple proceeds can best be visualized using Huygens' principle. In the central regions, where the dimple causes the wavefront to be concave, a focusing wave develops; in the outer regions, where the wavefront is convex, a diverging wave ensues. As the

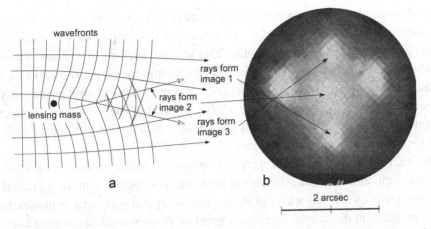

Fig. 2.6. Huygens' principle applied to gravitational lensing. (a) The distortion of the wavefront of a plane wave in the region of a massive body, causing a dimple on the axis, propagation of the dimpled wavefront, and the way in which multiple images result; (b) an example of the gravitationally distorted image of a quasar in the near infrared (courtesy of NASA).

wave progresses through the focus of the central region, a cusp develops. The result of all this is to create arcs and multiple images of a single source, an effect which has been observed astronomically on several occasion (figure 2.6). Gravitational lensing has become relevant to the extrasolar planet search (chapter 10), which is a major incentive for astronomical interferometry. When the lensing mass is very large, an image broken into distinguishable arcs or dots may be formed as shown in figure 2.6(b). If the mass is very small, considerably less than that of the Sun, these patterns are unresolved by atmospherically limited telescopes, but the gravitational lens effect still causes a larger area of the wavefront to be directed towards the observer than would be the case if the lensing mass were absent. As a result, a peak in flux occurring simultaneously at all wavelengths is expected. It can be quite considerable if the source, observer and lensing mass are well aligned (Paczynski, 1986). Such events, called "microlensing", have been observed and are monitored regularly. Moreover, the presence of a second body next to the lensing mass causes a double peak in intensity (Beaulieu *et al.* 2006), and from the ratio between the two, information about the relative masses and separation can be gleaned. This has led to a new way of looking for extrasolar planetary systems, and several observing programs of this type are currently being carried out.

2.2.3 Superposition

So far we have discussed individual waves, each having a sinusoidal profile. In fact, most waves are more complicated than this but fortunately, in a linear medium, the more complicated waves can always be expressed as a weighted sum of many individual simple waves. In most cases (including, in particular, electromagnetic waves) this summation is simple; one considers the propagation of each wave individually, and then sums the results. This idea is called *superposition*. The way in which a complicated wave can be broken into simple ones is *Fourier analysis*, and is discussed in Appendix A. When it applies, it considerably simplifies the analysis of complex problems. Fortunately, it applies to most of the situations we meet in astronomy, so that superposition is a great asset which can be used generously, as will be seen in the following chapters.

A rather surprising result of superposition is the "speckle pattern," which will be discussed much more seriously later in the book. The speckle pattern is created by the superposition of many waves of similar frequency and with random phases, traveling in different directions. They are important in astronomical imaging because such a situation is created when light waves from a star pass through the turbulent atmosphere. If all the waves have exactly the same frequency, the superposition is stationary with time; otherwise it changes at a rate proportional to the frequency spread.

To illustrate speckle patterns with water waves, consider a group of crayfish, standing at the bottom of a water tank, vibrating their tails and causing wavelets at the surface. Two of them would cause a moving fringe pattern since the wavelets interfere, but their frequencies would not be identical. Many cause a speckle pattern, whose amplitude would be indicated at each point by the vertical motion of floating dust particles. It has "bright" speckles, where the water surface has unusually large vibration amplitude and "dark" speckles where it is stagnant. These can only occur because of chance completely constructive or destructive interference between the various waves. The mean square of the amplitude indicates the energy or intensity of the pattern, and the speckle scale size is λ/α if λ is the wavelength of the crayfish vibration and α is the angular size of the crayfish flock, seen from the surface.

A classical way of representing the addition of vibrations is Fresnel's amplitude–phase vector diagram, where vectors represent each vibration; the length of a vector represents its amplitude and the angle its phase. Superposition means adding the vectors from the different waves. With randomly-phased vibrations, the diagram amounts to a two-dimensional random walk, or "drunkard's walk", a classical exercise in physics (although requiring unreasonable quantities of alcoholic beverage for practical verification). Indeed, it is usually associated with the motion of a drunk pedestrian shifting direction randomly at each step. As can be realized, even without drinking, the final distance from the starting point can amount to any value between zero and the sum of the step lengths, in any orientation (phase). These are the dark and bright speckles. Figure 2.7 shows a representation of this experiment, in which several plane waves coming from random directions overlap.

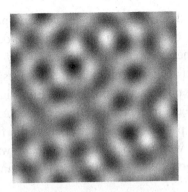

Fig. 2.7. Speckle pattern amplitude resulting from the superposition of 17 real-valued plane waves with random phases traveling in random directions. Black is most negative and white most positive.

2.3 Electromagnetic waves and photons

Electromagnetic waves are disturbances in equilibrium static electric and magnetic fields, which for convenience can be taken as zero in the absence of charges and currents. The waves which are of interest to astronomers are mainly electromagnetic, with frequencies ranging from radio frequencies of the order of 10^6 Hz to γ-rays with frequencies greater than 10^{18} Hz. Today, one could add two additional types of wave – neutrinos and gravitational waves – but these are not the subject of this book. The identification, by means of their velocity, of light and radio waves as manifestations of electromagnetic waves predicted by J. C. Maxwell was one of the greatest triumphs of nineteenth century physics. The basic derivation of electromagnetic waves from Maxwell's equations, as we express it in vector notation[†] today, is given in Appendix A.

The telescope is a device which concentrates the electromagnetic radiation from a distant source onto a detector. The detector is usually an imaging detector, i.e. its output is a position-dependent signal, which is interpreted eventually by the observer's brain as an image of the source. But there can be a lot of processing between the initially received signal and the final image! That is what this book is about. The interaction between the wave and the detector is basically one between the electromagnetic field and the atoms of the device, causing an output electric current which is the received signal.

In astronomy, the quantum nature of electromagnetic waves becomes relevant when we consider the detection of light, since the light signals are almost always very weak. Under these conditions, the energy transport appears not to be

---

[†] The vector notation, invented after Maxwell's days, makes the derivation seem almost trivial! But when you look at Maxwell's original derivation, you see what genius he had to see the waves through such a mass of component equations.

continuous, but has to be described in terms of *photons*, packets of radiant energy, which deliver energy in quantized packets with well-defined values of energy and momentum. The energy of a photon is $E_p = hf = \hbar\omega$, where $h = 2\pi\hbar$, is Planck's constant, and its momentum is E_p/c [†]. Despite the common belief that the existence of the photon was proved by Einstein's (1905) interpretation of the photo-electric effect, in fact this effect can actually be explained without the aid of quantized packets of radiant energy; the first phenomenon which could not be described without the photon concept was the Compton effect (1923), which involves the transfer of both energy and momentum from X-rays to electrons.

When our eyes are fully dark-adapted, they respond to a signal corresponding to about 10 photons, so we can roughly say that the individuality of photons starts to be important when the light is about the weakest that we can see. It is a good analogy to say that a light beam is like a rain shower. The photons are like raindrops: in heavy rain, we just appreciate their general effect in terms of mm rainfall per minute. However in very light rain, drops of approximately uniform size fall in uncorrelated positions at random times; we can certainly calculate average probabilities, but we can predict nothing about when or where a particular raindrop will appear. The "shadow" of an umbrella (observed as a dry patch on the ground) will be completely unrecognizable or very fuzzy until some minimum density of raindrops has fallen; the more detail we want about the edge of the umbrella, the more raindrops we need. Images taken with few photons have similar characteristics; their resolution is severely compromised by small photon counts, and so in astronomy every effort has to be made not to lose photons. In figure 2.8 we show a simulation of the way in which the image of a checkerboard appears out of the noise as the number of random photons used to image it increases.

But photons have more remarkable properties than purely geometrical ones. These arise when we ask how interference occurs in a photon world. To focus on this question we ask, as did G. I. Taylor in 1909, whether a interference pattern can be observed if only one photon can be found in the apparatus at a time. In this experiment Young's screen was replaced by photographic film. Obviously one photon cannot produce more than one black grain on the film, so that the film had to be exposed for three months for sufficient photons to pass through the system and record the pattern. But since the photon had to pass through one slit [‡] or the other – surely it could not pass through both – there was no other photon in the system to interfere with at the time, so no interference pattern should be observed. But the

[†] The latter relationship comes directly from special relativity for a particle with zero mass, so that it seems quite remarkable that it can also be proved for electromagnetic waves from Maxwell's equations.

[‡] Actually, in this experiment Taylor observed the diffraction pattern of a wire, similarly to Young (section 2.1.1), so we should say that the photon had to pass one side of the wire or the other; but the principle is the same.

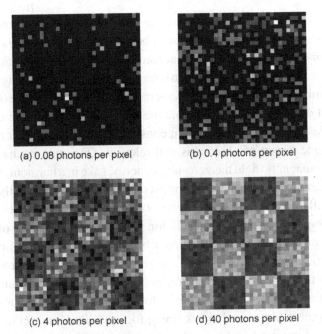

(a) 0.08 photons per pixel

(b) 0.4 photons per pixel

(c) 4 photons per pixel

(d) 40 photons per pixel

Fig. 2.8. Simulation of the development of an image out of noise as the number of photons in each white pixel increases.

interference pattern was indeed observed; although one cannot predict whether a particular grain in the film will turn black, wave theory gives a correct prediction of the *probability* whether the grain at a particular position will turn black or not, thus producing a clear interference pattern when enough photons have passed through the system. It appears that the photon must be considered as a delocalized particle, as delocalized as the corresponding wave is, thus allowing one photon to pass through both slits and interfere with itself; it only becomes localized when it is absorbed by the film. There are still many arguments on this subject centering around the question: at what point does the photon manifest itself as an event in the probability distribution? But in linear optical systems, the correct answer can always be obtained by considering the light as a wave phenomenon until the last moment, when it interacts with the film (or other detector). Then, it is considered as an electromagnetic perturbation to the molecular potentials of the film materials, which can cause an observable change (dissociation of a silver halide molecule, in the case of film) with a certain calculable probability. For this, the detector has to be described by quantum mechanics. If there is no electromagnetic field, such as at the zeros of the interference pattern, there is no dissociation and the film remains clear. But surely even weak light, which should only create an event with very small probability at a given point, might occasionally produce events simultaneously[†] at several points where the probabilities are not absolutely zero. The chance of this happening turns out to be exactly the same as the chance that several photons arrive simultaneously, according to the statistics of random events; so there is no way to distinguish between interpretations. In fact moving the quantum properties from the radiation to the detector makes no difference in the linear system. In the context of this book, it means that the description of the wave optics can be carried out classically, leaving quantum considerations to the detectors.

However, one aspect of photons is still relevant. Quantum mechanics predicts that the electromagnetic field has zero-point energy. Like in a harmonic oscillator, or any bound quantum system, there is a lowest energy (not zero) which the system can have, essentially arising from the uncertainty principle. This is called the *vacuum field* and corresponds to one-half of a photon in every localized mode of the system. Because it is only half a photon, it cannot be absorbed and thus eliminated! The electromagnetic fields associated with the zero-point energy can have exactly the same effect as those we want to measure, and as a result any measurement is bound to be noisy to a certain extent. This sets a lower limit to spontaneous dissociation in the photographic film, which gives rise to fog, or to dark current in a detector.

This chapter has introduced several basic ideas which will be used continuously throughout the book. In the following chapters, we shall amplify these discussions from a quantitative point of view.

References

Beaulieu, J.-P., D. P. Bennett, P. Fouqué, *et al.* (2006). *Nature*, **439**, 437.
Magie, W. F. (1935). *A Source Book in Physics*, New York: McGraw Hill.
Paczynski, B. (1986). *Astrophys. J.*, **304**, 1.
Schneider, P., J. Ehlers and E. E. Falco (1992). *Gravitational Lenses*, Berlin: Springer.
Udalski, A., M. Jaroszynski, B. Paczynski et al., (2005). *Astrophys. J*, **628**, L109.

[†] In this context "simultaneously" is used to mean "within a very short but finite time interval, short compared with any other time in the problem."

3

Interference, diffraction and coherence

3.1 Interference and diffraction

Following the qualitative introduction in the previous chapter, we continue with a more detailed discussion of interference. When two or more waves arrive at a point from different sources, or from the same source by different routes, they can interfere. In most cases, interference occurs when the waves originate from the same source (e.g. a star) but arrive at a detector or screen by different routes (e.g. via different telescopes, or different subapertures of the same telescope)[†]. An interference pattern then occurs as a result of superposition of the waves from the various sources or routes. Roughly, the term *interference* is used to describe the situation where the different routes are separate, and *diffraction* for the case where the different routes form a continuum, but there are cases (such as the diffraction grating) which don't support this categorization. In any case it is only semantic and doesn't have any real physical significance; it is the same as the difference between a sum and an integral.

As we pointed out in section 2.2.2, Huygens' construction allowed one to see qualitatively how a wave propagates in the far field, but seemed to miss features arising from interference. That is because it is not easy to see from the geometry how the *intensity* varies along a wavefront. With relatively little trouble, the approach can be made more quantitative by defining the wave propagation algebraically rather than figuratively. As a further refinement, it has been made almost rigorous by Kirchhof, who was able to formulate the same principle as a boundary value problem; this approach will not be used here but can be found in numerous books devoted to basic optical theory (e.g. Born and Wolf 2000; Goodman 1996; Lipson et al. 1995).

[†] However, we shall also come across cases where the interference is between waves from completely different sources, such as ISI (section 8.5.5) where it is between starlight and the radiation from a laser. This situation will be discussed further in section 4.4.1.

3.1.1 Interference and interferometers

Although sometimes interference occurs naturally, such as in a thin film of oil on a wet road or within the periodic structure of the feathers of a sun-bird, we generally construct an instrument to carry out an interference experiment and use it to measure something. This is called an *interferometer*. There are specialist texts on interferometry in general, for example Hariharan (2003) and Steel (1983), so we will limit our discussion here to one particular interferometer which is closest to the practical needs of astronomy. This is the Michelson interferometer, which amongst other things was instrumental in putting the theory of relativity on an experimental footing, and earned Albert A. Michelson the Nobel Prize in 1907. The Michelson interferometer, as Michelson constructed it, is shown in figure 3.1, and consists basically of two mirrors M_1 and M_2, a plane-parallel "beam-splitter" which reflects about 50% of the light falling on it, the rest being transmitted, and a glass "compensator" plate of the same glass and having the same thickness as the beam-splitter plate. The configuration shown in the figure is such that a light wave incident on the beam-splitter is half reflected and half transmitted. Each of these two waves is reflected back again by one of the two mirrors, and they are recombined at the same beam-splitter surface. There are two recombined waves, one of which continues out of the interferometer in the direction A and the second returns in the direction B which is approximately back in the direction of the source. The mirrors in the figure have been set at a slight angle to the axes so as to make the A and B exits clearer, but this is not necessary in practice. The compensator plate is introduced into one of the waves at the same angle as the beam-splitter so that the two interfering waves at either A or B have both passed three times through the same thickness of the same glass. Michelson and others did a great variety of experiments with this simple set-up, and it is the basis of several of the astronomical interferometers to be discussed later.

Let the amplitude reflection coefficient of the beam-splitter be R and its transmission coefficient T. If there are no losses, energy is conserved and $T^2 + R^2 = 1$. T and R may be different for the two polarizations of light incident on the beam-splitter; this is often so since the incidence is at about $45°$ to the normal. Then we denote the optical path between the point at which the waves separated when they first met the beam-splitter to the point at which they recombine on their second meeting with the beam-splitter as l_1 by one route through mirror M_1, and l_2 by the route through mirror M_2. The *optical path* is defined as the physical path multiplied by the refractive index of the medium in which it is propagating. Most of the path is in air but some parts are within glass. Now consider the recombined beam at A. If the original wave arriving at the beam-splitter was a plane wave with field $E_0 \exp[i\omega t]$, then the wave returning after the path through M_1 has field

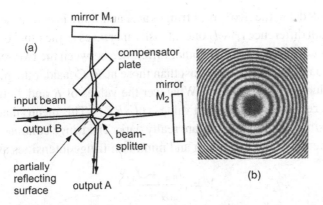

Fig. 3.1. The Michelson interferometer: (a) optical layout; (b) a typical fringe pattern from an extended source, when the configuration of figure 3.2 (b) is used.

$E_0R \exp[i(\omega t - k_0l_1)]$ and the wave through M_2 has field $E_0T \exp[i(\omega t - k_0l_2)]$. At exit A, the former wave has been further transmitted by the beam-splitter and the latter reflected, so that their respective fields are $E_0RT \exp[i(\omega t - k_0l_1)]$ and $-E_0TR \exp[i(\omega t - k_0l_2)]$[†], so that on superposition we have the sum of these fields with intensity

$$I_A = |E|^2 = E_0^2 R^2 T^2 |\exp(i\omega t - ik_0l_1) - \exp(i\omega t - ik_0l_2)|^2$$
$$= 2E_0^2 R^2 T^2 \sin^2[k_0(l_1 - l_2)/2] \,. \tag{3.1}$$

This gives us sinusoidal interference fringes, which can be investigated by sampling them as one of the mirrors is moved, which changes l_1 or l_2. At the other exit, B, the first wave is reflected a second time from the same side of the beam-splitter, and the second beam transmitted a second time, giving

$$I_B = E_0^2 |R^2 \exp(i\omega t - ik_0l_1) + T^2 \exp(i\omega t - ik_0l_2)|^2$$
$$= E_0^2 \{(R^2 - T^2)^2 + 2R^2T^2 \cos^2[k_0(l_1 - l_2)/2]\} \,. \tag{3.2}$$

Notice that the fringe patterns at the two exits are in antiphase; when one is weakest, the other is brightest. Summing the two outputs we get

$$I_A + I_B = E_0^2 \{(R^2 - T^2)^2 + 2R^2T^2[(\sin^2 + \cos^2)[k_0(l_1 - l_2)/2]]\}$$
$$= E_0^2(R^2 + T^2) = E_0^2 \,; \tag{3.3}$$

energy is conserved if the beam-splitter is lossless!

[†] The minus sign arises because one wave was reflected from one side of the beam-splitter surface and the other from the other side. This is a subtle point, discussed further in various texts such as Lipson et al. (1995). It arises because time reversal applies to the wave propagation equations if there is no absorption. It is necessary in order to conserve energy, as we shall see. An example would be reflection at a bare glass surface for which $R = (n_1 - n_2)/(n_1 + n_2)$ from one side, and $R = (n_2 - n_1)/(n_1 + n_2)$ from the other side.

As we pointed out, the two sets of fringes at A and B are in antiphase, so when we change the path difference $l_1 - l_2$ one intensity increases as the other decreases, and the sum stays constant. There is another difference between the two sets of fringes: those at A usually have better contrast than those at B. Consider the maximum and minimum values of intensity at A. Whatever the values of R and T, the minimum value of I_A is zero and its maximum value is $(E_0 RT)^2$. This means that the contrast is very high, since the dark regions are really dark. The fringe contrast, or *visibility*, is defined in terms of the maximum and minimum fringe intensities as

$$V = \frac{I_{max} - I_{min}}{I_{max} + I_{min}}, \tag{3.4}$$

which essentially expresses this point. At A, the fringe visibility is unity, whatever the values of R and T. However, at B the fringe visibility is less. Substituting the values of $I_{max} = (R^2 - T^2)^2 + 2R^2T^2 = (R^2 + T^2)^2$ and $I_{min} = (R^2 - T^2)^2$ into (3.4) we get

$$V = \frac{2R^2T^2}{R^4 + T^4}. \tag{3.5}$$

This is unity only if $R^2 = T^2$, which is true if the beam-splitter divides the energy equally between the two arms of the interferometer. For the lossless beam-splitter, this means $R = T = \sqrt{2}$. Otherwise the visibility is less than unity; the fringe contrast is reduced. The fringes at A are generally used in a Michelson interferometer for two reasons: first, that output is more convenient; second, the visibility is always high. Michelson pioneered the use of fringe visibility to get information on the source size in his stellar interferometer; but that story is told later (section 8.1).

The spatial fringe patterns obtained with a Michelson interferometer depend on the exact configuration of the mirrors. It is most easily explained by considering the relative positions of a particular point on the source as it is seen through the beam-splitter (twice) and the mirrors M_1 or M_2, respectively (figure 3.2, where the beam-splitter has been considered as infinitely thin and the compensator plate can therefore be ignored). This gives us two coherent point images I_{B1} and I_{2B} which then interfere. If the two points lie side-by-side, the interference pattern is like Young's fringes, and a set of parallel fringes is obtained. I_{B1} and I_{2B} are side-by-side when $l_1 = l_2$ and the mirrors are at slightly different angles to the axes. On the other hand, if $l_1 \neq l_2$, I_{B1} might lie behind I_{2B} and a circular pattern of fringes ("Newton's rings") is seen, as in figure 3.1(b). The scale of the pattern depends on the actual distance of the source from the interferometer. In astronomical interferometry, a configuration similar to this is often used (see section 8.3.7) and in order to get a uniform single fringe across the field of view (so that all photons contribute

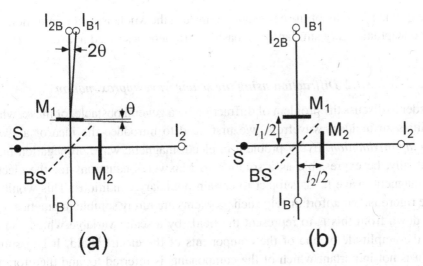

Fig. 3.2. The two virtual images I_{2B} and I_{B1} of a source point S as seen through the mirrors M_1, M_2 and beam-splitter BS of a Michelson interferometer. Image I_{2B}, for example, is formed by reflecting S first in M_2, giving image I_2, and then reflecting I_2 in BS. The fringe patterns result from the interference between the two virtual images. In (a) the two images are side-by-side, and equidistantly spaced straight fringes are seen; in (b) they are one behind the other, and the concentric ring interference pattern is like figure 3.1(b).

to the pattern with the same phase), the two interfering sources must be at the same effective distance from the beam-splitter. This is often called the "Michelson configuration" or "pupil–plane interference"; it is characterized by the fact that the output field is uniform in intensity at any given time, and the fringes are scanned by changing $l_1 - l_2$ dynamically. With appropriate optics, the whole output field can then be sensed with a single-pixel detector. In many cases, the beam-splitter has different values of R and T for different polarizations of light. Each polarization has to be treated separately and the results combined. Sometimes, poor fringe contrast is obtained for unpolarized light, and the contrast can be improved by polarizing the input either in the plane of incidence or normal to it. The use of polarizers, however, results in loss of precious light, and astronomical interferometers are usually designed carefully so that the interfering beams have the same polarization and polarizers are not necessary to get good contrast. Some examples are given in section 8.5.

There are many other types of interferometer in common use, but the above example will be sufficient for this book. The Michelson interferometer, as described, has many applications in fields such as optical testing and fundamental quantum optics and is the basis of Fourier spectroscopy, a very powerful method of measuring absorption and emission spectra. Although it is not directly used in astronomical interferometry, many of the existing systems use the Michelson configuration, and are conceptually very similar to the basic instrument described above.

3.1.2 Diffraction using the scalar wave approximation

In order to discuss the problem of diffraction by a general obstacle, or mask, which modifies an incident wavefront, we first need to introduce the idea of a *scalar wave approximation*. Every problem in electromagnetic wave propagation must, eventually, be expressible as a solution to Maxwell's equations for the electric and magnetic wave fields subject to certain boundary conditions. This would be quite rigorous, but unfortunately such problems are rarely soluble in practice. One step down from this is to represent the fields by a scalar variable which can be, say, the amplitude of one of the components of the electric field. It is assumed that it is not important which of the components is referred to, and therefore the approach implicitly ignores polarization-dependent effects. In practice, this limits the usefulness of the scalar wave approximation to situations where the size of the objects involved is $\gg \lambda$, since edges or discontinuities which are anisotropic are a major source of polarization modification, affecting the parallel and normal polarizations differently out to a range of order λ. Essentially, a criterion for "believing" the scalar wave approach should be that the value of edge-length $\times \lambda$ divided by the total area should be considerably less than unity. For example, it should not work well for diffraction by a grating with spacing comparable to λ; despite this, it does indeed give many qualitatively useful results even in this region. With the scalar wave approximation alone, Kirchhof created his rigorous interpretation of Huygens' principle. Here, we shall treat the problem less rigorously, and interpret Huygens almost literally, replacing his wavelets by mathematical spherical waves with appropriate phase and amplitude. This turns out to be a very fruitful method and can be used to derive most results of importance in the theory of diffraction by fairly large obstacles. One place, however, where the applicability of the scalar wave approximation has to be considered seriously is in apodization and coronagraphy (chapter 10), where small inaccuracies may be critically important.

Let us consider an example in two dimensions, in which an incident plane wave propagating in the z direction is limited by an aperture mask lying within some limiting region $-H < x < H$ symmetrical about the origin O, as shown in figure 3.3. A more complete version of this argument in three dimensions is given in Appendix A. Outside the region $2H$, no light is transmitted. Within it, the wavefront emerging has an amplitude $f(x)$ which can in principle be complex, $f(x) = |f(x)| \exp[i\phi(x)]$. In other words, the aperture mask might have some phase-changing mechanism. We observe the light reaching a general point Q on a screen at distance $L \gg H$; then angles such as ϕ are small. Now we assume that every point P at x on the emerging wavefront acts as a new source of a spherical wave, as taught by

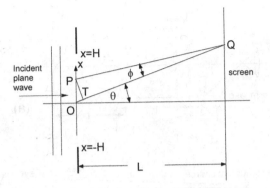

Fig. 3.3. Fraunhofer diffraction by an aperture, using Huygens' principle. When $|x| < H \ll L$, ϕ is small and $OQ - PQ = OT \approx x \sin \theta$.

Huygens. What are the propagated fields at the point Q? The propagation distances are PQ so the phase delays are $k_0 PQ$ and the wave amplitude for the spherical wave is proportional to $PQ^{-1} f(x_P) \exp[ik_0 PQ]$. Now from the figure you can see that when $|x| < H \ll L$, $PQ = OQ - OT \approx OQ - x \sin \theta$ and for smallish angles θ, $PQ \approx L$. Thus, integrating for all points x we have for the amplitude $A(\theta)$ at Q,

$$A(\theta) = \frac{1}{L} \exp(ik_0 OQ) \int_{-H}^{H} f(x) \exp(-ik_0 x \sin \theta) dx. \tag{3.6}$$

Remembering that $f(x)$ is zero outside $(-H, H)$, the limits of the integral can be replaced by $\pm\infty$ and the integral takes on a well-studied form known as the *Fourier transform*:

$$F(u) = \int_{-\infty}^{\infty} f(x) \exp(-iux) dx. \tag{3.7}$$

The observed amplitude $A(\theta)$ is therefore proportional to $F(u)$, where u, called the *spatial frequency*, is in the present case equal to $k_0 \sin \theta$, and thus corresponds directly to the position of Q. The dimensions of spatial frequency are [length]$^{-1}$. For readers not familiar with Fourier transforms, some of their properties are discussed in Appendix A. When we observe the intensity $|A(\theta)|^2$ of the field at Q, the phase term $\exp(ik_0 OQ)$ preceding the integral is irrelevant, because its modulus is unity. The *far-field diffraction pattern*, which has been calculated here, is usually known as the *Fraunhofer diffraction pattern* and has played an important part in the development of physical optics[†]. It is usual, in experimental work in optics, to take L

[†] The definition of the "far field" can be made more rigorous by using the cosine theorem on the triangle OPQ: $PQ^2 + x^2 + 2xPQ \sin \theta = L^2 \cos^2 \theta$ and calculating the path difference $PQ - L \cos \theta$ by a Taylor series expansion. Then you can see that the second-order term in the phase, which is neglected in this argument, is $k_0 x^2 \cos \theta / 2PQ$. The formal condition for the "far field" is that this second-order phase term be $\ll 1$ when x^2, $\cos \theta$ and PQ^{-1} have their maximum values, namely H^2, 1 and L^{-1}, respectively. This gives $L \gg k_0 H^2 / 2$.

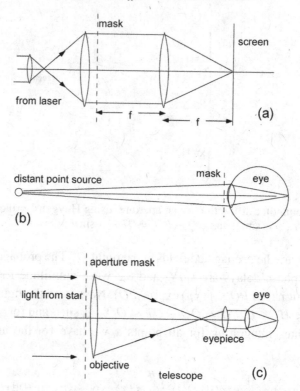

Fig. 3.4. Three experimental arrangements for observing Fraunhofer diffraction patterns: (a) with an expanded laser beam illuminating the mask, and a converging lens which gives the diffraction pattern in its focal plane; (b) visually, viewing a distant point source of monochromatic light and putting the mask directly in front of the eye pupil; (c) a point star observed by a telescope, where the mask is the telescope aperture.

effectively to ∞ and to observe in the focal plane of a converging lens of focal length F. Moreover, it can be shown (Appendix A) that the phase prefactor $\exp(ik_0 O Q)$ can be canceled by placing the lens a distance F after the aperture plane, but this is not necessary in order to observe the diffraction pattern intensities. The relationship is then exact, in the framework of the scalar wave approximation. As you can see, on the one hand it represents a very simple experimental situation, while on the other it can also be described mathematically by a Fourier transform. The relationship between the two has been very fertile ground for fundamental advances in fields such as crystallography, cryptography and analog image processing. Several experimental methods of observing or recording Fraunhofer diffraction patterns are illustrated in figure 3.4.

As a final point, we shall point out, without proof, what is added by Kirchhof's more rigorous treatment. The main addition is that (3.6) should be multiplied by a factor i/λ, indicating that the source of a Huygens wavelet is in fact phase-shifted by $\pi/2$ from the wavefront it represents. A second point is that there is an angular factor $(1 + \cos\theta)/2$, which is rarely of any great importance since the angles θ are generally small when the scalar wave approximation is valid. But the fact that this factor is zero when $\theta = \pi$ does explain why Huygens was correct in ignoring the backwards-propagated part of his wavelets!

3.1.3 Fraunhofer diffraction patterns of some simple apertures

We don't intend here to write a treatise on diffraction patterns; this subject is dealt with in detail in many texts (e.g. Goodman 1996; Lipson et al. 1995). But just in order to complete the picture, we shall give a few examples, which will be important in the context of astronomical interferometry. Apertures are usually two-dimensional (functions of x and y), and the extension from a one-dimensional to a two-dimensional Fourier transform is straightforward. In the following examples, we refer the reader to Appendix A for details of the Fourier transform mathematics.

Young's slits

The classic experiment of Young, with two slits each having width $2b$ separated by $2a$, can be described by an aperture transmission function in one dimension

$$f(x) = 1 \qquad \text{when } -a - b < x < -a + b,$$
$$f(x) = 1 \qquad \text{when } a - b < x < a + b,$$
$$f(x) = 0 \qquad \text{otherwise} \tag{3.8}$$

which is easily described by the convolution (denoted by $\star$; see Appendix A)

$$f(x) = [\delta(x + a) + \delta(x - a)] \star \text{rect}(x/b). \tag{3.9}$$

The Fourier transform $F(u)$ of this function and its intensity $|F(u)|^2$ are

$$F(u) = 4b \cos(ua) \, \text{sinc}(ub), \tag{3.10}$$
$$|F(u)|^2 = 16b^2 \cos^2(ua) \, \text{sinc}^2(ub), \tag{3.11}$$

where the function $\text{sinc}(x) \equiv \sin(x)/x$. Remember that $u = k_0 \sin\theta$ and corresponds to the position of the observation point Q in figure 3.3. The form of this function is a set of cosine fringes with period in u equal to $2\pi/a$. The fringes are modulated by a $\text{sinc}(ub)$ envelope (figure 3.5a). Since b must necessarily be less than a, the period of the cosine has to be smaller than that of the sinc, and the number of cosine fringes ($\cos(ua) = \pm 1$) within the central peak of the sinc is $2a/b$. Fringes of this type, showing both the fringes and the envelope, can be seen in figure 2.1. From the photograph in figure 2.3, from which one can judge that in figure 2.2 the ratio b/a between the slit width and the separation is 2–2.5.

Young's slits with a phase difference between them

It often occurs in astronomical interferometry that the atmosphere or object geometry causes the illumination at the two slits to have phases differing by an amount we shall define as 2Δ. The diffraction pattern is then modified by a fringe shift relative to the envelope as follows. The aperture transmission function can now be written in a symmetrical form:

$$f(x) = e^{-i\Delta} \quad \text{when } -a-b < x < -a+b,$$
$$f(x) = e^{i\Delta} \quad \text{when } a-b < x < a+b,$$
$$f(x) = 0 \quad \text{otherwise.} \tag{3.12}$$

This is described by the convolution

$$f(x) = [e^{-i\Delta}\delta(x+a) + e^{i\Delta}\delta(x-a)] \star \text{rect}(x/b). \tag{3.13}$$

The Fourier transform and its intensity are:

$$F(u) = 4b\cos(ua - \Delta)\,\text{sinc}(ub), \tag{3.14}$$
$$|F(u)|^2 = 16b^2\cos^2(ua - \Delta)\,\text{sinc}^2(ub), \tag{3.15}$$

indicating that the fringe phase has been shifted by Δ (figure 3.5 c). Michelson specifically referred to the fringe shift introduced by atmospheric fluctuations in his observations with his first stellar interferometer (section 5.1). When this diffraction pattern is used in stellar interferometry to determine the contrast and phase shift of fringes produced by interference between the waves illuminating two subapertures, it is generally referred to as the "Fizeau" or "image-plane" configuration, since the image of a distant source is also formed in the focal plane, and Fizeau was the first to suggest using the fringes for astronomical measurements.

A square aperture

A two-dimensional square aperture with side $2a$ is represented by the product $f(x, y)$ of two functions, $f_1(x)$ and $f_2(y)$:

$$f_1(x) = 1 \quad \text{when } -a < x < a,$$
$$f_2(y) = 1 \quad \text{when } -a < y < a,$$
$$0 \quad \text{otherwise.} \tag{3.16}$$

The transform $F(u, v)$ is the product of the transforms $F_1(u)$ and $F_2(v)$, namely $4a^2\text{sinc}(ua)\text{sinc}(va)$. The intensity of the diffraction pattern is the square of this; notice that its axial value increases with a like $16a^4$ because both the aperture

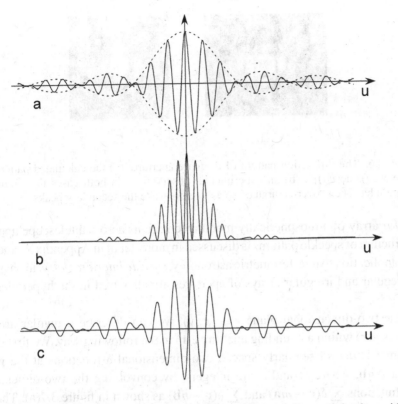

Fig. 3.5. The Fraunhofer diffraction pattern of a pair of slits each having width $2b$ separated by $2a$ when $a = 6b$: (a) amplitude; (b) intensity; (c) amplitude when there is a phase difference $2\Delta = 1$ rad between the slits.

area increases and the width of the pattern in both directions decreases. Another useful feature of this pattern, which will be employed in section 10.3 as the basis of a method to discover extrasolar planets, is that along the diagonals $u = \pm v$ the intensity falls off very fast, like $\mathrm{sinc}^4(ua)$, so that there is very little light along the diagonal except in the central peak (figure 3.6). For example, at the point $u = v = 3.5\pi/a$, which is a peak on the diagonal, the intensity is about 10^{-4} of that at the origin.

An array of apertures

The diffraction pattern of a regular array of identical objects or apertures was the starting point for the widespread application of Fourier theory to crystallography, since this object represents the electron density in a perfect crystal, and its X-ray diffraction pattern is the Fraunhofer pattern. We bring it up here as an example, not for that reason, but because it provides a simple example from whose extension we can understand features of a speckle pattern, which is the diffraction pattern of an

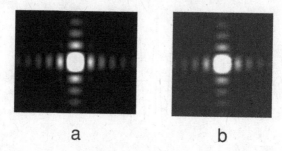

Fig. 3.6. The diffraction pattern of a square aperture: (a) the calculated pattern, $[\text{sinc}(ud)\text{sinc}(va)]^2$; (b) an experimental observation. In both cases the central region has been "over-saturated" so as to emphasize the secondary peaks.

irregular array of atmospherically modulated regions across a telescope aperture. The structure of speckle patterns is discussed in more detail in Appendix A, and their major application to interferometric astronomy, *speckle interferometry*, in chapter 6. Both regular and irregular arrays of apertures are also used in the hypertelescope (chapter 9).

In the two-dimensional plane $\mathbf{r} \equiv (x, y)$, we can describe a regular array of apertures $g(\mathbf{r})$ within a bounding aperture $c(\mathbf{r})$ in the following way. We first define an infinite lattice of regularly spaced two-dimensional δ-functions at the points $\mathbf{r} = m\mathbf{a} + n\mathbf{b}$, where m and n are integers, by convolving the two-dimensional comb functions $\sum \delta(\mathbf{r} - m\mathbf{a})$ and $\sum \delta(\mathbf{r} - n\mathbf{b})$ as shown in figure 3.7(a). The two vectors $\mathbf{a}$ and $\mathbf{b}$ are the *lattice vectors* and may have any non-zero angle γ between them. The two vectors define the *unit cell*, which is a parallelogram with edges $\mathbf{a}$ and $\mathbf{b}$. Next, we limit this infinite array to the region of the bounding aperture by multiplying by $c(\mathbf{r})$. Finally, we convolve the resulting finite periodic array with the unit aperture $g(\mathbf{r})$. These stages are shown in figure 3.7(b),(c), and are represented formally by

$$f(\mathbf{r}) = \left\{ \left[\sum_{-\infty}^{\infty} \delta(\mathbf{r} - m\mathbf{a}) \star \sum_{-\infty}^{\infty} \delta(\mathbf{r} - n\mathbf{b}) \right] \cdot c(\mathbf{r}) \right\} \star g(\mathbf{r}). \qquad (3.17)$$

The Fourier transform of this is

$$F(\mathbf{u}) = \left\{ \left[\sum_{-\infty}^{\infty} \delta(\mathbf{u} \cdot \mathbf{a} - 2\pi m^*) \cdot \sum_{-\infty}^{\infty} \delta(\mathbf{u} \cdot \mathbf{b} - 2\pi n^*) \right] \star C(\mathbf{u}) \right\} \cdot G(\mathbf{u}), \qquad (3.18)$$

where m^* and n^* are integers. Figure 3.8 illustrates the transform in a way which makes it clearer than the mathematics! Then the observed diffraction pattern is $|F(\mathbf{u})|^2$.

The diffraction pattern, as shown schematically in figure 3.8(c) and experimentally in figure 3.9, has the following features. First, it is based on the *reciprocal*

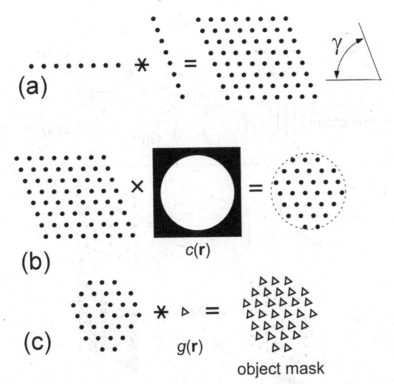

Fig. 3.7. Description of a limited periodic array of finite apertures by means of multiplication and convolution. (a) Two infinite vectors of δ-functions at angles 0 and γ are convolved to give a two-dimensional array of δ-functions. (b) This is multiplied by the bounding-aperture function $c(\mathbf{r})$ (a circle). (c) The resulting finite array of δ-functions is convolved with the individual aperture $g(\mathbf{r})$.

lattice which is the transform of the original infinite lattice and has unit cells[†] which have size inversely proportional to that of the original lattice, are the same shape as those of the lattice but are rotated by 90°. Second, the points of this reciprocal lattice are not ideal δ-functions, but are convolved with the diffraction pattern $|C(\mathbf{u})|^2$ of the bounding aperture. The reciprocal lattice convolved with $|C(\mathbf{u})|^2$ is called the *interference function*. And third, the intensities of the points are determined by the diffraction pattern $|G(\mathbf{u})|^2$ of the repeated apertures, called the *diffraction function*, which is therefore sampled at the reciprocal lattice points. In X-ray crystallography, it is this final point which is most important, because it provides a way of determining the details of the electron density of the unit cell contents, which are the molecules under investigation.

[†] The unit cell is based on reciprocal lattice vectors $\mathbf{a}^*$ and $\mathbf{b}^*$, with $a^* = 2\pi/a \sin\gamma$ and $b^* = 2\pi/b \sin\gamma$, where $\mathbf{a}^*$ is normal to $\mathbf{b}$ and $\mathbf{b}^*$ is normal to $\mathbf{a}$.

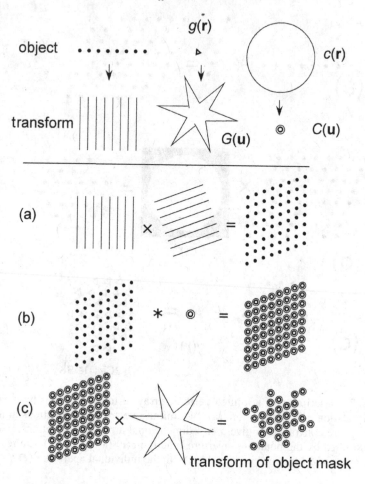

Fig. 3.8. Schematic description of the transform of the array in figure 3.7. The individual transforms of the vector of δ-functions, $c(\mathbf{r})$ and $g(\mathbf{r})$; then (a), (b) and (c) are the transforms of the corresponding processes in that figure.

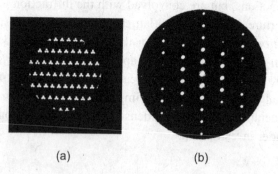

Fig. 3.9. (a) A finite array of apertures and (b) its diffraction pattern.

It is maybe surprising that most of these features go over from a periodic lattice to a random lattice, created for example by atmospheric fluctuations. The details are given in Appendix A, but we shall summarize the results here. In practice the entrance (bounding) aperture of a telescope is modulated by atmospheric disturbances, which can be looked at as patches of uniform but differing phase with typical size r_0 distributed on a random lattice of δ-functions in **r** space. The random lattice changes continuously, but can be "frozen" by a photograph with exposure of a few ms. It transforms to a new random lattice of δ-functions in **u** space. Each point in this random lattice, which is called a "speckle," is then convolved with the diffraction pattern of the entrance aperture of the telescope, and the array of δ-functions is multiplied by an envelope which is the transform of the typical uniform regions; this is a patch with dimensions inversely related to r_0. The atmospheric seeing determines r_0, typically 10 cm in the visible, but greater at longer wavelengths. The patch containing the random array therefore has an angular diameter λ/r_0, which is typically 1–2 arcsec: the "seeing limit." A long-exposure photograph only shows this, since the array of speckles is continuously changing and averages to a fairly uniform function. But a short-exposure photograph shows the individual speckles within this envelope, and as we saw, each one is convolved with the diffraction pattern of the entrance aperture (see figure 5.12). Because of this, individual speckles can give angular resolution corresponding to the whole telescope aperture, which is much better than the seeing limit.

3.1.4 The point spread function

Suppose that a lens, or other imaging instrument, is used to image a point test source. According to geometrical optics, if the lens were ideal, the image would be a point. However, from the point of view of wave optics, this is a Fraunhofer diffraction experiment (figure 3.4c); the point source ensures that the lens is illuminated coherently, and the observed far-field diffraction pattern is therefore that of the limiting aperture, which we represent by a two-dimensional complex transmission function $f(\mathbf{r})$. For optics with axial symmetry, the transform is that of a circular aperture of radius R, represented by $f(\mathbf{r}) = \text{circ}(r/R)$, which is shown in Appendix A to be the Airy disk function $2\pi R^2 J_1(\rho R)/\rho R$, where $J_n(x)$ is the n-th order Bessel function, and $\rho = k_0 \sin\theta = \sqrt{u^2 + v^2}$ is the radial distance in reciprocal space. Its intensity is $4\pi^2 R^4 [J_1(\rho R)/\rho R]^2$, which is illustrated in figure 3.10. In astronomy, where angles are usually small, $\sin\theta$ can be approximated by θ, giving intensity $4\pi^2 R^4 [J_1(k_0\theta R)/k_0\theta R]^2$. If the lens is not ideal, this function is modified; in general, the diffraction pattern of the imaging optics, which is the image of a point object, is called the *point spread function* (PSF). It usually has the form of a strong peak surrounded by some weaker detail, for example the rings in

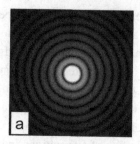 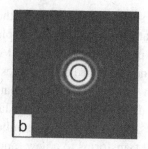

Fig. 3.10. The diffraction pattern of a circular aperture: (a) the calculated pattern, $[2\pi R^2 J_1(\rho R)/\rho R]^2$; (b) an experimental observation. In both cases the central region has been "over-saturated" so as to emphasize the rings.

figure 3.10; the width of the strong central peak determines the resolution of the optical system. For a telescope, the test point object is always at infinity, and so the relevant resolution is angular resolution. The function $J_1(x)$ has its first zero at $x = 3.83$ and so the width of the peak (out to its first zero) is $\rho R \approx k_0 \theta R = 3.83$, indicating an angular resolution $\theta = 3.83/k_0 R = 1.22\lambda/2R$, which is the well-known Rayleigh resolution limit for a telescope. According to this criterion, which is a little pessimistic, two point objects can be resolved only if the central maximum of the PSF around the image of one of them lies outside the first zero of the PSF around the image of the other one. Other criteria, based on the half-width of the peak, or on its differentials, give slightly different resolution limits which are closer to real observations (see, for example, Lipson et al. 1995).

Another point spread function which we shall need later is that due to a thin annular aperture of radius R and width t. This has been shown to have the best Rayleigh resolution of any real positive aperture of maximum size R. The function representing the annulus, $f(\mathbf{r}) = 2\pi R t \delta(r - R)$, has transform $2\pi R t J_0(\rho R)$ and intensity $4\pi^2 R^2 t^2 J_0^2(\rho R)$. This point spread function has a narrower peak, but considerably stronger rings, than has the Airy disk function (figure 3.11).

The concept of the point spread function applies to any imaging optical system, and there is extensive work on the effect of lens aberrations on it. In chapter 6, for example, we will use the concept in connection with a telescope with aberrations resulting from the atmospheric turbulence, which results in serious degradation of the resolution of the telescope. Knowing the point spread function of an optical system allows us to create an image of any incoherent object, by convolution. To demonstrate this, first consider the ideal image of the object, with a given magnification and orientation. Since each point on the object can be considered as an independent source, which does not interfere with any other neighboring point, it is clear that each point on the image will be a replica of the point spread function centered on the corresponding point in the ideal image. As a result, the observed

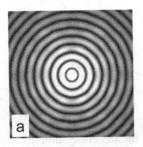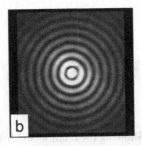

Fig. 3.11. The diffraction pattern of an annular aperture: (a) the calculated pattern $[\pi R t J_0(\rho R)]^2$, on the same scale as that of figure 3.10; (b) an experimental observation.

image is seen to be the convolution of the ideal image function with the point spread function.

3.1.5 The optical transfer function

Another concept which is widely used in optical imaging studies is the *optical transfer function* (OTF). Its absolute value is usually called the *modulation transfer function* (MTF). It is closely related to the point spread function, being its Fourier transform. The optical transfer function tells us how clearly we can image a periodic target with a given spatial frequency. It always has zero value when the spatial frequency is greater than the inverse of the resolution limit, since imaging a grating with period smaller than the resolution limit would require a resolution better than the instrument is capable of.

We can visualize the OTF by supposing that, instead of using a point source, we have an object which is an incoherently illuminated screen with intensity sinusoidally varying in space (like a sinusoidal fence) with a specified spatial frequency. The sinusoidal intensity at the source is assumed to have maximum value 1 and minimum zero. The image of this will also be sinusoidally varying, but may have different modulation and phase from the ideal image. The *modulation*, which is the same as the visibility of interference fringes (section 3.1.1), is defined by (3.4) in terms of the maximum and minimum intensities of the image. The phase Δ is measured relative to that of the the ideal image, so that the complex OTF is $M \exp(i\Delta)$. The object therefore had $M = 1$. Now, since the system aperture is represented by the transmission function $f(\mathbf{r})$, the PSF is the square modulus of its Fourier transform, $|F(\mathbf{u})|^2$. It follows that the OTF is the autocorrelation of $f(\mathbf{r})$. In the case of a simple aperture, this function is readily calculable, being the overlap area between the aperture and itself, shifted, as a function of the shift vector S (figure 3.12). If the aperture is more complicated, particularly if the function $f(\mathbf{r})$ is not real

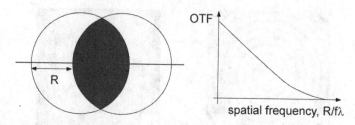

Fig. 3.12. Showing the relationship between the autocorrelation function (overlap area between the aperture and itself, shifted by R) and the optical transfer function. The spatial frequency is related to R by $u = R/f\lambda$, where f is the focal length, in the paraxial approximation.

(i.e. there are phase changes, from the atmosphere, for example), the function is more complicated, but can never be greater than what would be achieved for the same aperture without the phase changes. Thus aberrations always reduce the value of the OTF. Examples of such calculations are given by Born and Wolf (2000). In chapter 4, we meet the OTF under another guise. When we use interference between separated telescopes or subapertures to form an image using aperture synthesis, the spatial frequency vectors sampled are determined by the vectors between all pairs of subapertures. This is the autocorrelation function of a map of the subaperture positions; the OTF is then called the (u, v) diagram, and will be discussed in detail in section 4.1.2. The synthetic PSF is then its Fourier transform.

3.2 Coherent light

3.2.1 The effect of uncertainties in the frequency and wave vector

If we have an extended source of light or a wavefront on which all points in a given region oscillate in precisely the same way, i.e. with the same frequency and known phases, the source is said to be *spatially coherent*. If, moreover, the frequency is so accurately defined that after a long period of time the relative oscillation phases at the points have not changed, it is said to have *temporal coherence*. Clearly, the terms here have to be made quantitative: how accurately is implied in "precisely the same way," how long is a "long period," and how are these related to the size of the "given region?" This will be done in the next section. In general terms, coherent waves can interfere, because the phase difference between them is defined accurately, and incoherent waves cannot.

The diffraction calculations presented in the previous section assumed that the wave incident on the diffracting aperture is both spatially and temporally coherent, which means in practice that it is derived from a single monochromatic point source. Then, the field at every point has the same time variation $\exp(i\omega t)$. We know that this is not always the case; in fact, as we described very qualitatively in section 2.1.2, the whole idea of astronomical interferometry is that we can discover how different the source is from a point by inspecting the diffraction patterns produced using that source. The straightforward approach to this is to consider the source as a dense quasimonochromatic array of point sources; each one is independent and has its own value of ω (within a very small range – hence *quasi*monochromatic). Each one produces its own diffraction pattern centered on its own axis but, because of the frequency differences, the patterns do not interfere with one another. As a result, the observed intensity is just the sum of the intensities of the individual patterns. Now if it turns out that the diffraction patterns from the individual sources are insignificantly displaced one from another, the result of the multitude of sources is just to make the pattern brighter; but if the displacements are significant with respect to details in the pattern, the result is always some form of blurring of the original pattern. We shall treat this problem analytically later in section 3.3, but the following will give a qualitative perspective of the concept of coherence.

3.2.2 Coherent light and its importance to interferometry

The ideas of coherent light have become very prominent since the invention of lasers, which are almost ideal sources of coherent light. But the concept of coherence dates back to the work of Van Cittert in 1934 and Zernike in 1938. Two waves are mutually coherent if the phase difference between them at a particular point in space is constant during a long period of time. If this is so, then interference between the waves will be observed. But of course nothing in the real world can be exactly constant, so we should really ask how much this phase difference changes *during the period of an observation*. If the answer is that the change is much smaller than π, then clear interference effects will be seen. If not, then interference won't be seen. Suppose for example that we have two independent sources, with frequencies ω_1 and ω_2. Then the phase difference between them is changing as $(\omega_2 - \omega_1)t$. So if the observation time (say, the exposure time for a photograph of an interference pattern) is 0.01 second, the difference $(\omega_2 - \omega_1)$ has to be $\ll \pi/0.01 = 314\,\mathrm{s}^{-1}$ for fringes to be recorded. Since at optical frequencies ω is of order $10^{16}\,\mathrm{s}^{-1}$, it follows that no two independent sources could be stable enough (to one part in 10^{13}) for this to be possible; so in practice interference patterns can only be observed when the interfering waves derive from the same source. The situation is different at much lower frequencies; for example, acoustic interference between independent sources, such as beats between a tuning fork and a piano string, are commonplace.

3.2.3 Partial coherence

Even when the two interfering waves apparently have a common origin, the geometry of the situation may result in incomplete coherence. *Partial coherence* can arise when we observe interference from a source which is not really a single source of

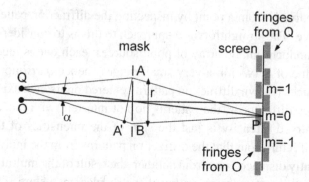

Fig. 3.13. The phase difference between the waves from a point source Q reaching the pinholes A and B depends on their separation r. Drawing A' such that $QA = QA'$, the phase difference is seen to be $k_0 BA' \approx k_0 r\alpha$ for small α. On the screen, the zero-order fringe is at P, where QP passes through the mid-point of the two pinholes. The fringes from O and Q as shown have π phase difference, so that r is about equal to r_c.

fixed frequency. This may be because the source has a fluctuating frequency itself: for example, emission from a molecule in a high-pressure gas, where the random collisions between molecules cause randomly changing Doppler shifts of the frequency. It may also arise if the source is an ensemble of many molecules, each one emitting at a slightly different frequency, depending on its surroundings and velocity. The result is partial coherence, in both the temporal and spatial domains. We shall discuss spatial and temporal coherence separately, in order to make things clearer, but one should remember that both can occur together.

3.2.4 *Spatial coherence*

Partial coherence between neighboring points in space (we'll visualize them as two pinholes separated by distance r) can arise because the illuminating source is not a point, but is actually an array of sources at slightly different spatial positions. Two such points are shown in figure 3.13. The difference between the optical paths from a given source point to the two pinholes results in a phase difference between the waves emerging from them. As we saw in section 3.1.3, the result of this phase difference (there we called it 2Δ) is a shift in the positions of the fringes observed on the screen. The zero-order bright fringe for this element is defined by the locus of the points on the screen where the total path difference between the source and screen via the two pinholes is zero. Similarly, the locus of the points on the screen where this difference is $m\lambda$ defines the m-th order fringe. From the simple geometry of the figure, assuming small angles, one sees that the zero-order fringe has to be on the line from the source going through the mid-point between the pinholes. When there are many sources in the array, what we actually see on the screen is a superposition of fringes from all the sources, each one giving a pattern with shifted fringes; for a large enough source, this will clearly result in a blurring of the pattern. Now a shift of half a fringe corresponds to a path difference of $\frac{1}{2}\lambda$ between the waves received at the two pinholes, so that the criterion for noticeable blurring of the fringes is that the variation in this path difference coming from the various source points should be about $\frac{1}{2}\lambda$. The larger the source, or the larger the separation between the pinholes, the more serious is the blurring. The separation between the pinholes at which the fringes become very unclear is called the *coherence distance*, r_c, of the wavefront, and is clearly dependent on the source size. When the pinholes are separated by a distance larger than r_c, the waves emerging from the two pinholes are incoherent and no interference is observed between them. This picture was envisaged by Michelson, who built the first stellar interferometer, with a distance between the pinholes (apertures in his case) which could be varied up to about 6 m. His analysis was on the lines of the above arguments (see Michelson 1927), and allowed him to measure the angular diameters of several stars by finding at what value of r the fringes disappeared.

The quantitative relationship between the source size and structure and the degree of partial coherence which we use today is summarized in a theorem discovered independently by Zernike and van Cittert; we shall prove this theorem in section 3.3.3. But at this stage, we can estimate the relationship between r_c and the size of the source by considering two incoherently radiating points separated by angle α in the sky which illuminate the plane containing two pinholes A and B. Suppose that one source, O, is on the axis of the optics (normal to the plane) and the second, Q, is at angle α to it (figure 3.13). O illuminates the two pinholes with a plane wave at normal incidence, and so the phase difference is zero. Q gives a plane wave incident at α and so the phase difference is $2k_0r\sin(\alpha/2) \approx k_0r\alpha$. This has the value π (path difference $\frac{1}{2}\lambda$) when $r = r_c = \lambda/2\alpha$, for which situation no fringes will be seen if the sources have equal intensity. This is an estimate based on just two points, but gives the order of magnitude of r_c. We'll see below that for a filled source, r_c is about twice this value, since the intermediate sources are closer than α.

3.2.5 Temporal coherence

As remarked earlier, coherence in general can be measured in both the spatial and temporal domains. Temporal coherence is related to the frequency bandwidth of a source, i.e. its spectral purity. Suppose that a source of nominal frequency ω_0 actually consists of several emitters with slightly different frequencies, all in the range $\omega_0 \pm \delta\omega$. Light from this source is received at a distant point at time $t = 0$. The waves from the various emitters arrive with various phases which are essentially random; what we see is the superposition of these waves. Now we wait a short time τ and look at the superposition again. Any given wave has its phase advanced by

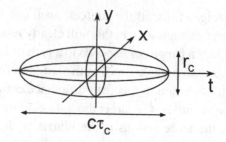

Fig. 3.14. A schematic picture of the coherence region; interference can be observed between points separated in space and time by a vector lying within this region.

$\omega\tau$, so that the various phases in the ensemble are advanced by amounts in the range $(\omega \pm \delta\omega)\tau$. If this range is very small compared to π, what we see will be essentially a continuation of the wave we saw at $t = 0$, as if we had a monochromatic wave of frequency ω_0. But if the range $\delta\omega\tau$ is of the order π or larger, the superposition will be quite different, and in particular its average phase will be unrelated to that of the wave at $t = 0$. If the waves at $t = 0$ and $t = \tau$ have unrelated phases, we can't perform an interference experiment between them, and so they are incoherent. The largest time difference for coherence to exist is about $\tau_c \equiv \pi/\delta\omega$, because after this time the range in phase differences is approximately π and this time is called the *coherence time*. The concept is very relevant to interferometric measurements; if one superimposes two beams with a path difference greater than $c\tau_c$, no interference fringes will be seen even if the two waves are spatially coherent. They also have to be temporally coherent. We can in fact define a region in x, y, ct space within which the two interfering points have to be situated in order to observe interference between them; this *coherence volume* has lateral dimensions r_c in the x, y directions and length $c\tau_c$ in the direction of propagation. In figure 3.14 we picture this as an ellipsoid, but of course its real shape might be much more complicated, as we shall see later. As was pointed out by Hanbury Brown and Twiss (1956), the physical significance of this coherence volume is that, since they can interfere, photons measured within that volume are indistinguishable.

Temporal coherence has other effects too, the main one being to result in fluctuations in the intensity of the wave on a time-scale of τ_c. These will be discussed in section 3.4.1, and are very relevant to intensity interferometry, which is the subject of chapter 7.

3.3 A quantitative discussion of coherence

The various ways in which interferometry has contributed to astronomy, whether radio or optical, are closely linked with coherence theory, which was originally formulated by Van Cittert (1934) and Zernike (1938). Coherence theory leads to a quantitative relationship between the signals received and the angular structure of the stellar source.

3.3.1 Coherence function

The coherence function (Zernike 1938) relates the wave fields received at two neighboring points in space. The word "neighboring" must be considered as relative to the distance from the source, so that in astronomy, stations on opposite sides of the Earth, or even at opposing points on the Earth's orbit, might be called "neighboring." As mentioned above, if the vector joining the points is normal to the propagation vector, or parallel to it, we quantify respectively the concepts of spatial coherence (section 3.2.4) or temporal coherence (section 3.2.5). But in principle, the vector can be in any direction. The *complex degree of coherence* or, more briefly, the *coherence function* γ is a normalized correlation function between the complex wave fields $f(\mathbf{r}, t)$ at the two points $\mathbf{r}_1$ and $\mathbf{r}_2$:

$$\gamma(\mathbf{r}_1, \mathbf{r}_2) = \frac{\langle f(\mathbf{r}_1, t) f^*(\mathbf{r}_2, t) \rangle}{[\langle |f(\mathbf{r}_1, t)|^2 \rangle \langle |f(\mathbf{r}_2, t)|^2 \rangle]^{\frac{1}{2}}}, \tag{3.19}$$

where the averages $\langle ... \rangle$ are taken over time. Note that $|\gamma(\mathbf{r}_1, \mathbf{r}_2)| \leq 1$. If the vector $\mathbf{r}_1 - \mathbf{r}_2$ is parallel to the propagation direction, the two fields are in fact the same field sampled at times differing by $\tau = |\mathbf{r}_1 - \mathbf{r}_2|/c$, and instead of $\gamma(\mathbf{r}_1, \mathbf{r}_2)$ we write $\gamma(\tau)$. We shall assume, although there may be cases where this is not exactly correct, that γ is only a function of the vector difference $\mathbf{r} = \mathbf{r}_1 - \mathbf{r}_2$, because this makes life a lot easier. The physical significance of the coherence function is that it tells us how good an interference pattern we can get between the waves emerging from two pinholes at $\mathbf{r}_1$ and $\mathbf{r}_2$. Putting the pinholes at these positions is, of course, symbolic; there are many different ways in which we can interfere the two signals, but the idea of two pinholes makes it easier to imagine.[†]

3.3.2 The relationship between the coherence function and fringe visibility

The importance of fringe visibility as a measurable parameter was appreciated by Michelson in his early experiments. Now we shall show that visibility of the interference fringes, as defined in section 3.1.1, is directly related to γ. Assume the two waves emerging from the pinholes have mean intensities A_1^2 and A_2^2, and that the functions $f(\mathbf{r}_1)$ and $f(\mathbf{r}_2)$ are normalized. The wave fields are therefore

---

† For example, we could have two optical fibers of equal length, with one end of the first at $\mathbf{r}_1$ and one end of the second at $\mathbf{r}_2$. The other two ends would be side by side, and separated by about a millimeter, radiating onto a screen where the fringes would be seen.

$A_1 f(\mathbf{r}_1)$ and $A_2 f(\mathbf{r}_2)$, so that the coherence function becomes

$$\gamma = \frac{\langle A_1 f(\mathbf{r}_1) \cdot A_2 f^*(\mathbf{r}_2) \rangle}{A_1 A_2} = \langle f(\mathbf{r}_1) f^*(\mathbf{r}_2) \rangle. \qquad (3.20)$$

The interference pattern is a superposition of the two fields with a phase difference ϕ between them, in which ϕ varies from point to point in the fringe pattern. The total field is

$$g(\phi) = A_1 f(\mathbf{r}_1) + A_2 f(\mathbf{r}_2) \exp(i\phi). \qquad (3.21)$$

The instantaneous intensity of the interference pattern is

$$\begin{aligned}
I(\phi) = |g(\phi)|^2 &= |A_1 f(\mathbf{r}_1) + A_2 f(\mathbf{r}_2) \exp(i\phi)|^2 \qquad (3.22) \\
&= A_1^2 |f(\mathbf{r}_1)|^2 + A_2^2 |f(\mathbf{r}_2)|^2 \\
&\quad + A_1 A_2 [|f(\mathbf{r}_1) f^*(\mathbf{r}_2)| \exp(i\phi) \\
&\quad + |f^*(\mathbf{r}_1) f(\mathbf{r}_2)| \exp(-i\phi)]. \qquad (3.23)
\end{aligned}$$

Taking time averages, this can be written

$$\begin{aligned}
I(\phi) &= A_1^2 + A_2^2 + A_1 A_2 [\gamma \exp(i\phi) + \gamma^* \exp(-i\phi)] \qquad (3.24) \\
&= A_1^2 + A_2^2 + 2A_1 A_2 |\gamma(\mathbf{r}_1, \mathbf{r}_2)| \cos(\phi + \Delta), \qquad (3.25)
\end{aligned}$$

where $\gamma \equiv |\gamma| \exp(i\Delta)$. This represents the interference pattern; the cosine gives fringes, because ϕ varies from point to point on the screen, and Δ specifies the position of the central fringe with respect to the axis. But notice that the fringes may not be of high contrast, either because the minimum value of I is not zero, or because $A_1 \neq A_2$. Substituting the maximum and minimum values of $I(\phi)$ from (3.25) into the definition (3.4), we find that the visibility of the fringes has value

$$V = \frac{2A_1 A_2}{A_1^2 + A_2^2} \cdot |\gamma(\mathbf{r}_1, \mathbf{r}_2)|. \qquad (3.26)$$

If the two interfering waves have equal intensities, i.e. $A_1 = A_2$, $V = |\gamma|$. But in general, if the intensities A_1^2 and A_2^2 are measured simultaneously with the fringe pattern, the complex value of γ can be found by measuring V and Δ (see figure 8.11). Figure 3.15 shows three examples of interference patterns observed between sources with different degrees of coherence.

3.3.3 Van Cittert–Zernike theorem

The value of the coherence function in interferometric imaging comes from its relationship to the source structure. We saw in section 3.2.4 that if the source is not a point source, the fringe contrast is reduced, because the coherence function is less

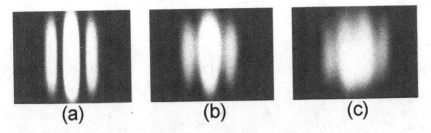

Fig. 3.15. Fringes observed between sources with degrees of coherence (a) $\gamma = 0.97$, (b) 0.50 and (c) -0.07. Notice in (c) that there is minimum intensity on the center line, indicating that $\Delta = \pi$.

than unity. The exact quantitative relationship between the fringe visibilty and the source structure is called the *Van Cittert–Zernike theorem*. It is the cornerstone of aperture synthesis using multiple apertures or telescopes, in any wavelength region. We define the optical axis of a telescope system as the z-axis. Suppose that we have a stellar source situated close to the z-axis. It can be described by its intensity $I(\theta_x, \theta_y)$ as a function of angle with respect to that axis. Although the source probably has a wide spectrum, we consider now only the component at wavelength $\lambda = 2\pi / k_0$; other wavelengths can be added by superposition. It is natural to use angular coordinates (θ_x, θ_y) to describe the source structure, but angles are not very good coordinates because they do not add like vectors. It is more correct to use *direction cosines* $\vec{\ell} \equiv (\ell, m, n)$ to indicate a direction in space[†] (figure 3.16). Since the source is very close to the z-axis, $\ell = \sin\theta_x \approx \theta_x$ and $m = \sin\theta_y \approx \theta_y$; then $n \approx 1$ since $\ell^2 + m^2 + n^2 = 1$. We shall write the intensity $I(\ell, m)$ of the source in terms of its amplitude $a(\ell, m)$, despite the fact that all we know is that $I = \langle |a|^2 \rangle$. The wave reaching the observatory and measured in the (x, y) plane $z = 0$ is a superposition of plane waves originating from each point on the source and traveling in the direction $-\vec{\ell}$ (i.e. wavevector $-k_0\vec{\ell}$). The elementary source $a(\ell, m)\mathrm{d}\ell\,\mathrm{d}m$ gives a contribution in the $z = 0$ plane (figure 3.17)

$$\mathrm{d}^2 f(k_0 x, k_0 y) = a(\ell, m) \exp[\mathrm{i}(\omega t + k_0 \vec{\ell} \cdot \mathbf{r})]\mathrm{d}\ell\,\mathrm{d}m , \qquad (3.27)$$

which is integrated to give

$$f(k_0 x, k_0 y) = \iint a(\ell, m) \exp[\mathrm{i}(\omega t + k_0 \vec{\ell} \cdot \mathbf{r})]\mathrm{d}\ell\,\mathrm{d}m \qquad (3.28)$$

$$= \exp[\mathrm{i}\omega t] \iint a(\ell, m) \exp[\mathrm{i}(k_0 \ell x + k_0 m y)]\mathrm{d}\ell\,\mathrm{d}m. \qquad (3.29)$$

The integrals can be considered as infinite although in practice one only needs to integrate over the source region where $a(\ell, m) \neq 0$. Then (3.29) is the

[†] In the astronomical case, the angular regions of interest are so small that this change is really not important.

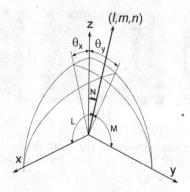

Fig. 3.16. Direction cosines (ℓ, m, n) of a vector. The components ℓ, m and n are the cosines of the angles shown as L, M and N.

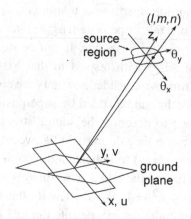

Fig. 3.17. Geometry of the proof of the Van Cittert–Zernike theorem.

two-dimensional Fourier integral

$$f(k_0 x, k_0 y) = \exp[i\omega t]A(-k_0 x, -k_0 y). \qquad (3.30)$$

Now consider the spatial coherence function γ in the (x, y) plane. To simplify the mathematics, we again normalize the time-averaged fields to unit intensity $\langle|f(\mathbf{r})|^2\rangle_t = 1$, where $\langle...\rangle_t$ indicates average over time. Then $\gamma(\mathbf{r})$, (3.19), can be written simply as the time-averaged convolution (autocorrelation)

$$\gamma(\mathbf{r}) = \langle f(\mathbf{r}) \star f^*(-\mathbf{r})\rangle_t , \qquad (3.31)$$

and in terms of the dimensionless coordinates $k_0\mathbf{r}$

$$\begin{aligned}\gamma(k_0\mathbf{r}) &= \langle f(k_0\mathbf{r}) \star f^*(-k_0\mathbf{r})\rangle_t \\ &= \langle A(-k_0\mathbf{r}) \star A^*(k_0\mathbf{r})\rangle_t .\end{aligned} \qquad (3.32)$$

Now, one can see the reason why it was wise to define γ with the second f as the complex conjugate; the term $\exp(i\omega t)$ implicit in f cancels with $\exp(-i\omega t)$ in f^* – otherwise the time-average would result in zero. The two-dimensional Fourier transform of (3.32) is

$$\Gamma(\ell, m) = \langle a(\ell, m) \cdot a^*(\ell, m) \rangle_t$$
$$= \langle |a(\ell, m)|^2 \rangle_t = I(\ell, m). \tag{3.33}$$

This is the *Van Cittert–Zernike theorem* which states that, **for a monochromatic incoherent source, the Fourier transform of the complex spatial coherence function is the angular intensity distribution of the source,** $I(\ell, m)$. It follows that by measuring the spatial coherence in the observatory plane, as a function of the vector separation between the "pinholes," the source intensity image can be deduced by a Fourier transform. Because of the wavelength scaling implied in the use of the dimensionless coordinates $(k_0 \mathbf{r})$, each wavelength has to be transformed independently. The theorem is the spatial analog of the Wiener–Khinchin theorem for a time-dependent wave, which states that its power spectrum is the Fourier transform of its temporal autocorrelation function.

The Van Cittert–Zernike theorem is so important to interferometric astronomy that we should see how it works with several important specific examples. In these examples we quote results in the astronomers' coordinates u and v mentioned in section 3.1.5 and defined by $(u, v) = (x, y)k_0/2\pi = (x, y)/\lambda$, with $w \equiv \sqrt{u^2 + v^2}$, which will be used extensively in chapter 4. The physical meaning of (u, v) is the projection of the baseline vector on the plane normal to the direction of the star origin, measured in units of the wavelength.

A circular star with angular diameter α

This uniformly bright circular star can be represented by

$$I(\vec{\ell}) = \text{circ}(2\ell/\alpha). \tag{3.34}$$

The Fourier transform is

$$i(k_0 x, k_0 y) = \frac{\pi \alpha^2}{4} \frac{2 J_1(k_0 r \alpha/2)}{k_0 r \alpha/2}. \tag{3.35}$$

Normalizing to unity at $x = y = 0$, we have the complex coherence function in terms of the coordinates (u, v):

$$\gamma(w) = \frac{2 J_1(\pi w \alpha)}{\pi w \alpha} \tag{3.36}$$

which has zero value on the circle $w = 1.22/\alpha$ (figure 3.18).

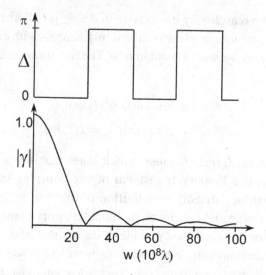

Fig. 3.18. Phase and value of the coherence function $\gamma(w)$ for a circular star of angular diameter $\alpha = 10^{-3}$ arcsec.

A circular star with a monotonic decrease of intensity with radius

This is an important model for real stars, since the emitted light passes through an absorbing atmosphere whose optical thickness depends on the radius because of increasing angle of incidence. This effect is called *limb darkening*. Using data from the Sun, where this effect can be measured accurately, it is usual to represent the intensity function as

$$I(\vec{\ell}) = \text{circ}(2\ell/\alpha)(1 - 4\ell^2/\alpha^2)^{n/2} \tag{3.37}$$

where n has value up to about 3. The Fourier integral can be evaluated numerically and some examples are shown in figure 3.19(a). In experimental work, it is important to determine the diameter α from such observations, when the value of n is not known. In that case, measurements of $\gamma(r)$ have to be fitted to the model. If these are made only out to the first zero in γ, it is virtually impossible to determine α and n independently, as can be seen from figure 3.19(b), where the curves of (a) have been scaled (by varying α only) to have their first zeros at the same value of r. The three curves overlap almost exactly within the central peak. But after the first zero, the different values of n give distinctly different curves and so, if γ is measured in this region of r also, the value of n can also be determined. This procedure is very important in measuring stellar diameters accurately, and will be discussed further in terms of actual measurements in section 7.5.3 and section 11.2.1. The problem is that, even for $n = 0$, a uniform disk, the maximum value of $|\gamma|$ in the second lobe is 0.13; usually $|\gamma|^2$ is measured, and values less than 0.13^2 are not too easy to measure accurately unless the star is bright.

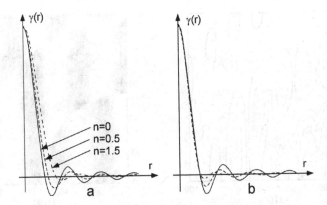

Fig. 3.19. Coherence function for limb-darkened circular disks. (a) shows $\gamma(r)$ for three degrees of limb-darkening, and (b) shows the same data when scaled so that the first zeros of the three curves coincide.

A binary pair of point stars

A binary pair of point stars (δ-functions) with intensities I_1 and I_2, separated by small angle β is represented by

$$I(\ell, m) = I_1 \delta(\ell - \beta/2)\delta(m) + I_2 \delta(\ell + \beta/2)\delta(m), \tag{3.38}$$

when we put the z-axis half-way between them. Then its Fourier transform is

$$i(k_0 x, k_0 y) = I_1 \exp(ik_0 x \beta/2) + I_2 \exp(-ik_0 x \beta/2), \tag{3.39}$$

$$= (I_1 - I_2) \exp(ik_0 x \beta/2) + 2I_2 \cos(x k_0 \beta/2). \tag{3.40}$$

Normalizing to unity at $x = y = 0$, we have the complex coherence function in terms of the coordinates (u, v):

$$\gamma(u, v) = \frac{I_1 - I_2}{I_1 + I_2} \exp(2\pi i u \beta/2) + 2\frac{I_2}{I_1 + I_2} \cos(2\pi u \beta/2), \tag{3.41}$$

Notice that there is no variation with v since the stars are considered as point sources. If the two stars have equal intensity, $\gamma(u, v)$ is real and goes through zero at the values $u = (\frac{1}{2} + m)/\beta$.

A binary pair of disk-like stars

Now suppose that the individual stars are large enough to be resolved, and can each be represented a circular disk with angular diameter α. Then the source is the convolution of (3.38) with the function $\mathrm{circ}(2\rho/\alpha)$. The transform of $I(\ell, m)$ is the product of the transform of this circ function, namely $2J_1(\pi w\alpha)/(\pi w\alpha)$, with the value of $\gamma(u, v)$ in (3.41). Now indeed the coherence function falls off in both u and v directions as is shown in figure 3.20. This is essentially the form of

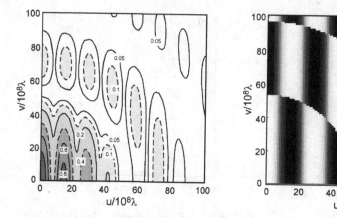

Fig. 3.20. Value and phase of the coherence function $\gamma(u, v)$ for a pair of disk-like stars with angular diameter 0.5 mas, separated by 1.5 mas and with intensity ratio 1:2. (a) shows $|\gamma(u, v)|$ as a contour plot with contours at 0.05, 0.1, 0.2, 0.4, 0.6, and 0.8. (b) shows $\cos \Delta$ in gray scale (1 = white to −1 = black); in both figures u and v are in units of $10^8\lambda$.

the coherence function which gave the first two-dimensional image of an extended object obtained by infrared interferometry, the image of Capella obtained by the COAST group in 1997, shown in figure 3.21. The techniques behind this and other measurements of $I(\ell, m)$ will be described in later chapters.

3.4 Fluctuations in light waves

Fluctuations are an inherent property of classical light waves, and are related to their coherence. In this section, we'll assume that the fluctuations occur as a function of time in a spatially coherent wave, but the same approach can be applied to a monochromatic wave which is spatially incoherent, and gives rise to the phenomenon of speckle which is a major theme of chapter 6. The temporal mode considered here is one-dimensional (time), which simplifies the algebra. The analysis sheds light on the ideas of correlation in quasimonochromatic light, and the results will also be directly applicable to intensity interferometry, the subject of chapter 7.

3.4.1 A statistical model for quasimonochromatic light

Suppose that we have a quasimonochromatic source which radiates simultaneously several waves $j = 1, \ldots, N$, each having frequency ω_j in a narrow band $\omega_0 \pm \delta\omega$. The picture represents, for example, a collection of atoms in a mercury discharge lamp, all radiating the same basic frequency ω_0, but Doppler shifted differently

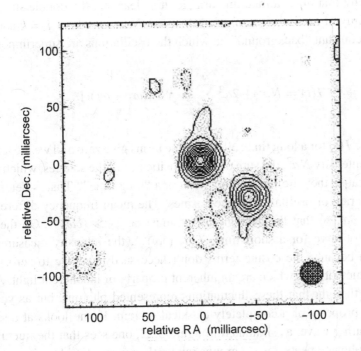

Fig. 3.21. Image of the double star Capella, obtained by the COAST group in 1997 at 1.29 μm (Young 1999). The circle at $(-100, -100)$ indicates the resolution limit.

because of their random thermal motion. The frequencies ω_j are assumed to be randomly distributed in the band, and their phases are random. However, we won't mention the phases specifically, because since the frequencies are random, the choice of the origin of time $t = 0$ to be at some time in the past will automatically randomize the phases $\omega_j t$ after a time $t \gg \delta\omega^{-1}$ (which is really rather short). We shall also assume the sources to have equal amplitudes a, to simplify the algebra. What interests us is the form of the composite wave $f(t)$ and its instantaneous intensity $I(t) = |f(t)|^2$:

$$f(t) = a \sum_{j=1}^{N} \exp(i\omega_j t) \tag{3.42}$$

$$I(t) = a^2 \left| \sum_{j=1}^{N} \exp(i\omega_j t) \right|^2 . \tag{3.43}$$

The latter can be understood if we write it as a double sum, in the form:

$$I(t) = a^2 \sum_{j=1}^{N} \sum_{k=1}^{N} \exp(i[(\omega_j - \omega_k)t]) . \tag{3.44}$$

Now, recalling that ω_j is a random variable, it's clear that the double sum (3.44) consists mainly of oscillating terms; but there are N terms where $j = k$ and these add up to a constant "background" on which the oscillations are superimposed:

$$I(t) = Na^2 + 2a^2 \sum_{j=1}^{N} \sum_{k=1}^{j-1} \cos[(\omega_j - \omega_k)t]. \tag{3.45}$$

If we average $I(t)$ for a long time, all the cosine terms give zero and we are left with the average intensity Na^2, the sum of the intensities of all the sources, which do not interfere because they are incoherent. What is a "long time"? That is determined by the mean rate of oscillation of the cosines. The mean frequency difference in the range is $\delta\omega$, so that the averaging time must be $T_1 \gg (\delta\omega)^{-1}$. On the other hand, if we observe for a short time, $T_0 < (\delta\omega)^{-1}$, the intensity measured will be uncertain because the cosine terms don't necessarily average to zero. These are natural fluctuations, which are an inherent property of incoherent light. At one time, it was thought that these fluctuations represented photons, but as you see, they are the property of a completely classical system. If one looks at a sample of the fluctuating wave, as simulated in figure 3.22, one sees that the fluctuations are like complicated beats – wave groups randomly occurring and having a typical duration. This duration is in fact approximately $(\delta\omega)^{-1}$, and each beat contains $\omega_0/\delta\omega$ periods of the wave. Within one beat, the phase is more or less coherent (i.e. given just that one beat, you could determine the phase and frequency); but the phases of adjacent beats are not correlated. Therefore, from the concepts of coherence, we identify $(\delta\omega)^{-1}$ as the coherence time τ_c; interference between two samples taken within one beat would give stable fringes, whereas if the samples were in different beats, no stable fringes would be seen. It is interesting to notice in figure 3.22 that the phase shifts from beat to beat are mainly restricted to the regions where the intensity is very small.

We shall now show that the fluctuations don't go away, as maybe you might have expected, when the number N of samples becomes very large. To show that the fluctuations remain significant, we calculate the mean square fluctuation of the intensity. First, the fluctuation at a given time is the deviation from the mean; from (3.44)

$$I(t) - Na^2 = a^2 \sum_{j=1}^{N} \sum_{k=1}^{N} \exp[it(\omega_j - \omega_k)] - Na^2 \tag{3.46}$$

$$= a^2 \sum_{j=1}^{N} \sum_{k \neq j=1}^{N} \exp[it(\omega_j - \omega_k)]. \tag{3.47}$$

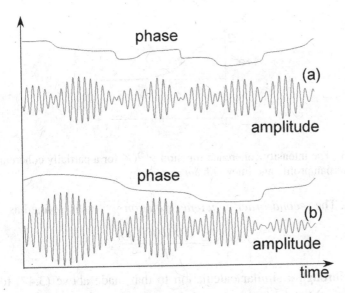

Fig. 3.22. Incoherent waves simulated by adding 20 components with unit ampli-tude and randomly chosen frequencies within the band $\omega_0 \pm \delta\omega$. (a) $\omega/\delta\omega_0 = 6$; (b) $\omega/\delta\omega_0 = 16$. In both cases the phase, relative to the phase at the start of the example, and the amplitude measured during periods T_0 are shown. The coherence time $\tau_c = (\delta\omega)^{-1}$ is the length of a typical wave group.

Now the square of (3.47) is, using two more suffices l and m for the summations:

$$[I(t) - Na^2]^2 = a^4 \sum_{j=1}^{N} \sum_{k \neq j=1}^{N} \sum_{l=1}^{N} \sum_{m \neq l=1}^{N} \exp[it(\omega_j - \omega_k + \omega_l - \omega_m)]. \quad (3.48)$$

Once again, on taking an average for a long period T_1 above, the oscillating terms average out, and what remains are only the non-oscillating terms given by $j = m$ and $k = l$. There are N^2 such combinations which each contribute a^4 to (3.48), and so its long-time average is

$$\langle [I(t) - Na^2]^2 \rangle_{T_1} = N^2 a^4. \quad (3.49)$$

The root-mean-square fluctuation is therefore Na^2 **which is equal to the mean value**. The result is that there are *macroscopic fluctuations*, even in the limit $N \to \infty$, which is a somewhat unexpected conclusion. One can see this in the examples in figure 3.22; there are occasions when the combined intensity is close to zero.

3.4.2 Intensity correlation – the second-order coherence function

In the same way as the coherence function $\gamma(\tau)$ represents the amplitude correlation in a partially coherent wave, we can define a coherence function to describe intensity

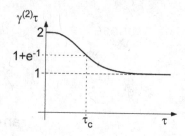

Fig. 3.23. The intensity coherence function $\gamma^{(2)}(\tau)$ for a partially coherent wave with Gaussian profile and linewidth $\delta\omega = \tau_c^{-1}$.

correlations. The *second-order coherence function* $\gamma^{(2)}(\tau)$ is defined as

$$\gamma^{(2)}(\tau) \equiv \frac{\langle I(t)I(t+\tau)\rangle}{\langle I(t)\rangle^2}. \tag{3.50}$$

We can go through a similar calculation to that made above (3.47) to evaluate $\langle I(t)I(t+\tau)\rangle$. What we find is that when $\delta\omega\tau \ll 1$, the number of nonoscillating terms is $2N^2$, whereas when $\delta\omega\tau \gg 1$ it is N^2, giving a numerator for $\gamma^{(2)}(\tau)$ which varies from $2N^2 a^4$ at $\tau \ll \tau_c$ to $N^2 a^4$ when $\tau \gg \tau_c$. The denominator is $N^2 a^4$. Thus $\gamma^{(2)}(\tau)$ varies from 2 at $\tau = 0$ to 1 at $\tau \to \infty$, the change taking place around τ_c (figure 3.23).

Intensity fluctuations and their correlation were first explored by Hanbury Brown and Twiss (1956), for emission from a Hg discharge lamp. They also showed (which can be done by writing $\gamma(\tau)$ also in terms of the model we discussed) that $\gamma^{(2)}(\tau) = 1 + |\gamma(\tau)|^2$, which suggested that measurements of correlation between intensity fluctuations could replace interferometric measurements of $\gamma(\tau)$ in astronomy. The important point to notice is that because the waves from two receivers do not have to interfere in order to make this measurement, many of the stability and atmospheric problems inherent to large-scale amplitude interferometry are absent in this method. However, these problems are replaced by those of fast electronics, which in the 1960s was only just developing; one should recall that τ_c is of order 10^{-9} s for a spectral line, and considerably shorter for starlight even after filtering it through a narrow-band filter. Subsequently, Hanbury Brown and his colleagues developed this type of measurement considerably, with the intensity interferometer at Narrabri in Australia, and the important achievements of this method are discussed in chapter 7 (Hanbury Brown 1974).

3.4.3 *Photon noise*

When we observe light, the electromagnetic energy is converted into a signal, usually electrical, which we record. For example, when we look at a star, light energy at the retina induces a chemical reaction resulting in an electric potential change which is transmitted by nerves to the brain by ionic conduction. In a photo-electric device, light falling on a photosensitive surface releases bound electrons which flow as a current to a charge-measuring device. These processes are quantum processes, since the electric charge of the electron or the ion is an integral number of units. The quantum process is statistical; given a certain electric field, the number of charges n released is statistically related to the exposure W (wave power times time) by Einstein's relation, $W = n_0 \hbar \omega$, n being an integer having mean value n_0 and obeying Poisson statistics. Poisson statistics indicate that the distribution of values of n has a variance n_0 or mean error $n_0^{\frac{1}{2}}$. In practice, photo-electric conversion may compete with other processes of absorption of the wave power, and the number n_0 might be less than the above by a factor called the *quantum efficiency*, η. Astronomical detectors today usually have $\eta > 90\%$.

However, we saw in section 3.4.1 that for classical reasons, the power in a light beam undergoes macroscopic fluctuations (i.e. the intensity varies on a scale comparable with the mean intensity). As a result, there are two components to the fluctuations in the photo-electron current: first from the fluctuations in the light beam intensity, and second from the statistics of conversion from light into electrons. The two fluctuations are additive: if in a certain interval of time T (less than the coherence time τ_c) the expected number of photo-electrons is n_0, n has a mean square fluctuation (3.49)

$$\langle (n - n_0)^2 \rangle = n_0^2 \tag{3.51}$$

from the wave fluctuations and

$$\langle (n - n_0)^2 \rangle = n_0 \tag{3.52}$$

for the Poisson statistics, so that

$$\langle (n - n_0)^2 \rangle_{T < \tau_c} = n_0^2 + n_0. \tag{3.53}$$

If the measuring time $T > \tau_c$, the Poisson fluctuations in counting are unchanged, but the intensity fluctuations are smoothed and reduced in proportion to τ_c / T. It follows that

$$\langle (n - n_0)^2 \rangle_{T > \tau_c} = \frac{\tau_c}{T} n_0^2 + n_0. \tag{3.54}$$

These statistics are called *super-Poisson* and the light is said to be *bunched*. The origin of this term follows from generating a typical sequence of events with this character. This can be done by taking a wave of the form of figure 3.22 and generating events with Poisson statistics whose probability is proportional to the instantaneous intensity. Comparing this with the same process for a wave of constant intensity

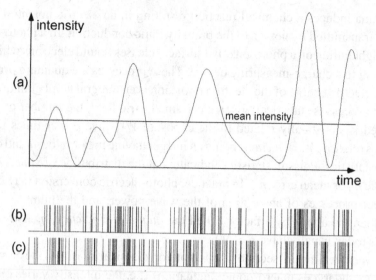

Fig. 3.24. Super-Poisson statistics. (a) Typical intensity fluctuations in a wave, generated as in figure 3.22; (b) corresponding photo-electron sequence; (c) photo-electron sequence for a steady wave with the same mean intensity as (a).

makes the difference clear (figure 3.24)[†]. It is interesting that the fluctuations in intensity can also be calculated using the quantum theory of light (Mandel and Wolf 1995). The result is identical to what we find from the classical theory above, but arises there from the Bose–Einstein statistics of photons. In quantum optics, there also exists light with sub-Poisson statistics, but this is a nonlinear phenomenon and hardly relevant to astronomy.

Most astronomical measurements are made by instruments which average over the longer period T_1 and so the super-Poisson aspects are irrelevant, and what concerns us are mainly the Poisson statistics of photo-electron conversion, for which the mean error in a count of electrons is equal to the square root of its number. However, the technique of intensity interferometry relies on the correlation of the intensity fluctuations (section 7.2), where the super-Poisson noise is relevant.

3.4.4 Photodetectors

The flux of photons from stars is very small, and it is imperative to collect and analyze them as efficiently as possible. Astronomers have always been in the forefront of the development and application of new light-measuring and imaging devices

---

[†] Super-Poisson statistics are well-known to us in everyday life as applying approximately to the statistics of traffic on a narrow road, where overtaking is difficult. When there is little traffic, cars arrive with a Poisson distribution, but in heavy traffic they arrive in groups, each being headed by a slower than average vehicle.

and concepts, and a good up-to-date account of detectors currently in use is given by Léna et al. (1998). In this section we shall concentrate on some of the basic aspects of light detection relevant to optical interferometry; details of actual detectors used in working systems will be given in the appropriate places.

What is actually important is always the *signal-to-noise ratio* (SNR) of a given detector; that is what determines the smallest signal that can be reliably detected. When the SNR has unit value, the signal is then barely detectable, and the value of that signal is called the *noise equivalent power* (NEP). The units in which this is expressed generally show the way in which it depends on measurement parameters. For example "Watt cm^{-1} $Hz^{-\frac{1}{2}}$" indicates that the NEP is proportional the square roots of the detector area and the electronic measurement bandwidth. In practical terms, since the power reaching a detector is proportional to its area, and the noise only to its square root, this means that the signal-to-noise ratio of single detectors increases as the square root of the area, and as the square root of the time during which the signal is integrated (which is inversely proportional to the electronic bandwidth).

In general, there are two types of noise: signal-independent, and signal-dependent. Processes such as a receipt of light from an unwanted background, or intrinsic events in the photodetector, are examples of signal-independent noise. Statistical counting fluctuations are an example of signal-dependent noise. A photodetector eventually converts incident photons and noise events into electronic pulses, which may be individually measured or whose rate may be measured as an electric current. Ideally, we should like every photon to produce a pulse (quantum efficiency $\eta = 1$), and there to be no noise events, in which case the counting fluctuations would be those of the photons, but this is rarely the case, although the best photodetectors today in the optical region are close to achieving this. If the number of incident photons in a given period is n_0, then the expected number of pulses is

$$n = \eta n_0 + n_n \qquad (3.55)$$

where n_n is the signal-independent noise count in the same period. The number n_0 depends not only on the source, but also on radiation from the surroundings. In the visible region, thermal radiation from the surroundings at ambient temperature is rarely a problem, but this is not so in the infrared, so we shall deal with the subject briefly here to justify this statement. Suppose the detector surface to be situated within an enclosure having black walls at temperature T. The radiation energy density $u(\nu)\delta\nu$, in terms of photons of frequency between ν and $\nu + \delta\nu$ traveling in all directions, is given by Planck's formula:

$$u(\nu)\delta\nu = \frac{8\pi\nu^2}{c^3} \frac{1}{\exp(h\nu/k_B T) - 1} \delta\nu. \qquad (3.56)$$

The number of photo-electrons generated by this radiation depends on the quantum efficiency and its dependence on ν. Generally a good photodetector will have a high quantum efficiency $\eta(\nu)$ in the region where it is being used, and as low as possible elsewhere; in particular, most photodetectors have a cut-off ν_0 below which the efficiency is zero – the celebrated photoelectric effect. The total number of background events is therefore given by integrating (3.56) multiplied by $\eta(\nu)$ to represent the rate at which the photons impinge on a detector of surface area A (which sees half of the total solid angle of the cavity) in units of photons per second:

$$n_b = cA \int_0^\infty \tfrac{1}{2} u(\nu)\eta(\nu)\,d\nu$$

$$= A \int_0^\infty \frac{4\pi \nu^2}{c^2} \frac{\eta(\nu)}{\exp(h\nu/k_B T) - 1}\,d\nu. \tag{3.57}$$

Now since this background signal consists of random events, with a Poisson distribution, we can calculate its fluctuations within an electronic band δf, which is equivalent to a counting time $\tau = \delta f^{-1}$. During this time, the signal count is τn_s and the noise count is τn_b. For Poisson statistics, the fluctuation of the latter is its square root, $\sqrt{\tau n_b}$. As a result, the signal equals the noise (limit of detectability) when $n_s = \sqrt{n_b/\tau} = \sqrt{n_b \delta f}$. The NEP is this value n_s multiplied by the photon energy $h\nu$:

$$\mathrm{NEP} = h\nu \sqrt{A\delta f \int_0^\infty \frac{4\pi \nu^2}{c^2} \frac{\eta(\nu)}{\exp(h\nu/k_B T) - 1}\,d\nu}. \tag{3.58}$$

It is important to see when the NEP becomes relevant. For a detector with area $1\,\mathrm{mm}^2$, and $\delta f = 1\,\mathrm{Hz}$, typical values of $\mathrm{NEP}/h\nu$ are shown in table 3.1.

One sees that cooling the background is necessary for observing in the infrared but not in the visible. In fact, it is clear that in the 8–12 μm region, even cooling the detector surrounds from 77 K to 63 K, which can be achieved by pumping on liquid nitrogen until it solidifies, reduces the noise count by almost an order of magnitude. But to remove background noise completely, $h\bar{\nu}/kT$ must be at least of the order of 100.

Once background noise has been reduced or eliminated by working with the detector at a suitable temperature, the remaining noise comes from the source itself and the fact that counting a random sequence introduces Poisson noise. We can consider the detector as situated within a cavity of the size of the coherence volume for the source in question which therefore classically supports one mode (of each polarization) within the range of directions of the incoming radiation and its bandwidth. Provided that the detector is smaller than the cross-section of this volume, it

Table 3.1. *Noise equivalent photon counts in 1 sec. for 1 mm square detectors in four spectral regions with backgrounds at different temperatures*

Background temp. (K)	Wavelength range (μm)	Counts in 1 sec.	$h\bar{v}/kT$
300	8–12	1.9×10^8	5
77	8–12	2.1×10^5	18
63	8–12	3.6×10^4	22
4	8–12	0	350
300	3–5	1.8×10^7	11
77	3–5	10	45
300	1.0–1.5	1000	36
77	1.0–1.5	0	140
300	0.4–1.0	0.5	78

will sample the number of photons in the cavity, which is given by (3.56) multiplied by the volume of the cavity. Now for a source subtending solid angle Ω (essentially $\pi\alpha^2/4$ for a star of angular diameter α), and within frequency band $\delta v = c\delta\lambda/\lambda^2$, the coherence volume (figure 3.14) has lateral area λ^2/Ω and length $c/\delta v$, so its volume is $V_c = \lambda^4/\delta\lambda\Omega$. According to the quantum theory, the number of photons occupying this one mode is $\delta \equiv [\exp(h v/k_B T) - 1]^{-1}$, which is called the "degeneracy factor." The energy of the mode is therefore $E = h v \delta$. Now, from statistical mechanics, a thermal system with mean energy E has mean square fluctuation (Landau and Lifshitz 1980):

$$\langle \Delta E^2 \rangle = \langle E^2 \rangle - \langle E \rangle^2$$

$$= k_B T^2 \frac{dE}{dT} \tag{3.59}$$

$$= h v E (1 + \delta) \tag{3.60}$$

in this case. In terms of photon counts, $n = E/hv$,

$$\langle \Delta n^2 \rangle = n(1 + \delta). \tag{3.61}$$

When the source temperature is much higher than that of the detector, $\delta \approx \exp(-h v/k_B T) \ll 1$, so that we confirm the Poisson-counting result (3.52) that $\langle \Delta n^2 \rangle = \langle n \rangle$. This will essentially be true for any source which can be detected without appreciable background noise, since this is the same condition as required there. The detector then works in the *photon noise limit*.

Most of the astronomical interferometry systems work in the visible to near infrared regions of the spectrum, so detector cooling is not a serious problem; only for work in the mid-infrared region (for example, ISI, section 8.5.5 and the space interferometer Darwin, section 12.2.2) is it absolutely necessary. However, now that some photon detectors have dark counts of tens of counts per second, and detection frequencies may be as high as kHz, some cooling is necessary for the near infrared region too.

It is interesting to estimate a photon counting limitation to the magnitude of stars which can be observed interferometrically in the visible region. Because of atmospheric turbulence, to be discussed in chapter 5, the fringe sweep rate must be at least 50 fringes per second and an entrance aperture larger than about $0.2 \, m^2$ cannot be used without adaptive optics correction (section 5.8). This implies that several photons must be detectable within a time of 0.01 sec, i.e. at least 1000 photons per sec. Further, we assume that a bandwidth of $\lambda/\delta\lambda = 10$ is needed to give a fringe pattern with about 20 fringes. From Appendix B, we see that the photon flux received from a zero-magnitude star at $0.55 \, \mu m$ wavelength is about $1.1 \cdot 10^{11}$ photons.$s^{-1} \, m^{-2} \, \mu m^{-1}$, implying $1.2 \cdot 10^9$ photon/sec within the limited bandwidth of $0.055 \, \mu m$. An allowance for absorption, reflection and other losses discussed in chapter 8, which can also result in poor fringe visibility, reduces the usable photon flux by at least one order of magnitude. During measurement, either interference fringes have to be swept (Michelson configuration) or observed with a multichannel imaging detector (Fizeau configuration), either of which probably reduces the mean useful photon arrival at an individual detector by another order of magnitude, giving about $1.2 \cdot 10^7$ photon/sec (for a zero-magnitude star, we recall). Comparing this to the detector requirement of 1000 photon/sec, we have a factor of $1.2 \cdot 10^4$ in hand, which is about 10 magnitudes. We can therefore expect that, under the above conditions, unresolved stars as weak as 10th magnitude might be accessible to amplitude interferometry. However, for interferometry to produce good science, and in order to measure interesting features on resolved stars, it will be necessary to measure the visibility of fringes down to less than 0.1, which increases the limit by about 3, to 7th magnitude. Of course, improvements such as adaptive optics, which have already increased the useful telescope area at Keck and VLTI to about $50 \, m^2$, and employing dispersed fringes (section 8.3.10) to increase the optical bandwidth, will add several magnitudes to this estimate. It can be compared to the practical achievements of some of the interferometers described in chapter 8.

References

Born, M. and E. Wolf (2000). *Principles of Optics*, 7th edn, Pergamon, Oxford.

Van Cittert, P. H. (1934). *Physica*, **1**, 201.

Goodman, J. W. (1996). *An Introduction to Fourier Optics*, 2nd edn, New York: McGraw-Hill.

Hanbury Brown, R. and R. Q. Twiss (1956). *Nature*, **177**, 27.

Hanbury Brown, R. (1974). *The Intensity Interferometer*, London: Taylor and Francis.
Hariharan, P. (2003). *Optical interferometry*, 2nd edn, San Diego, CA: Academic Press.
Landau, L. D. and E. M. Lifshitz (1980). *Statistical Physics*, 3rd edn, Oxford: Pergamon.
Léna, P., F. Lebrun and F. Mignard (1998). *Observational Astrophysics*, Berlin: Springer.
Lipson, S. G., H. Lipson and D. S. Tannhauser (1995). *Optical Physics*, 3rd edn, Cambridge: Cambridge University Press.
Mandel, L. and E. Wolf (1995). *Optical Coherence and Quantum Optics*, Cambridge: Cambridge University Press.
Michelson, A. A. (1927), (1962). *Studies in Optics*, University of Chicago Press, reprinted in Phoenix Science Series.
Steel, W. H. (1983). *Interferometry*, 2nd edn, Cambridge: Cambridge University Press.
Young, J. S. (1999). *Infrared Imaging with COAST*, Ph.D. thesis, University of Cambridge.
Zernike, F. (1938). *Physica*, **5**, 785.

4

Aperture synthesis

4.1 Aperture synthesis

Aperture synthesis is the way that the Van Cittert–Zernike theorem (section 3.3) is used in practice to get higher resolution images than a single large aperture will allow. Although Michelson's stellar interferometer was the first implementation of the concept, and occurred before the formalization of coherence theory, its real application began in the late 1940s in radio astronomy, where today it is responsible for almost all high-resolution images. At radio frequencies the problem of getting high angular resolution is very acute because of the long wavelengths involved, and Michelson's stellar interferometer inspired Martin Ryle (1952) to use the same idea in radio astronomy. And so, while optical aperture synthesis languished for 60 years for lack of suitable electronics, the radio astronomical applications of the technique blossomed. There are several excellent texts on the theory and practice of aperture synthesis, mainly directed to the radio regime, such as Thompson (2001), Perley and Schwab (1989) and Rohlfs (1996), which will give the reader insight not only into the principles but also the techniques involved.

4.1.1 The optics of aperture synthesis

Suppose we measure the complex spatial coherence function due to a distant source using two receivers separated by a vector $\mathbf{r}$ lying in a plane normal to its direction in inertial space. Following this observation, we can change the vector and make another measurement, and so on, until a sufficiently large bank of data for $\gamma(\mathbf{r})$ is accumulated. From this data, the Fourier transform can be calculated to give the stellar image, provided that we assume that the image is not changing significantly with time. Clearly, since $\gamma(\mathbf{r})$ is only sampled at a finite number of points, the image may have missing details; for example it would miss a sinusoidal feature at

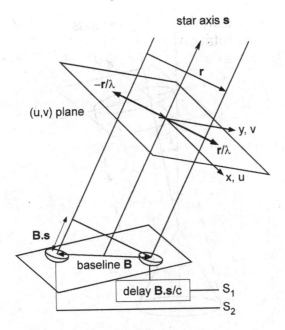

Fig. 4.1. The (u, v) plane and time-difference compensation.

a periodicity which would give a strong contribution to γ at a value of $\mathbf{r}$ which was not sampled. But, since such strongly periodic features are unlikely components of stellar images, this is unlikely to happen, and since $\gamma(\mathbf{r})$ is a continuous function, algorithmic methods can be used to fill in the missing details. In the astronomical literature, the spatial vector $\mathbf{r}$ is generally normalized to the wavelength by defining $\mathbf{u} \equiv (u, v) \equiv k_0 \mathbf{r}/2\pi = \mathbf{r}/\lambda$ (section 3.3.3). This dimensionless variable is the Fourier conjugate to angular position (ℓ, m) in the sky[†]. Then the angular resolution limit is about equal to the reciprocal of the maximum value of u or v.

Fortunately, it is not actually necessary that the physical baseline vector $\mathbf{B}$ joining the two receivers be in the plane normal to the star direction in inertial space, because for a source whose angular extent is limited, the path difference from the star to the two receivers can be compensated by introducing an appropriate delay into the signal from the closer receiver (figure 4.1). This allows the receivers to be located at convenient points in the observatory, although it should be pointed out that both the Michelson stellar interferometer and the system used by Hanbury Brown and colleagues for intensity interferometry (chapter 7) did in fact keep the receivers equidistant from the source.

[†] In this case the Fourier transform would be defined by the integral $I(\vec{\ell}) = \int_{-\infty}^{\infty} \gamma(\mathbf{u}) \exp[-2\pi i \mathbf{u} \cdot \vec{\ell}] \mathrm{d}^2 \mathbf{u}$.

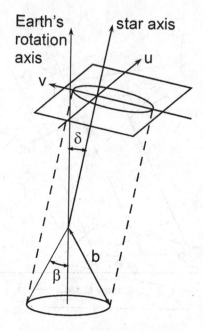

Fig. 4.2. Geometry of aperture synthesis.

4.1.2 Sampling the (u, v) plane

Ideally, we should like to fill the (u, v) plane completely and uniformly, but this would require an infinite number of observations, so we have to concentrate on getting a good enough coverage of the (u, v) plane with an appropriately arranged finite set of receiver positions. We shall discuss this in terms of the set of receiver separation vectors $\mathbf{r} = (u, v)\lambda$, which we represent by drawing them relative to an origin in inertial space. Notice that each separation gives two vectors $\pm\mathbf{r}/\lambda$ relative to the origin of (u, v). Suppose we have two receivers (radio antennas, or optical telescopes) separated by $\mathbf{B}$ in the Earth's frame. The vector $\mathbf{r}$ is the projection of $\mathbf{B}$ on the (u, v) plane, which is fixed in inertial space because it is normal to the star axis $\mathbf{s}$. The (u, v) plane is sampled by two processes: first, the diurnal changes in $\mathbf{r}$ as the Earth rotates and second, as a result of deliberate changes in $\mathbf{B}$.

The diurnal changes in $\mathbf{r}$ result from simple geometry. First, we transform the Earth-bound vector $\mathbf{B}$ to $\mathbf{b}(t)$ in inertial space. Then we project $\mathbf{b}$ onto the (u, v) plane. For example, if the two receivers are at the same latitude, $\mathbf{b}$ describes during the day a circle with radius B about the origin. If they are not at the same latitude, the vector $\mathbf{b}$ describes a cone with apex at the origin and circular base with radius smaller than B. The projection of $\mathbf{b}$ onto the (u, v) plane is shown in figure 4.2. For a non-polar star, we clearly get an elliptical locus in both cases; in the second case

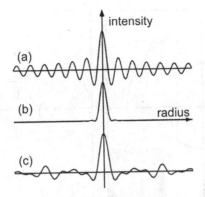

Fig. 4.3. Synthetic point spread functions for a polar star: (a) single baseline B compared to (b) the optical point spread function for a circular aperture of diameter B and (c) sum of baselines $0.5B$, $0.75B$ and B with equal weights.

the center of the ellipse is displaced from the origin. In addition, we must take into account the fact that part of it may be missing because of obscuration by the Earth.

The simplest situation to envisage is when $\mathbf{B}$ has east–west orientation and the star is close to the Earth's axis; then the projection is a circle of radius B and there is no obscuration! In this case, the set of data acquired describes the sample of $\gamma(\mathbf{r})$ on this circle. This sample is described mathematically by the product $\gamma(\mathbf{r}) \cdot \delta(r - B)$, where $\delta(r - B)$ describes an annular ring of radius B and zero thickness. When it is transformed to give the image intensity, we get the synthesized image I_{obs}:

$$I_{\text{obs}}(\vec{\ell}) = I(\vec{\ell}) \star J_0(k_0 \ell B) \tag{4.1}$$

since the Fourier transform of the annulus is the zero-order Bessel function (see Appendix A). In the terminology of optics, the *point spread function* (PSF, section 3.1.4) of the system, which is the image that would be calculated for an unresolvable point star, is $J_0(k_0 \ell B)$. This has a bright center point surrounded by quite strong rings of alternating sign, and does not give a very acceptable image. However, it can be improved by algorithmic deconvolution; this type of approach will be discussed further in section 4.3.1 and is illustrated in a simple example in section 4.6. However, no algorithm can restore information which has never been measured, so clearly this approach is limited.

We must make two remarks about the synthesis procedure. The first relates to the fact that since $I(\vec{\ell})$ is a real function, its Fourier transform satisfies $\gamma(-k_0\mathbf{r}) = \gamma^*(k_0\mathbf{r})$ so that only half of the (u, v) plane need be measured (taking 12 hours). Second, we must make a formal distinction between the optical point spread function and that from aperture synthesis. The former is positive definite, since it is the intensity of the image of a point and intensities cannot be negative. However the

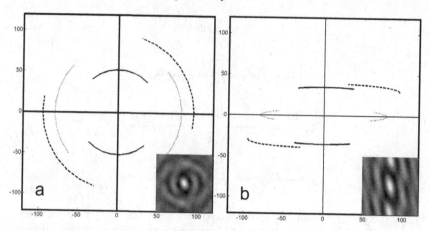

Fig. 4.4. Two examples of (u, v) plane coverage (arbitrary units) and calculated equally-weighted point spread functions for a group of three receivers observing sources (a) on the Earth's axis and (b) at 6° to the equator. The receivers are arranged in a 3-4-5 triangle with the 4-unit side EW, situated at latitude 60°.

latter has no restriction of sign, since it is a calculated function. Actually it is the square root of the equivalent optical PSF, where the (u, v) plane is considered as a transmitting mask representing $\gamma(\mathbf{r})$ (figure 4.3a). So synthesized point spread functions are qualitatively different from optical ones.

Once the data have been measured for one value of $\mathbf{B}$, the receivers are moved so that a new value can be investigated, and so on, day after day. In the simple case of a source near the pole star, the result is to gradually build up a denser region of sampled circles, up to the maximum value of B. The resultant point spread function is the sum of the transforms of all the rings measured. Because it is the result of a digital calculation, mathematical processing such as variable weighting c_n of the contributions from different baselines at different times can be used to improve the PSF quality $\sum_1^N c_n J_0(k_0 \ell B_n)$ (figure 4.3c). However, if the source is not close to the pole, the circles become nonconcentric ellipses, and they may not be complete because of obscuration. The point spread function becomes more complicated and loses its simple symmetry. An example of how the (u, v) plane becomes filled with these curves is shown in figure 4.4. In radio astronomy this is essentially a geometrical problem, which will be discussed further in section 4.1.3, since clouds and daytime do not limit observing; but in optical aperture synthesis, these also have to be taken into account when deriving the point spread function. In very long baseline radio interferometry, where the receivers may be on different continents, the problem of obscuration may become very acute, since calculation of the coherence function requires that both receivers be unobscured at the same time, which may be true only for a short part of the day.

4.1.3 *The optimal geometry of multiple telescope arrangements*

When an interferometric telescope array is being planned, one has to decide on a geometric arrangement for the telescope positions. These may be fixed positions, or may be a set of well-defined optional positions where all the necessary connections are available, or the telescopes may be continuously movable. The decision involves many things; one of them is the desire to achieve the best point spread function. Other considerations, such as the need for the minimum total length of pumped vacuum tubes for the interfering light beams, may also be important. In fact, as Mozurkewich has pointed out (p. 242 in Lawson 2000), since there are many *almost* optimal arrangements, the other considerations are often predominant. In addition, the data treatment method must be considered. Phase closure (section 4.2.1) can be used with a nonredundant array of telescopes or apertures, but leads to practical limitations to the smallest fringe visibilities that can be measured. If a certain amount of redundancy is introduced, baseline bootstrapping (section 8.3.12) can be employed to measure smaller visibilities reliably, and this may be important for getting interesting astronomical results.

In the absence of a completely defined optimization problem, this section therefore deals with the theoretical problem of choosing the best arrangement of N receiver positions (x_i, y_i) to fill the (u, v) plane as uniformly as possible at a given time, without taking into account the diurnal rotation of the Earth. This problem was first discussed in a short paper by Golay (1970) and a more detailed discussion, on which we have based this section, was given more recently by Keto (1997). The results will also be directly relevant to optical interferometry using aperture masking (section 6.2). We represent each receiver position by a δ-function at its center. Each *pair* of receivers gives two vectors $\mathbf{r} = \pm(x_i - x_j, y_i - y_j)$, and so the array of values of $\mathbf{r}$ from a given array of positions is the autocorrelation function of the receiver position map. The point spread function of the synthesized image is the Fourier transform of this autocorrelation function. Ideally we should like the point spread function to be as sharp as possible and to have side-lobes as weak as possible, but these are somewhat antagonistic considerations. One one hand, the width of the central peak of the point spread function is determined by the outer dimensions of the region (the *support*) where the cross-correlation is nonzero, i.e. by the largest distance between receivers, and it is narrowest when the autocorrelation function is strongest at its outer edges. On the other hand, the requirement of weak side-lobes suggests that the density of points should be maximum in the center and tail off gradually toward the edge of the support region; best that it should approximate a Gaussian (which would give no wings at all). Usually, the sharpness criterion is more important because the wings can be cleaned off algorithmically.

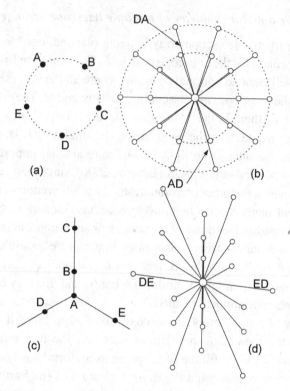

Fig. 4.5. Annular and "Y" receiver arrays, and the corresponding autocorrelation functions. (a) A circular array of five receivers and (b) its autocorrelation function; (c) five receivers in a "Y" array and (d) their autocorrelation. The black circles *A* to *E* represent receiver positions and the open circles peaks in the autocorrelation function. The lines represent the construction vectors.

To see where the problems lie, let us take two well-known examples. One is a set of telescope positions arranged uniformly around a circle (an annular array). The other is a "Y"-shaped telescope array with positions at intervals along the bars. The autocorrelation functions for these two cases for five telescopes are shown in figure 4.5. The autocorrelation function as a function of the vector $\mathbf{r}$ can easily be visualized by drawing all the interposition vectors, in both directions; selected vectors are labeled in the figure as examples. Ideally, to get a uniform density of points within the support we first want no value of $\mathbf{r}$ to be a vector joining more than one pair of receivers (except for $\mathbf{r} = 0$; there we have no choice because vector $\mathbf{0}$ joins every receiver to itself). Otherwise some of the δ-functions in the autocorrelation will have values greater than unity, and in total there will be less distinct values of $\mathbf{r}$ where the function is nonzero. This is called a *nonredundant array* (see section 6.2). From figure 4.5 one can see that a uniformly distributed annular array with an odd

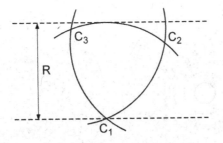

Fig. 4.6. The Reuleaux triangle.

number of receivers is nonredundant, and gives a distribution of points which does indeed get stronger toward the edge[†] and therefore has good resolution but rather strong side-lobes. If the array is to have the same resolution in all orientations (i.e. the support is circular) it is necessary that the maximum separation of receivers in all directions be the same. This can be obtained not only with the annular array, but with any of the "smooth roller" shapes which are essentially curved-edged polygons with an odd number of edges each of which is replaced by an arc centered on the opposite vertex. For example, the Reuleaux triangle (figure 4.6) is the simplest example. Using the Reuleaux triangle with uniform antenna spacing gives better uniformity but still a strong central region; in general, the autocorrelation functions tend to be dense in the center and get weaker toward the edge of the support, and it is difficult to avoid this (see particularly the examples given by Golay 1970).

Keto (1997) used an approach based on the classical "traveling salesman" problem[‡] of computer science lore to optimize the arrangements of receivers. He showed that perturbations of the antennas from the simple geometrical arrangements of the circle and the Reuleaux triangle break down the symmetry in the autocorrelation pattern and lead to more uniform coverage; some results are shown in figure 4.7. The angular point spread function in each case can then be calculated as the Fourier transform of the point array.

4.2 From data to image: the phase problem

The practical methods by which $\gamma(\mathbf{r})$ is measured vary from system to system, and will be described in chapters 6, 7 and 8. In the optical domain it is determined from recordings of the fringes obtained when signals from a pair of receivers are mixed using a detector which measures the intensity of the light, i.e. the square

---

[†] Compare this to imaging through an annular aperture, which gives the narrowest point spread function of any aperture mask; see for example section 3.1

[‡] A salesman must serve each of several fixed towns once, but can choose the order in which he visits them; find the shortest route he has to travel.

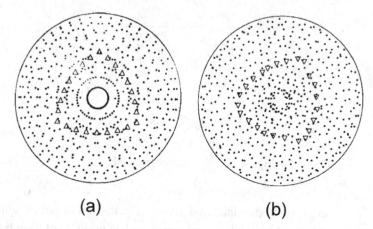

(a)   (b)

Fig. 4.7. Autocorrelation functions for 24 receivers around a Reuleaux triangle: (a) on the triangle, but spaced non-uniformly around it; (b) with deviations from the triangle to optimize autocorrelation uniformity. The triangles show the receiver positions, and the dots the autocorrelation points. From Keto (1997).

of the combined fields from the two receivers, as a function of a controlled phase difference between them. The controlled phase shift can be provided in several ways, such as the Young's slit set-up (image-plane interferometry, also called the "Fizeau" configuration, since he was the first to use it for astronomy), or phase-sweeping optics (aperture-plane interferometry, also called the "Michelson" configuration from its similarity to the Michelson interferometer, section 3.1.1[†]); phase-sweeping may either occur naturally as the Earth rotates or be introduced optically.

Once the coherence function $\gamma(u, v)$ has been measured using (3.26), the image can in principle be calculated by a Fourier transform. However, to do this means that both the absolute value and the phase of γ must be known. The visibility of the interference fringes, however they are obtained, gives $|\gamma|$, but finding the phase is a more serious problem, common to all inverse diffraction problems. In principle, in the Fizeau configuration, the offset of the central fringe from the optical axis gives Δ (section 3.3.1), but of course, as Michelson remarked (section 5.1), the fringes are also displaced laterally in a random manner by differential atmospherically induced phase changes at the two apertures. As a result, while the visibility $|\gamma|$ can be well-measured, Δ is elusive.

Much useful work has been done using the visibility alone. This is often based on a model, such as a circular or elliptic star, either of uniform intensity or with limb darkening, or a binary pair (section 3.3.2; see figure 3.20). Once the model

[†] This statement is bound to be confusing because the original "Michelson stellar interferometer" used the Fizeau configuration, although Michelson did indeed consider the possibility of using a beam-splitter in the paper proposing the instrument, where he called it the "refractometer configuration."

has been chosen, the visibility can be correlated with that expected for the model, and various free parameters found. Since the work of Michelson and Pease, this has been the accepted method for determining the diameters of stars and binary separations with maybe unequal component intensities. When visibilities alone are measurable for subaperture separations larger than the first value at which $|\gamma| = 0$, Δ is assumed to be π, in which case limb-darkening can be calculated (sections 7.5.3 and 11.2.1). However, it will never be possible to distinguish between an image and its centrosymmetric counterpart (enantiomorph) from visibility data alone, and true imaging cannot be carried out.

4.2.1 What can be done to measure phases? Phase closure

If three or more observations are made at different receivers simultaneously, information can be obtained about the relative phases of γ for the various vectors between pairs of these receivers by a technique called *phase closure* (Jennison 1958; Cornwell 1989), in which the phase fluctuations cancel out. We should emphasize that only *relative* phases are important, since changing the phases of all the components by an amount linear in u and v just moves the calculated position of the source to a new angular origin without changing its character (Appendix A). Therefore, in fact, the phases of any two values of γ can be assumed to be zero. Moreover, $\Delta(0, 0) = 0$ from the definition of γ.

Although the phase closure technique applies to any number of receivers greater than three, a larger number can always be broken down into independent groups of three ($i = 1, 2, 3$). The apparent coherence function is measured simultaneously for each pair in such a loop. We assume that the amplitude and phase signals from each receiver contain a slowly changing atmospheric[†] (or other) phase error of unknown values ϕ_{i0}. The signals will then be $|f_i(t)| \exp[i(\phi_i(t) - \phi_{i0})]$ where $|f_i(t)| \exp[i\phi_i(t)]$ is the true value. Three apparent coherence coefficients can be calculated and will have the values (assuming $\langle |f_i|^2 \rangle$ to be normalized in each case):

$$\gamma_{ij}^a = \langle |f_i| \exp[i(\phi_i - \phi_{i0})]| f_j| \exp[-i(\phi_j - \phi_{j0})] \rangle_t, \qquad (4.2)$$
$$= \gamma_{ij} \exp[i(\phi_{i0} - \phi_{j0})] \qquad (4.3)$$

since ϕ_{i0} and ϕ_{j0} are independent of time during the measurement period. As a result, the product of the coherence functions for a closed loop is independent of the atmospheric phases; for a loop of three:

$$\gamma_{12}^a \gamma_{23}^a \gamma_{31}^a = \gamma_{12}\gamma_{23}\gamma_{31} = |\gamma_{12}\gamma_{23}\gamma_{31}| \exp(i\phi_c) \qquad (4.4)$$

[†] Atmospheric phase changes occur at frequencies not greater than about 100 Hz, so that the fringes must be swept or recorded at a rate faster than this.

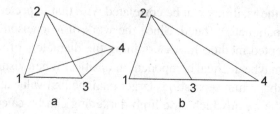

Fig. 4.8. (a) A nonredundant array of four receivers; (b) a redundant array, in which vectors 13 and 34 are equal.

and ϕ_c is called the *closure phase*. In terms of the phases Δ, we have, for baseline vectors $\mathbf{r}_{12}$, $\mathbf{r}_{23}$ and $\mathbf{r}_{31}$ satisfying $\mathbf{r}_{12} + \mathbf{r}_{23} + \mathbf{r}_{31} = 0$, $\Delta_{12} + \Delta_{23} + \Delta_{31} = \phi_c$. Note that, formally, $\Delta_{ij} = -\Delta_{ji}$, since the intensity is a real function.

The closure phase can be used in several ways. First, it can be used for equipment checks, by confirming that $\phi_c = \Delta_{12} + \Delta_{23} + \Delta_{31} = 0$ for an unresolved source, for which all three γ_{ij}s must be positive. Second, it can be used in a simple manner for model confirmation. Suppose, for example, we were fitting data to a model for a double star of unknown separation, whose coherence function is shown by figure 3.20, and ϕ_c is found to be π for a certain triplet. Then this tells us that one (or all) of the three algebraic values of γ_{ij} are negative, which leads to a lower bound for the separation between the components.

It is not possible to determine the phase Δ_{ij} deterministically by phase closure if the array is nonredundant, but the method does reduce enormously the number of degrees of freedom available. Suppose that signals from N receivers are measured. Then there are $N(N-1)/2$ values of Δ_{ij}. The number of independent closure phases is found as follows. Given any four receivers, 1, 2, 3 and 4 (figure 4.8a), we can define four closure phases $\phi_{1hj} \equiv \Delta_{1h} + \Delta_{hj} + \Delta_{j1}$, etc. Note that $1, h$ and j appear cyclically. Remembering that $\Delta_{ij} = -\Delta_{ji}$, we then find that $\phi_{234} = \phi_{134} + \phi_{124} - \phi_{123}$. Thus, any closure phase ϕ_{234} can be expressed in terms of the closure phases of triangles including the node 1. The number of *independent* closure phases is therefore equal to the number of triangles which can be constructed including node 1, and this is $(N-1)(N-2)/2$. As a result, there are always $\frac{1}{2}[N(N-1) - (N-1)(N-2)] = N - 1$ more values of Δ than there are closure equations. Given the fact that two of the phases can arbitrarily be set to zero, there is a deterministic solution only for $N = 3$, which is obvious because if two sides of the triangle are defined to have zero phase, the third must have the closure phase. However, it is clear that the number of variables which have to be adjusted to get an acceptable image (positive definite, restricted in support) is now much smaller than would be the case if all values of Δ could be adjusted independently. This in fact leads to the conclusion that it is simply necessary to give to each receiver signal an adjustable phase in model-fitting, since changing the phase of a single receiver signal will not alter the closure phase of any of the triangles in which it is involved. This is called "autocalibration."

If the array does contain enough redundant elements, the phases can be found deterministically, since then several of the values of Δ_{ij} will be known to be the same. In fact, this will be the case if, of the $N(N-1)/2$ baselines, $N-3$ are repeated. In the development of speckle masking, a single-aperture technique described in section 6.4.2, phase closure was rediscovered; but since the "baselines" there are on a computational grid, the redundancy is very high and the accuracy of the phase determination is improved by averaging the closure phase results over many different triangles. Consider the following example in the redundant case. We recall that $\Delta(0) = 0$ and values of Δ for two other noncollinear baselines can be defined as 0. Let these two other baselines be 12 and 31. Then, when ϕ_{123} is measured, we know $\Delta_{23} = \phi_{123}$. Now define a point 4 such that one of the baselines is repeated (figure 4.8b), say $\mathbf{r}_{34} = \mathbf{r}_{13}$ so that $\Delta_{34} = \Delta_{13}$. Then

$$\phi_{124} = \Delta_{12} + \Delta_{24} + \Delta_{41}$$
$$= \Delta_{12} + \Delta_{24} - 2\Delta_{13} = \Delta_{24}. \tag{4.5}$$

In this way, working with a sufficiently redundant array of receivers, the phases of all of the coherence functions between them can be measured.

The COAST interferometer provided the first optical images which used phase closure with interference between light from three separate telescopes (Baldwin et al. 1996). The data on the paradigm binary star Capella (which has two almost identical components) in this experiment is appropriate to the model in section 3.3.3 and shows clearly the way in which the fringe visibilities and closure phase vary with the baseline, whose projection varies with time. This is shown in figure 4.9. One sees, for example, that until approximately 1.5 hrs, all three values of γ_{ij} have zero phase, so that $\phi_c = 0$. Then, between 1.5 and nearly 3 hrs, γ_{31} alone has phase π, making $\phi_c = -\pi$, and between 3.2 and 4 hrs all three phases are $\pm\pi$, while at 4 hrs the phase of γ_{31} returns to 0, making ϕ_c zero once again. This, of course is a "direct" example, where the phases are rather obvious; the real problem is to deduce the individual phases from the measured closure phases in a manner which is consistent with an image whose intensity is positive definite.

4.3 Image restoration and the crowding limitation

A basic question which applies to any aperture synthesis imaging system asks how good an image can one really obtain, under ideal conditions, of a complicated object. Using aperture synthesis optics, one samples the Fourier transform of the

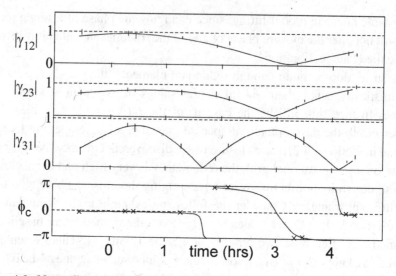

Fig. 4.9. Normalized fringe visibilities and phases determined by phase closure for Capella at 830 nm (Baldwin et al. 1996).

image intensity, the sampling function being the (u, v) plot of the synthesis optics. The synthetic point spread function is then the transform of this plot. One can envisage trying to create a perfect image from the raw aperture synthesis image by deconvolving with the point spread function. But, as is well known, this is quite a difficult problem, and there is no way of restoring information at spatial frequencies where γ was never measured, represented by the regions where the (u, v) plot is zero. In the Fourier plane, deconvolution corresponds to dividing by the (u, v) plot, and therefore to dividing zero by zero. So in fact, what can be done? Below we briefly mention algorithmic methods, and then construct an estimate of the image limitations which arise when observing objects which are clusters of points.

4.3.1 Algorithmic image restoration methods

Two algorithmic methods, developed for aperture synthesis imaging in radio astronomy, are commonly used in optical interferometry. Details of the procedures can be found in Perley and Schwab (1989) and useful summaries are given by Rohlfs (1996) and Monnier (2003). One is called CLEAN (Högbom 1974), in which the raw (or "dirty") image is reproduced by an iterative method as the convolution between the known point spread function and a two-dimensional array of δ-functions. Essentially, the brightest pixel is chosen, and is represented by a δ-function. The associated point spread function is then subtracted from the image and the process repeated until only noise is left. The resulting array of δ-functions is the basis of the "clean" image, which is then smoothed. A second algorithmic method widely used is the maximum entropy method (MEM), which looks for the smoothest image which is consistent with the raw data and the given point spread function, using information entropy as the quantity to be optimized. The image is constrained to be limited in size and to be positive definite, and to have no Fourier components outside the region measured. As might be expected, CLEAN is better for images consisting of clusters of point sources, while MEM is better for extended sources. A third method, which is easy to implement and sometimes gives good results for simple images, is Wiener deconvolution; this is used in the demonstration of aperture synthesis in section 4.6.

4.3.2 The crowding limitation

The crowding limitation attempts to answer the basic question: how many individual sources can be simultaneously imaged with an aperture synthesis system? Suppose there are N subapertures between which interference fringes are measured. We consider them simply as point receivers, arranged within a circular region of diameter D according to a certain pattern. In one limiting case, the receivers are optimally arranged (section 4.1.3) for nonredundancy, so that γ is measured on $N(N-1)$ baseline vectors (including both positive and negative values). Because of the optimal arrangement, we can envisage these vectors as being distributed uniformly within a circle of radius D, so that the shortest one has length approximately $D_{min} = D/\sqrt{N(N-1)} \approx D/N$ for large N. The unique field of view is limited by this *shortest* baseline; point sources separated by distance larger than λ/D_{min} would give rise to spatial oscillations in γ too fast to be sampled correctly by the array of receivers, and therefore would be reconstructed at the wrong points. Basically, this can be understood if we consider just two apertures separated by D/N. The phase difference between the signals received by them from a source at angle θ to the axis is $Dk_0\theta/N$. Since a phase difference can only be measured modulo 2π, only if it is assumed to be between $\pm\pi$ can the phase measurement define θ uniquely, i.e. $|\theta| < \lambda N/2D$. A source outside this range will give a false reconstruction within the range, a process called "aliasing" (Appendix A). As a result, the unique field of view subtends a solid angle of diameter $\lambda N/D$. Now, consider the angular resolution limit for imaging a point source, which is given by the *longest* baseline D as λ/D. The number of point sources that can be distinguished within the unique field of view is therefore $S \approx [(\lambda N/D)/(\lambda/D)]^2 = N^2$. This result is approximate for several reasons: N is not always large, the optimal nonredundant baseline distribution is not uniform, and the receivers are not point receivers.

Now consider the case of the most redundant baseline distribution, a periodic array – say a square lattice, for simplicity. If there are N receivers within the limiting aperture of diameter D, there are only about N distinct baseline vectors, many of them being repeated. Following through the same argument as above, the number of distinguishable points in the field of views $S \approx N$. These points will be intrinsically cleaner, because of the repeated measurements of the same baselines, and so if the number of sources is known to be small, as in astrometric measurements (section 8.4) there are advantages in using redundant or semiredundant arrays.

The result that the number of distinguishable point sources varies from N to N^2 as the array goes from redundant to nonredundant, is called the "crowding limitation;" it seems physically reasonable, and is independent of the type of imaging system, as we shall see in the case of the hypertelescope (chapter 9). The redundant crowding limit above also agrees with the limit of a simple optical imaging system. For an ideal system, the unique field of view extends (in theory) to the whole hemisphere on the imaging side of the lens, i.e. has solid angle 2π. The angular resolution limit is about λ/D, defining a solid angle resolution element ("resel") of $4\pi\lambda^2/D^2$. So the number of resels in the field of view is about $S = D^2/2\lambda^2$. But the number of distinguishable receivers in the aperture (i.e. the number of regions which could, in principle, receive waves with significantly different phases) is its area divided by one square wavelength, so that $N \approx D^2/\lambda^2$. This array is definitely redundant, since similar vectors joining pairs of points can be placed almost anywhere in the aperture. Thus, also for this case, $S \approx N$.

The nocturnal rotation of the Earth extends the number of baseline vectors considerably, and improves the quality of the image reconstruction. The crowding limitation is affected, since the field of view is increased. Aliasing is avoided because there is no longer a degeneracy between the signals from a source outside the unique field of view and a single aliased source within it, independent of the orientation of the shortest baseline. Thus, point sources outside the unique field of view can also be located. The argument has been extended by Koechlin and Pérez (2003) to take into account noise in the interferometric measurements.

4.4 Signal detection for aperture synthesis

4.4.1 Wave mixing and heterodyne recording

An interferometric signal is recorded when two (or more) signals are received simultaneously by a detector, and the detector has a nonlinear response. The simplest nonlinear response is a square-law, and this typifies most optical detectors which respond to the intensity (the square) of the incident amplitude. This is essentially so obvious that we have not thought to discuss it in detail before: if we have two real signals $s_1(t)$ and $s_2(t)$, the detected quantity is $e(t) = [s_1(t) + s_2(t)]^2$. When the signals are represented by complex quantities, we can write the detected quantity as $e(t) = |s_1(t) + s_2(t)|^2$. Even if the detector is not really a square-law detector, such as a diode (which detects zero if the combined signal is negative), the square-law approximation is often assumed as a lowest order. The detector output $v(t)$ is the received signal, filtered by some response function – for example a low-pass filtering mechanism which outputs only changes that occur below some cut-off frequency $\omega_{max} \equiv 2\pi f_{max}$.

Suppose that both signals have frequencies ω considerably above this cut-off: $s_i(t) = a_i(t) \cos(\omega t + \phi_i)$. Then

$$
\begin{aligned}
e(t) &= [a_1(t) \cos(\omega t + \phi_1) + a_2(t) \cos(\omega t + \phi_2)]^2 \\
&= \tfrac{1}{2} a_1^2(t)[1 + \cos(2\omega t + 2\phi_1)] + \tfrac{1}{2} a_2^2(t)[1 + \cos(2\omega t + 2\phi_2)] \\
&\quad + a_1(t) a_2(t)[\cos(\phi_1 - \phi_2) + \cos(2\omega t + \phi_1 + \phi_2)] \quad (4.6) \\
v(t) &= \tfrac{1}{2}[a_1^2(t) + a_2^2(t) + 2a_1(t)a_2(t) \cos(\phi_1 - \phi_2)] \quad (4.7)
\end{aligned}
$$

since the frequency 2ω is too high to pass the filtering. The term $a_1(t)a_2(t) \cos(\phi_1 - \phi_2)$ is the required interference term; note that it preserves the phase difference between the signals. If its average is zero, as in the case of incoherent signals, the output $v(t)$ simply gives the sum of the intensities of the two incident waves.

In most interferometric instruments, the above represents in principle the way in which the interference signal from two receivers is detected. However, consider the case of long-baseline radio interferometry in which the receivers are too far apart for the signals to be brought together to interfere in real time. In this case they are recorded on tape and the coherence function $\gamma(\mathbf{r})$ is calculated from the two recordings at a later time. Actually, in the radio case the signals recorded are electric current oscillations derived from the wave fields sensed by the antennas. These oscillations have both amplitude and phase, both of which are needed in calculating the coherence function. There is no obvious problem in recording the amplitude of the oscillations, but the phase is a different matter. In order to record phase information, it is necessary to record simultaneously with the signal a time-base which is common to both of the signals being measured, so that when the time comes to calculate the coherence function, the product term will have the right phase. This is a nonnegligible problem. Instead of recording the signal and the time-base separately on the same tape, it is much more efficient to record the signals by the heterodyne technique in which they are mixed with a very stable local oscillator (LO) signal having frequency close to the wave frequency, which plays the part of the time-base, using a square-law detector. This creates beats at the difference between the two frequencies, and the beats preserve the amplitude and phase. Then it is only necessary to employ recording techniques working at an intermediate frequency (IF) much lower than that of the signal itself. In optical interferometry, the same technique is used at infrared wavelengths in the ISI interferometer (section 8.5.5).

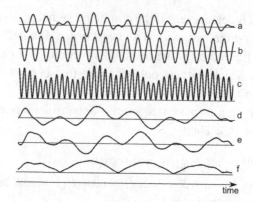

Fig. 4.10. Illustrating the principle of heterodyne detection: (a) the signal, as a function of time; (b) the local oscillator; (c) the square of the sum of the amplitudes of (a) and (b), which is the instantaneous intensity measured by the detector; (d), (e) and (f) the detector output after filtering through a filter which passes frequencies between f_{min} and f_{max} ((d) – real part, (e) – imaginary part and (f) – modulus). The filtering is illustrated in figure 4.11. The observer is interested in the envelope of the signal (a), which is retrieved in (f); its phase can also be found from (d) and (e).

Continuing from (4.7), we now let $s_2(t)$ represent the local oscillator with frequency ω_0, phase ϕ_0 and constant amplitude A, $s_2(t) = A\cos(\omega_0 t + \phi_0)$, and $s_1(t)$ the signal at ω with phase ϕ and amplitude $a(t)$. The squared signal $v(t)$ is then:

$$e(t) = [A\cos(\omega_0 t + \phi_0) + a\cos(\omega t + \phi)]^2 \tag{4.8}$$

$$= \tfrac{1}{2}A^2[1 + \cos(2\omega_0 t + 2\phi_0)] + \tfrac{1}{2}a^2[1 + \cos(2\omega t + 2\phi)]$$
$$+ aA\cos[(\omega + \omega_0)t + (\phi + \phi_0)]$$
$$+ aA\cos[(\omega - \omega_0)t + (\phi - \phi_0)]. \tag{4.9}$$

Now since $2\pi f_{max} \ll \omega$ or ω_0, all the above terms are averaged except the last, and that also unless $(\omega - \omega_0)/2\pi < f_{max}$. Thus the actual IF output is

$$v(t) = \tfrac{1}{2}(A^2 + a^2) + aA\cos[(\omega - \omega_0)t + (\phi - \phi_0)]. \tag{4.10}$$

By including in the output filter a lowest frequency of response f_{min} the first two zero-frequency terms can also be filtered out[†], giving

$$v(t) = aA\cos[(\omega - \omega_0)t + (\phi - \phi_0)] + \text{noise} \tag{4.11}$$

which is the IF heterodyne signal. The principle is illustrated in figures 4.10 and 4.11.

[†] In practice f_{min} should be higher than the frequency of typical fluctuations in the intensities A^2 and a^2 caused by the atmosphere, although we have to be remember that *noise* in the terms A^2 and a^2 at frequencies between f_{min} and f_{max} will still contribute to the measured $v(t)$.

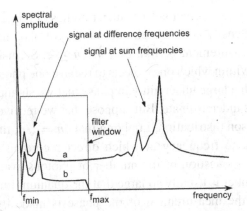

Fig. 4.11. The spectra of the wave (c) in figure 4.10, (a) before, and (b) after filtering through the band-pass filter window shown. Note that the signal shown contains two basic frequencies, so that the sum and difference spectra each contain two peaks. Fourier synthesis based on the filtered spectrum (b) returns the demodulated signals (d), (e) and (f) in figure 4.10.

Nothing is lost in this recording, since $a(t)$ and $\phi(t)$ are there explicitly. But now we remember that the two signals are to be recorded at well-separated points, and each one has its own local oscillator. Suppose that the two local oscillators have very slightly different frequencies, ω_{01} and ω_{02}. We can represent the phase difference between them as $\phi_{01} - \phi_{02} = (\omega_{01} - \omega_{02})t$. This will appear as an error in the difference between the apparent values of $\phi(t)$ in the two recordings, and will result in errors in the calculated $\gamma(\mathbf{r})$. So it is of crucial importance to have exactly synchronized local oscillators, of identical frequencies; in radio astronomy these are nowadays based on atomic clocks or hydrogen masers. The way this is done in optical interferometry will be described in section 8.5.5.

4.5 A quantum interpretation of aperture synthesis

Although classical electromagnetic theory is sufficient for understanding the methods used in aperture synthesis, they must also have a complete quantum description. Several interesting questions arise. At radio frequencies it is usually claimed that there is no problem in recording both amplitude and phase; the electric wave field induces an electric current with related amplitude and phase in the antenna – and that's it. But at optical frequencies, photo-electric detection is a quantum phenomenon, and the output is a series of electrons whose properties are determined by the intensity of the waves alone (section 2.3). In what frequency range does the transition occur from one description to the other? Is it possible to conceive of direct phase measurement at optical frequencies? This question can be discussed in terms of the uncertainty principle. The uncertainties of phase ϕ and intensity are related, when the intensity is translated into number of photons n received during the experimental measurement period, by $\delta\phi \cdot \delta n \geq \frac{1}{2}$. So, using a measurement lasting for the time within which one expects to receive one photon alone, for which $\delta n = 1$, there is such a large uncertainty in phase that we should say that the phase of a single photon is indeterminate. But suppose that we receive a large number of photons with a Poisson distribution, implying that $\delta n = \sqrt{n}$; then $\delta\phi \approx 1/(2\sqrt{n})$. So the question of the frequency at which direct recording of phase becomes impossible is really a question of the number of photons available, and at radio frequencies this number is usually so large that the quantum limit is not relevant.

We assume that the measurement of the phase is made by a method which introduces no noise or uncertainty itself, such as heterodyne mixing with a very intense and coherent local oscillator. For light intensity I collected by a detector with area A for time T, the number of photons is $n = IAT/\hbar\omega$. Now A can be no larger than the coherence area, otherwise different regions of the detector will see different phases. Likewise, T can be no longer than the coherence time, during which the phase is stable. Substituting these for A and T, n becomes the average number of photons within one coherence volume at a given moment (see section 3.3). For light originating from a source of diameter α at frequency ω with bandwidth $\delta\omega$, $A = \lambda^2/\alpha^2$ and $T = 1/\delta\omega$:

$$n = \frac{I\lambda^4}{\alpha^2 c^2 \hbar} \cdot \frac{\omega}{\delta\omega}. \tag{4.12}$$

For the Sun or a typical star at visible wavelengths (I/α^2 does not depend on the distance except as a result of of interstellar scattering and absorption) n is of order 10^{-3} even for broad-band light, for which $\delta\omega/\omega \approx 1$. But, because of the λ^4 dependence, n is very large at radio frequencies, even when the bandwidth is narrow. Of course, when *relative* phases of two waves originating from the same source are measured, the coherence time no longer limits the measurement time and n can be $\gg 1$ for much weaker sources.

Another quantum problem refers to the observation of interference *post factum* in long baseline interferometry. As we discussed above (section 2.3), a consistent description of interference phenomena is obtained if we consider the fields as classical electromagnetic fields until the moment of detection when, on interaction with the detector, their particle behavior becomes dominant. Interference takes place between waves, not between particles. But in long-baseline interferometry the signals are recorded on tape, which is a detection process, and the interference is only apparent when two tapes are compared at a central processing station, maybe months later. The problem is resolved when the time-base or common local oscillator is specifically included. We must remember that without one of these, the

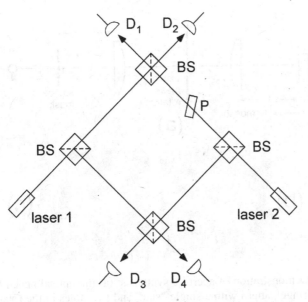

Fig. 4.12. An experiment in which two lasers interfere, and four output signals are obtained. BS is a beam-splitter and D is a detector. The individual signals from detectors D_1 to D_4 consist of randomly arriving photons and contain no signs of the interference (i.e. dependence on the phase shifter P) but correlation between the signals shows the expected sinusoidal dependence on the phase.

phases recorded on the tape are arbitrary and therefore the *post factum* interference between the two signals is meaningless. It is easiest to look at the case of a shared local oscillator, say a hydrogen maser. Then we have two "coherent" sources: the star and the hydrogen maser, which are both sources of delocalized photons. At each station we record the interference between them, which results in a certain output signal from their superposition. This optical signal is converted to data on the tape. The data contains fluctuations and, because the two tapes recorded the fluctuations caused by interference between the same two photons at the same time but at different places, there is correlation between the fluctuations and this is what is discovered when the two tapes are compared at the processing station. The situation is similar to a laboratory experiment (figure 4.12) in which two independent lasers are allowed to interfere at two different places, and the output signals are correlated (Noh et al. 1991).

4.6 A lecture demonstration of aperture synthesis

The idea of recording interference patterns between groups of receivers in different places at different times, and subsequently superposing them to get an image, seems

Aperture synthesis

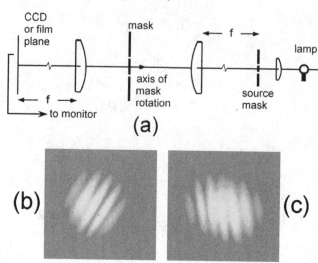

(a)

(b)   **(c)**

Fig. 4.13. Demonstration of aperture synthesis: (a) the optical bench layout; (b) stationary fringe pattern with a single "star" and two holes in the rotating mask; (c) as (b), but with a double star.

somewhat unintuitive, despite the many proofs from radio astronomy that it works. Although the patterns were obtained at different times, we make them interfere with one another to give the equivalent result to what would be obtained from all the receivers simultaneously. Donald Wilson and John Baldwin of Cambridge University devised several years ago a lecture demonstration to show that this indeed works, and we shall describe it in this section in the form that we use it as a student laboratory experiment (Lawson et al. 2002).

At one end of a long optical bench (figure 4.13a) we put an incoherent source consisting of one or more point objects (stars). To take them to astronomical distance, a converging lens with long focal length L is introduced so that the stars are in its focal plane. This source illuminates a mask consisting of a pair of holes in an opaque substrate, representing a pair of telescopes. The dimensions are arranged so that the illumination at the two holes in the mask is partially coherent. Interference between the light from the two holes is observed on an imaging detector (CCD) using the Fizeau configuration, i.e. by putting a converging lens after the mask and observing in its focal plane. At this stage only a set of fringes can be seen (b), with orientation depending on the orientation of the pair of holes; if there are several sources (stars) the fringe pattern may be more complicated (c). The mask is then changed by rotating it in its own plane, and the images recorded on the detector are watched on a monitor. The eye or the phosphor of the monitor sums the images appearing sequentially and, lo and behold! an image of the source appears on the

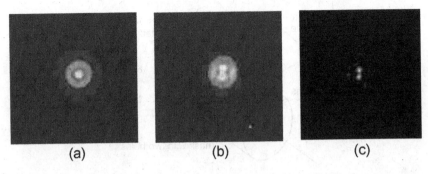

Fig. 4.14. In (a) and (b) we see integrated images when the mask rotates, corresponding to figure 4.13(b) and (c). Deconvolution of (b) using (a) as the point spread function gives the "clean" image (c).

screen (figure 4.14a and b). This has clearly been produced by summing the fringes sequentially, but appears similar to the image that would be obtained had the aperture consisted of all the sequential hole positions simultaneously (an annulus). This shows the basic idea behind aperture synthesis.

To carry out the demonstration successfully, several points have to be considered. First, to obtain partial coherence at the holes in the mask, we require that the individual stars have small enough angular diameter to illuminate both holes in the mask coherently, but that the combined group of stars should produce at most partial coherence between the points (this is the calculation in section 3.2.4 for a pair of disk-like stars). Second, the holes in the mask must be sufficiently small to produce a diffraction pattern with several fringes (see section 3.1.3). Third, the source needs to be monochromatic only to the extent that its bandwidth does not significantly affect the number of fringes visible.

We assume a wavelength $\lambda = 0.5\,\mu$m. If an individual star has diameter s, its coherence distance r_c at this mask, at distance L, is approximately $\lambda L/s$. When $s = 0.2$ mm and $L = 2$ m, we find $r_c \approx 5$ mm. If the pair of holes in the rotating mask have diameter 0.5 mm and are separated by about 2 mm, the coherence of the light transmitted by them will be sufficient to give reasonably good interference fringes. If the distance from the lens after the rotating mask to the detector is 1 m, a pair of holes separated by 2 mm will give Young's fringes separated by 0.25 mm. The central peak of the envelope of the fringes contains only about five fringes. This means that λ only needs to be determined to about $1/5 = 20\%$, which is why only a very broad-band colour filter is sufficient. An observation of the fringes is shown in figure 4.13(b) and (c), for single and double sources. Actually, the results shown were recorded on photographic film, and not a CCD, to facilitate the integration.

As the mask rotates, the fringes rotate with it. If we allow the CCD to record the fringes and project them on the monitor continuously while the mask rotates,

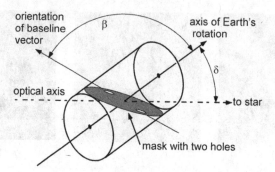

Fig. 4.15. Mask holder to simulate diurnal rotation of two antennas at different latitudes observing a non-polar star.

we are essentially doing Fourier analysis of the coherence function, since each orientation creates a sine wave with the appropriate amplitude and **k** vector, and the persistence of the monitor and the observer's eye sums the sine waves[†]. So the integrated images (figure 4.14a and b) for a complete rotation are the star image convolved with the effective point spread function, which is in this case like that for an annular aperture (figure 3.11). It is interesting to deconvolve the images, since the point spread function is known. The deconvolution in this case was done by the Wiener method as follows. If $f = g * h$ where f, the image, and h, the point spread function, are the measured data, then the Fourier transform of g, the "clean" image, is evaluated as $G = F * H^*/(|H|^2 + n)$. The number n is of the order of the noise in $|H|^2$ and is adjusted to give a clean image. This avoids the "zero-by-zero" division which would otherwise corrupt the calculation at points where $H = 0$. The result is shown in figure 4.14(c).

Although, as described by Lawson et al. (2002), it is easy to extend the experiment to masks containing several holes representing various nonredundant aperture groups, it loses its conviction because the groups are then observed simultaneously and only the rotation sequentially. However, as it has been described above, the geometry represents observation using pairs of antennas at the same latitude observing the pole star. It is easy enough to change the geometry of the rotation to simulate pairs at different latitudes observing stars at different elevations; this is shown in figure 4.15. Also, by putting the CCD immediately after the rotating mask, the (u, v) diagram can be recorded directly.

The demonstration was originally performed without lenses, which emphasizes the point that the telescope optics are irrelevant to aperture synthesis. This works

---

[†] Since the pattern is an intensity pattern, which is positive definite, there is also a constant term added, which creates an unacceptable background if the star cluster is too complex. But for a few sources it can be tolerated.

because the pair of pinholes produce Young's fringes also in the near-field (Fresnel diffraction) domain. However, in the near-field case the fringe positions are dependent on that of the origin of the mask, so that it is important to ensure that the mask rotates around the center-point between the pinholes. For Fraunhofer diffraction this is not so, which makes the alignment a little easier.

References

Baldwin, J. E., M. G. Beckett, R. C. Boysen *et al.* (1996). *Astron. Astrophys.*, **306**, L13.
Cornwell, T. J. (1989). *Science*, **245**, 263.
Golay, M. J. E. (1970). *J. Opt. Soc. Am.*, **61**, 272L.
Högbom, J. A. (1974). *Astron. Astrophys. Suppl.*, **15**, 417.
Jennison, R. C. (1958). *Mon. Not. R. Astron. Soc.*, **118**, 276.
Keto, E. (1997). *Astrophys. J.*, **475**, 843.
Koechlin, L. and J.-P. Pérez (2003). *Interferometry for Optical Astronomy II*, ed. W. A. Traub, Proc SPIE, **4838**, 411.
Lawson, P. R. (2000) ed. *Principles of Long Baseline Stellar Interferometry*, NASA-JPL.
Lawson, P. R., J. E. Baldwin and D. Wilson (2002). *Proc. SPIE*, **4838**, 404.
Monnier, J. D. (2003). *Rep. Prog. Phys.*, **66**, 789.
Noh, J. W., A. Fougères and L. Mandel (1991). *Phys. Rev. Lett.*, **67**, 1426.
Perley, R. A. and F. R. Schwab, (1989). edrs *Synthesis Imaging in Radio Astronomy*, San Francisco: Astronomical Society of the Pacific.
Rohlfs, K. (1996). *Tools of Radio Astronomy*, 2nd edn, Berlin: Springer.
Ryle, M. (1952). *Proc. Roy. Soc.*, **A 211**, 351.
Thompson, A. R. (2001). *Interferometry and Synthesis in Radio Astronomy*, 2nd edn, New York: Wiley.

5

Optical effects of the atmosphere

5.1 Introduction

The atmosphere behaves like a very thick bad piece of glass in front of your tele-scope, a piece which is constantly changing. The result of this bad optical element is that the image of a point star is not what the simple physics would lead us to expect, namely the diffraction pattern of the geometrical entrance aperture, but a much more complicated and diffuse image. The image has two general properties: an envelope, which is the image recorded in a long-exposure photograph (longer than, say, one second) and, within it, an internal speckle structure which is con-tinuously and rapidly changing and can only be photographed using a very short exposure (less than about 5 ms, as in figure 5.1). The angular diameter of the enve-lope, which is called the "seeing," has a value between 0.5 and 2 arcsec at a good observing site; this is determined by the averaged properties of the atmosphere, which are the subject of this chapter. On the other hand, although the speckle struc-ture is changing continuously, the angular diameter of its smallest distinguishable features correspond to the diffraction limit of the complete telescope aperture. For comparison, figure 5.2 shows the same effect in the laboratory when imaging a point source through the bad optics of a polythene bag.

Michelson (1927), in his book *Studies in Optics*, refers to the effect of the "seeing" on his 1890 stellar interferometer. The stellar interferometer at that time consisted of two apertures, whose separation could be varied, masking the objective of the telescope. He writes as follows:

In order to test the effect of atmospheric disturbances, trials with the 40-inch Yerkes refrac-tor and the 60-inch and 100-inch reflectors at Mount Wilson were made which, even with relatively poor "seeing" gave excellent results, showing that up to these distances and inferentially to much greater, the effects of atmospheric disturbances were not to be feared.

Fig. 5.1. Image of a point star through a 5-m telescope with an exposure of a few ms.

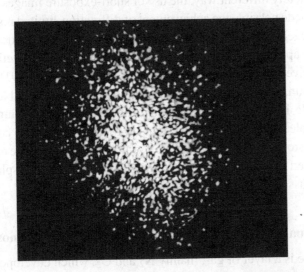

Fig. 5.2. Laboratory image of a point source through a polyethylene sheet.

To this he adds a footnote:

Doubtless this rather unexpected result may be explained as follows: The confusion of the image in poor seeing is due to the integrated effect of elements of the incident light waves, elements which are not in constant phase relation in consequence of inequalities in the atmosphere due to temperature differences; the optical result being a "boiling" of the image, closely resembling the appearances of objects viewed over a heated surface.

In the case of the two elements at the opposite ends of a diameter of the objective, the same differences in phase produce a motion of the (straight) interference fringes (and not a confusion) and if, as is usually the case, this motion is not too rapid for the eye to follow, the visibility of the fringes is quite as good as in the case of perfect atmospheric conditions.

Today, it has been realized that the "confusion" of the image mentioned by Michelson, and which he contrasts with the (straight) interference fringes from two apertures, is in fact also an interference effect which can be exploited like the fringes. In fact, since it is generated by the full aperture of a telescope, it contains more information than the interference produced by the pair of apertures.

In astronomical interferometry, as opposed to direct imaging, we find that the atmosphere affects measurements in several ways. On the one hand, it adds noise to multiple-aperture interferometry, which, as Michelson remarked, he was able to live with in 1890 when observing bright stars; today, using electronic fringe recording, its main effect is to limit the precision to which fringe visibility and phase can be measured for a given stellar magnitude because of photon statistics (section 8.3.11–12). In a completely different way, the use of short-exposure images of the speckle structure has made possible the achievement of diffraction-limited resolution with a single-aperture telescope, even in the presence of the seeing limitation, by the technique of single-aperture speckle interferometry (Labeyrie 1970) (section 6.3) and its various derivatives such as triple correlation imaging (Weigelt 1991) (section 6.4).

In the long run, when very weak sources are being observed and long exposures are unavoidable in order to collect enough photons, understanding atmospheric optics has enabled adaptive optical systems to be built in order to compensate the atmospheric distortions and create long-exposure images which are almost as good as those obtained with a space telescope situated outside the atmosphere. Adaptive optics are discussed briefly in section 5.8.

5.2 A qualitative description of optical effects of the atmosphere

The atmosphere is a layer of gas, mainly N_2 and O_2, which envelopes the Earth to a thickness of about 8 km; that is to say, it has pressure which is 1 bar (standard atmosphere) at the Earth's surface and drops off approximately exponentially by a factor of 10 every 18 km. The density, which is a function of both pressure and temperature, drops in a similar fashion. If all the gas were compressed to a layer of constant pressure equal to 1 bar at 300 K, it would be about 8 km thick.

For a given gas or homogeneous mixture, the refractive index difference from unity, $n - 1$, is proportional to the density. If the density were completely uniform, the atmosphere would have no effect on seeing, since it would simply result in a uniform phase retardation over the whole telescope aperture, and would therefore change the phase of the transmitted light uniformly by a given amount. However,

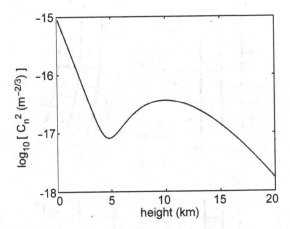

Fig. 5.3. Typical height profile of atmospheric turbulence.

the refractive index is not exactly uniform because of fluctuations in the density and thus, proportionally, in $n - 1$. Fluctuations have structures which vary in both space and time, which we shall discuss presently in more detail. The origins of these fluctuations are local variations of humidity, pressure and, most importantly, temperature, caused by hydrodynamic wind turbulence. The fluctuations also change within times of the order of milliseconds, which are very long compared with the wave period of order 10^{-15} s, and so from the optical point of view the atmosphere at each moment appears frozen; in other words, atmospheric fluctuations have no noticeable effect on the wave frequency, only on its phase, averaged over very many periods.

Quantitatively, the importance of fluctuations in refractive index is measured by a quantity C_n^2, which is the mean square of the difference in n between points separated by 1 m. At ground level C_n^2 may have a value of the order of $10^{-14} \mathrm{m}^{-\frac{2}{3}}$ (the units are explained in section 5.3.2). Since the fluctuations in density are in the first place proportional to the density itself, so is C_n^2, which therefore decays with height generally like the density.

When meteorological facts are taken into account, it appears that a rough approximation to the atmospheric turbulence can be made by assuming it to have two active layers. One layer is close to the ground and has a thickness of about 3 km within which C_n^2 falls about one order of magnitude; the second is at the tropopause, about 10 km above the ground, and is somewhat thicker (figure 5.3). The effect of atmospheric nonuniformity is to produce random phase changes in the waves reaching the telescope aperture. The phase change is an integral effect, having the value $\delta\phi = \int k_0 \delta n(z) \mathrm{d}z$ integrated along the line of sight, where δn is the deviation of n from its mean value. One can see immediately that if δn is approximately achromatic, $\delta\phi$ is smaller at longer wavelengths, giving an

Optical effects of the atmosphere

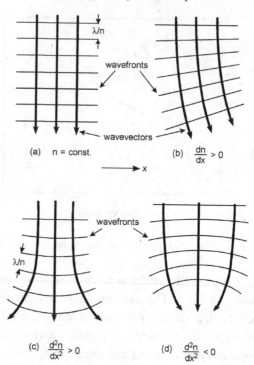

Fig. 5.4. Effects of inhomogeneous refractive index on light rays.

advantage to infrared, and also at the zenith where the length of the integration path through the atmosphere is shortest. This latter fact is clear when we compare the twinkling of stars at the zenith with those near the horizon.

The two layers mentioned above have different effects on the light reaching the telescope. A light ray passing through a layer of inhomogeneous medium suffers two effects: it is delayed (relative to the mean homogeneous medium) by an amount depending on the integrated value of δn, and it is deviated angularly by an amount depending on the lateral gradient of the delay. The latter effect is illustrated by figure 5.4 where we construct successive wavefronts separated by λ/n by using Huygens's principle. Remembering that the geometrical light ray propagates normal to the wavefronts, the figure shows that there are also focusing and defocusing lensing effects resulting from second derivatives like d^2n/dx^2, in addition to the angular deviation resulting from the lateral gradient in δn, dn/dx. Since the observer is at ground level, the lensing effects of the upper turbulence layer are distant enough to develop into intensity variations[†].

[†] The intensity variations can often be seen when sunlight illuminates a wall after passing a distant heated region. They can be seen briefly during solar eclipses when most of the solar disk is masked so that the Sun is like a point source, and are also visible on a telescope's primary mirror pointed toward a bright star when the eyepiece is removed.

If a star is observed through a converging region, $d^2n/dx^2 < 0$, it seems to be a little brighter, whereas if through a diverging region, $d^2n/dx^2 > 0$, it will seem a little weaker; this is one origin of twinkling, since the derivatives change with time. If the change in n is more abrupt, there can also be diffractive effects, resulting in the intensity variation characteristic of the corresponding diffraction pattern being observed at the ground. On the other hand, although the lower turbulence layer changes the phase of the waves reaching the telescope considerably more than does the upper layer, because n and δn are larger, its effect on twinkling is very small because there is no long path to develop the intensity variations. Both turbulent layers affect multiple-subaperture astronomical interferometry. The changes in phase (mainly originating in the lower layer) cause fringe movement or distortion, depending on how the interference takes place, and can result in uncertainty of the phase of the measured coherence function. The intensity variations (mainly due to the upper layer) result in a loss of contrast.

Another effect to which the upper and lower turbulence layers contribute differently is to the dimensions of the *isoplanatic patch* (section 5.4.2), which is of particular importance in adaptive optics (section 5.8). If we compare the effects of turbulence on observations of two different stars, at slightly different positions in the sky, clearly we have to take into account the fact that each star is observed through a different sample of the atmosphere. Since the lines of sight diverge as the height grows, they progressively sample regions of atmosphere with greater separation, and so there is increasing importance to the spatial correlation between fluctuations at higher levels. The isoplanatic patch is the angular region, of order a few arcseconds, within which the integrated effect of fluctuations at all heights can be considered the same function of x, y and t; the size of this region is mainly affected by the turbulence at greater heights. Its importance in adaptive optics arises because the atmospheric correction made according to a point reference star can be used to correct a second unknown star only if the two are within the same isoplanatic patch.

5.3 Quantitative measures of the atmospheric aberrations

5.3.1 Kolmogorov's (1941) description of turbulence

The history of turbulence analysis begins with two famous papers by Kolmogorov (1941a, b) – generally referred to as K41 – which described for the first time a simple model for the spatial and temporal behavior of turbulent motion of a fluid. The tests of Kolmogorov's predictions are today still a field of active research. Turbulent motion arises when motion is dominated by inertial rather than viscous effects, so that kinetic energy of flow may be transferred from one direction to another, or from one scale to another, but is not damped out and converted to heat by viscosity.

The basic idea of K41 is that turbulence is created by heat input to a fluid, at a rate ϵ per unit mass and time. This creates kinetic energy of motion of the gas per unit mass $\sim v_l^2$ in a coherent region (a turbulence cell) of dimension l with typical velocity v_l. The time taken to create this motion is $\tau = l/v_l$, and so $\epsilon \sim v_l^2/\tau \sim v_l^3/l$. On the other hand, since the velocity changes randomly from one cell to the next, the velocity gradient ∇v has size about v_l/l. A velocity gradient in a viscous medium with kinematic viscosity v converts the kinetic energy to heat with a rate of dissipation of energy $|\nabla v|^2 v \sim v v_l^2/l^2$. It follows that the input energy cannot be dissipated as heat unless $v_l^3/l < v v_l^2/l^2$ or $l v_l/v < 1$. The quantity $\mathrm{Re}_l \equiv l v/v$ is called the *Reynold's number* for scale l; this number is the most basic nondimensional quantity in hydrodynamics. Since all the numerical factors have been omitted in this argument, it shows simply that turbulence energy cannot be dissipated by viscosity unless the scale is such that $\mathrm{Re}_l < \mathrm{Re}_{\mathrm{crit}}$, which is some number that has to be determined experimentally and depends in fact on the geometry of the system.

When the whole system is in a steady dynamic state, with a constant energy input, motion is transferred from large scales to smaller and smaller scales, until it is eventually absorbed by viscosity at the smallest scale. The smallest typical velocities v_0 and scales l_0 of the turbulence are determined by the energy input; since $v_0 \sim (\epsilon l_0)^{\frac{1}{3}}$ and $\epsilon = v v_0^2/l_0^2$, it follows that

$$l_0 \sim (v^3/\epsilon)^{\frac{1}{4}}; \quad v_0 \sim (v\epsilon)^{\frac{1}{4}}. \tag{5.1}$$

At the other end of the scale, the largest eddies, with dimension L_0 and velocities v_{L_0}, satisfy $\epsilon \sim v_{L_0}^3/L_0$. We can then calculate the ratio between the largest scales L_0 and the smallest scales l_0:

$$l_0 = \left(\frac{v^3}{\epsilon}\right)^{\frac{1}{4}} = \left(\frac{v^3 L_0}{v_{L_0}^3}\right)^{\frac{1}{4}} \tag{5.2}$$

$$\frac{L_0}{l_0} = \left(\frac{v_{L_0}^3 L_0^3}{v^3}\right)^{\frac{1}{4}} \tag{5.3}$$

$$= \mathrm{Re}_{L_0}^{\frac{3}{4}}, \tag{5.4}$$

and the related velocities

$$\frac{v_{L_0}}{v_0} = \mathrm{Re}_{L_0}^{\frac{1}{4}}. \tag{5.5}$$

The two scales, l_0 and L_0 are called the inner and outer turbulence scales, and their ratio is a function of the Reynolds number of the turbulence. Typically, under atmospheric conditions (remember that everything depends on ϵ), Re_{L_0} is of order 10^6, so that the ratio L_0/l_0 is about 30,000. Actual values might be $L_0 = 15$ m and $l_0 = 0.5$ mm.

5.3.2 Parameters describing the optical effects of turbulence: Correlation and structure functions, B(r) and D(r).

As we saw in section 5.1 it is the fluctuations in refractive properties which lead to optical aberrations. In physical optics, correlation functions are widely used to describe statistically varying fields, so we might find it natural to express the fluctuations in such terms; however, since the mean value of the refractive index can vary slowly with position and time, it becomes uncertain which changes should be considered as drifts of the mean and which as fluctuations. Kolmogorov (1941a, b) devised a slightly different way of expressing his results which overcomes this difficulty. He defined a *structure function* to relate the values of a function $f(\mathbf{r})$ at neighboring points $\mathbf{r}_1$ and $\mathbf{r}_2$: $D_f(\mathbf{r}_1, \mathbf{r}_2) \equiv \langle [f(\mathbf{r}_1) - f(\mathbf{r}_2)]^2 \rangle$, where $\langle \ldots \rangle$ indicates a time-averaged value. In a homogeneous region, this is a function only of $\mathbf{r} = \mathbf{r}_1 - \mathbf{r}_2$ and not of their individual values, so that we can write

$$D_f(\mathbf{r}) = \langle [f(\mathbf{r}_1 + \mathbf{r}) - f(\mathbf{r}_1)]^2 \rangle . \tag{5.6}$$

Of course, the structure function can be related to a correlation function for the same variable. The correlation function $B_f(\mathbf{r})$ for normalized $f(\mathbf{r})$ is defined as $B_f(\mathbf{r}) = \langle [f(\mathbf{r}_1 + \mathbf{r}) - \bar{f}][f(\mathbf{r}_1) - \bar{f}] \rangle$, where $\bar{f}$ is the mean value of f. It can then be shown easily that, for a homogeneous region,

$$D_f(\mathbf{r}) = 2[B_f(0) - B_f(\mathbf{r})] . \tag{5.7}$$

The structure function for refractive index n is in fact very appropriate for studying the optical properties of the turbulent atmosphere, since the behavior of light rays is indeed governed by the refractive index *differences* between neighboring points on the wavefront, and not just by deviations from the mean. Tatarski (1961) described most of the basic concepts more than 40 years ago in his book *Wave Propagation in a Turbulent Medium*. Most significant is to see how the structure function for the refractive index $D_n(\mathbf{r})$ depends on $\mathbf{r}$. We assume the turbulence to be isotropic so that the vector separation $\mathbf{r}$ can be replaced by its magnitude r. When r lies between the inner and outer turbulence scales, the structure function for velocity fluctuations $D_v(r)$ can only be a function of the energy input ϵ and r itself. By dimensional arguments, it then follows from the definition (5.6) that in

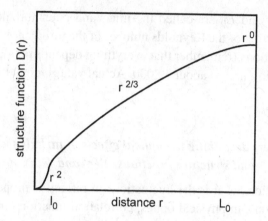

Fig. 5.5. Schematic diagram of the structure function $D_n(r)$. A typical value of C_n^2 is $10^{-17}\mathrm{m}^{-\frac{2}{3}}$.

the region $l_0 \ll r \ll L_0$

$$D_v(r) = C \cdot (\epsilon r)^{\frac{2}{3}}, \tag{5.8}$$

where C is a numerical constant. The spatial structure of the turbulence having been described this way, it is argued that other derived functions such as temperature, pressure, density and refractive index will have similar structures, so that one can write for the refractive index fluctuations:

$$D_n(r) = C_n^2 r^{\frac{2}{3}}, \tag{5.9}$$

where C_n^2 is an experimentally determined parameter which characterizes the strength of the turbulence. The relationship to $r^{\frac{2}{3}}$ has been quite well confirmed experimentally. For small separations, when $r < l_0$, Taylor expansion of $n(r)$ to first order gives $D_n(r) \propto r^2$, which joins smoothly to (5.9) as shown in figure 5.5. When $r > L_0$, the functional form of D_n has no ubiquitous form, but in view of (5.7), and the fact that $B_n(\infty) = 0$, $D_n(r)$ must approach the constant value $2B_n(0)$.

5.4 Phase fluctuations in a wave propagating through the atmosphere

The structure function $D_n(r)$ for the refractive index fluctuations contains enough information to build up a fairly complete picture of the atmospheric degradation of an optical telescope, which can also be confirmed experimentally. The main effect of the turbulence is to disturb the relative phases of the incident wave at different points in the aperture. Following the methods used in an important review by Roddier (1981), we can calculate the integrated effect of the whole atmosphere at the ground plane. We assume that the phase fluctuations are relatively small, so that when a plane wave is incident from outside the atmosphere we can calculate its phase properties at the ground by integrating along columns of air determined by the geometrical path it would have if the atmosphere were completely uniform. To make the arguments simpler, in the case of homogeneous turbulence we can limit ourselves to the (x, z) plane without losing generality and describe the complex wave amplitude at $z = h$ by amplitude and phase

$$\Psi(x, h) = |\Psi(x, h)| \exp[i\phi(x, h)] \tag{5.10}$$

and assume for the moment that only the phase is affected, i.e. $|\Psi(x, h)| = 1$. The spatial correlation function $\Gamma(r)$ at h has the same meaning as the correlation function $B_\Psi(r)$ in Kolmogorov's notation, which we shall continue to use here:

$$B_\Psi(r) = \langle \exp[i\phi(x)] \exp[-i\phi(x + r)] \rangle \tag{5.11}$$
$$= \langle \exp[i(\phi(x) - \phi(x + r))] \rangle . \tag{5.12}$$

Assuming that the value of ϕ is small, this can be expanded to give

$$B_\Psi(r) = 1 + \langle i(\phi(x) - \phi(x + r)) \rangle \tag{5.13}$$
$$- \tfrac{1}{2} \langle (\phi(x) - \phi(x + r))^2 \rangle + \ldots \tag{5.14}$$
$$\approx \exp[-\tfrac{1}{2} D_\phi(r)] \tag{5.15}$$

since for homogeneous turbulence $\langle \phi(x) \rangle = \langle \phi(x + r) \rangle$.

To relate $D_\phi(r)$ to the refractive index structure function we integrate the phase $\phi(x, z)$ along two columns of height h at $x = 0$ and $x = r$:

$$\phi(x, 0) = \int_0^h \frac{d\phi}{dz} \, dz = k_0 \int_0^h n(x, z) dz , \tag{5.16}$$

from which

$$D_\phi(r) = k_0^2 \left\langle \left(\int_0^h [n(0, z) - n(r, z)] \, dz \right)^2 \right\rangle . \tag{5.17}$$

We write the squared integral as a double integral by introducing a second variable z'

$$\int_0^h \int_0^h [n(0, z) - n(r, z)][n(0, z') - n(r, z')] dz dz' \tag{5.18}$$
$$= \int_0^h \int_0^h [n(0, z)n(0, z') + n(r, z)n(r, z')$$
$$- n(r, z)n(0, z') - n(0, z)n(r, z')] dz dz' . \tag{5.19}$$

On taking the average of each term, (5.19) can be written

$$2 \int_0^h \int_0^h [B_n(z - z') - B_n(\sqrt{r^2 + (z - z')^2})] \mathrm{d}z \mathrm{d}z', \qquad (5.20)$$

in which we have used the fact that the turbulence is homogeneous and isotropic and therefore B_n only depends on the scalar distance between the points of comparison. We can rewrite this in terms of the structure function $D_n(r)$ using the general relationship (5.7) and, with a variable $z'' \equiv z - z'$,

$$D_\phi(r) = k_0^2 \int_0^h \int_{z-h}^z [D_n(\sqrt{r^2 + z''^2}) - D_n(z'')] \mathrm{d}z'' \mathrm{d}z \qquad (5.21)$$

$$= h k_0^2 \int_{z-h}^z [D_n(\sqrt{r^2 + z''^2}) - D_n(z'')] \mathrm{d}z''. \qquad (5.22)$$

For the functional form $D_n(x) = C_n^2 x^{\frac{2}{3}}$ and integration limits replaced by $\pm\infty$, we then have the definite integral

$$\int_{-\infty}^\infty [(r^2 + x^2)^{\frac{1}{3}} - x^{\frac{2}{3}}] \mathrm{d}x = 2.91 r^{\frac{5}{3}}. \qquad (5.23)$$

It follows that

$$D_\phi(r) = 2.91 k_0^2 h C_n^2 r^{\frac{5}{3}}. \qquad (5.24)$$

The $\frac{5}{3}$ power law could have been obtained from dimensional arguments, but here we also have the prefactors. When the value of C_n^2 varies with the height z, clearly the term $h C_n^2$ is to be replaced by $\int_0^h C_n^2(z) \mathrm{d}z$, and if the light is incident at an angle γ to the zenith, this is multiplied by $(\cos \gamma)^{-1}$ for the longer slant path.

The final result for the phase correlation, or coherence function, is then

$$\Gamma(r) \equiv B_\Psi(r) = \exp[-\tfrac{1}{2} D_\phi(r)] \qquad (5.25)$$

$$= \exp \left[-\frac{1.46}{\cos \gamma} r^{\frac{5}{3}} k_0^2 \int_0^h C_n^2(z) \mathrm{d}z \right]. \qquad (5.26)$$

To insert approximate numbers, we note from figure 5.3 that $C_n^2(h)$ can be reasonably well approximated up to 5 km by $10^{-15} \exp(-h/1000)\ \mathrm{m}^{-\frac{2}{3}}$, where h is in meters. Thus $\int_0^\infty C_n^2(z) \mathrm{d}z$ is about $10^{-12}\ \mathrm{m}^{\frac{1}{3}}$. The bump at height 10 km adds approximately $3 \cdot 10^{-13}\ \mathrm{m}^{\frac{1}{3}}$ to the integral. For $\lambda = 0.5\ \mu\mathrm{m}$ at $\gamma = 0$, $D_\phi(r)$ is approximately $600 r^{\frac{5}{3}}\ \mathrm{rad}^2$. At a separation of $r \approx 8.5\ \mathrm{cm}$ this gives π^2, which is in reasonable agreement with observed phase correlation distances at ground level. The phase decoherence mainly results from turbulence close to ground level (i.e. the integral is dominated by the first km); we shall see later that other turbulence effects contain height weighting which can emphasize turbulence at higher levels (figure 5.7a).

5.4.1 Fried's parameter r_0 describes the size of the atmospheric correlation region

If a telescope has an aperture diameter D which is small enough that the atmospherically distorted phase front from a point star is well correlated over the whole of it, the resolution of the telescope is determined by D as if the atmosphere had no effect. On the other hand, if D is very large, the resolution is determined by the atmospheric properties (the seeing), and the large telescope diameter is only important in collecting more light. A physically intuitive parameter which describes the phase correlations was defined by Fried (1966). Fried's parameter r_0 is the value of D for which the two limitations are equivalent.

For a telescope with infinite aperture, the point spread function depends only on the atmospheric fluctuations and is the Fourier transform of the correlation function of the wave $B_\Psi(r) = \exp(-ar^{-\frac{5}{3}})$, which was calculated (5.26) on the assumption that $|\Psi|$ is constant. Because the correlation function is axially symmetric, its two-dimensional Fourier transform is a function only of the radial Fraunhofer diffraction variable $u = k_0\theta$ for small θ. The transform can be calculated analytically, but all we need here is the fact that, since $B_\Psi(r)$ is a function of the variable $a^{-\frac{3}{5}}r$, its transform scales as $a^{\frac{3}{5}}$ (Appendix A). On the other hand, without the atmosphere, the telescope with aperture radius R has a point spread function $|F(u)|^2 = 4\pi^2 R^4 [J_1(uR)/(uR)]^2$ whose dependence on u scales as R^{-1}. Numerically, we can compare profiles and find corresponding values R_a and a which give the same resolution; the criterion of "same resolution" is a bit fuzzy, because the profiles are different, but we can compare, for example, the "full width at half maximum" (FWHM) of the two point spread functions. Then we know, from the scaling relations of the Fourier transforms, that for the same resolution, $a^{\frac{3}{5}} \sim R_a^{-1}$, i.e. $R_a \propto a^{-\frac{3}{5}}$. The result of a numerical comparison using the FWHM is that

$$r_0 \equiv 2R_a = \left[0.6(\cos\gamma)^{-1}k_0^2 \int_h^\infty C_n^2(z)dz \right]^{-\frac{3}{5}}. \tag{5.27}$$

Using a different resolution criterion, Fried (1966) obtained 0.45 instead of 0.6 in the above expression. The wavelength dependence of r_0 follows, assuming a nondispersive atmosphere, as $\sim \lambda^{\frac{6}{5}}$. A typical value for r_0 at $\lambda = 0.5\,\mu m$, using again $1.3 \cdot 10^{-12}\,m^{\frac{1}{3}}$ for the integral, is 5.6 cm (6.5 cm using Fried's expression). In terms of r_0, the phase structure function (5.24) can be written

$$D_\phi(r) = 6.88(r/r_0)^{\frac{5}{3}}. \tag{5.28}$$

Actually, one should notice that the power $\frac{5}{3}$ is close to $\frac{6}{3} = 2$, so that qualitative observations might suggest that the point spread function is the Gaussian $\exp(-u^2/2b)$, which would correspond to a Gaussian correlation function $B_\Psi(r)$

for the atmospheric turbulence. This could be explained as resulting from random uncorrelated atmospheric fluctuations statistically distributed uniformly in space. However, careful measurements of the long-exposure (i.e. time-averaged) point spread function profiles for large telescopes ($D \gg r_0$) show that (5.27) is a significantly better description than the Gaussian (King 1971). $B_\Psi(r)$ has also been measured directly (Roddier 1976) by interferometry and the $\frac{5}{3}$ law confirmed. These are further experimental indications of the basic correctness of K41!

There is another intuitive way, described by Tatarski (1961), of looking at the phase correlations in terms of scattering. We consider the fluctuations to be the superimposed result of many weak events scattering an incident plane wave, and occurring at points randomly distributed throughout the atmosphere, with density indicated by the value of $C_n^2(h)$. This means that most of the scatterers are located within the first km or two of the ground (figure 5.3). A scattering event at height H radiates a spherical wave which interferes with the propagating plane wave, creating an on-axis region of approximately uniform phase (first Fresnel zone) of radius $\sqrt{\lambda H}$. Outside these regions the interference phase oscillates increasingly rapidly. The sum of the many scattering events is then dominated by such regions of uniform phase, the oscillating parts, on the average, canceling out. This picture also agrees qualitatively with the known seeing. A uniform domain of size $r = \sqrt{\lambda H}$ diffracts light to angles of order $\theta = \lambda/r = \sqrt{\lambda/H}$, and for $\lambda = 5 \cdot 10^{-7}$m and $H \approx 10^4$ m, the result is about 1.5 arcsec, which is of the order of a typical experimental seeing value. The wavefront from a point star at the entrance to a telescope can thus be described as having fluctuating phase and amplitude with a distance scale of the order of $\sqrt{\lambda H}$ which is about 7 cm.

5.4.2 Correlation between phase fluctuations in waves with different angles of incidence: the isoplanatic patch

Another aspect of interest is the correlation between the phase field $\phi(x)$ for waves incident in different directions. This affects the speckle patterns seen for stars at different points in the sky. It is particularly important for adaptive optics (section 5.8), because the adaptive system calculates the $\phi(x)$ for a reference source observed at a particular place in the sky, and uses the calculated values to correct the image of an unknown star which may be in a slightly different place; but it assumes the same function. The question is, how does the function vary with angular separation between the reference and unknown stars?

The phase correlation can be calculated, in a manner very similar to that used in the previous section, by assuming that the two columns used in the analysis meet at the ground $z = 0$ and have a small angle θ between them. The distance r between the two columns at height z is then $z\theta$ and (5.17) becomes:

$$D_\phi(r) = k_0^2 \left\langle \left(\int_0^h [n(0, z) - n(z\theta, z)] \, dz \right)^2 \right\rangle, \tag{5.29}$$

from which we continue

$$\int_0^h \int_0^h [n(0, z) - n(z\theta, z)][n(0, z') - n(z'\theta, z')] dz dz' \tag{5.30}$$

$$= \int_0^h \int_0^h [n(0, z)n(0, z') + n(z\theta, z)n(z'\theta, z')$$

$$-n(z\theta, z)n(0, z') - n(0, z)n(z'\theta, z')] dz dz'. \tag{5.31}$$

The averages of the terms now become

$$\int_0^h \int_0^h [B_n(z - z') + B_n(\sqrt{(z - z')^2(1 + \theta^2)} - B_n(\sqrt{(z\theta)^2 + (z - z')^2})$$

$$- B_n(\sqrt{(z'\theta)^2 + (z - z')^2})] dz dz'. \tag{5.32}$$

If θ is small, θ^2 can be neglected with respect to unity, so that (5.32) can be simplified to:

$$2 \int_0^h \int_0^h [B_n(z - z') - B_n(\sqrt{(z\theta)^2 + (z - z')^2})] dz dz'. \tag{5.33}$$

In terms of the structure function $D_n(r)$ (5.7) with $z'' \equiv z - z'$ again, and using the form $D_n(x) = C_n^2 x^{\frac{2}{3}}$,

$$D_\phi(\theta) = k_0^2 \int_0^h \int_{z-h}^z [D_n(\sqrt{(z\theta)^2 + z''^2}) - D_n(z'')] dz'' dz \tag{5.34}$$

$$= 2.91 k_0^2 \theta^{\frac{5}{3}} \int_0^h C_n^2(z) z^{\frac{5}{3}} dz. \tag{5.35}$$

The two important features of (5.35) are the dependence on $\theta^{\frac{5}{3}}$ and the way in which $C_n^2(z)$ appears multiplied by $z^{\frac{5}{3}}$ in the integral. This gives increasing importance to C_n^2 with height; at ground level, the degree of turbulence does not affect $D_\phi(\theta)$, but the effect increases quickly with z (figure 5.7b). Using the data shown in figure 5.3, we find that D_ϕ reaches a value of π^2 rad^2 for about $\theta \approx 4$ arcsec in the visible. This defines the dimension of what is called the *isoplanatic patch*, within which angular region the spatial phase fluctuations can be considered to be correlated. The value estimated above compares well with experimental determinations (Nisenson and Stachnik 1978). From (5.35) one immediately sees that since $D_\phi(\theta)$ is proportional to $k_0^2 \theta^{\frac{5}{3}}$, the angular size of the isoplanatic patch is proportional to $\lambda^{\frac{6}{5}}$, assuming that C_n^2 is not strongly wavelength dependent.

5.5 Temporal fluctuations

5.5.1 The wind-driven "frozen turbulence" hypothesis

The relationship between temporal and spatial fluctuations in the atmosphere can be understood by a simple hypothesis due to G. I. Taylor. We assume that in a frame moving with the local wind velocity the turbulent structure changes very slowly, and what we see as temporal fluctuation is in fact the spatial structure being blown across the field of view. The scaling argument which justifies this assumption, in the spirit of section 5.3.1, goes as follows (Tatarski 1961).

The wind velocity normal to the line of sight is v_n. For an optical path length L the transverse correlation distance of amplitude fluctuations is $\sim \sqrt{\lambda L}$ (section 5.4.1) and thus the correlation time τ_0 will be, according to the frozen-turbulence hypothesis, $\tau_0 \sim \sqrt{\lambda L}/v_n$. Now we recall from section 5.3.1 that the typical lifetime of a fluctuation of scale l is $\tau_l \sim l/v_l \sim l/(\epsilon l)^{\frac{1}{3}}$, where v_l was the typical random velocity in a coherent region and ϵ the heat input. For correlation distance $l \sim \sqrt{\lambda L}$, it follows that the time-scale for changes is $\tau \sim \sqrt{\lambda L}/(\epsilon \sqrt{\lambda L})^{\frac{1}{3}}$. For the frozen-turbulence hypothesis to be acceptable, τ must be considerably larger than τ_0, i.e.

$$v_n \gg \left(\epsilon \sqrt{\lambda L}\right)^{\frac{1}{3}}. \tag{5.36}$$

However, v_n is essentially the random velocity at the outer scale of turbulence L_0, i.e. $v_n \sim (\epsilon L_0)^{\frac{1}{3}}$, from which it follows that (5.36) becomes

$$(\epsilon L_0)^{\frac{1}{3}} \gg (\epsilon \sqrt{\lambda L})^{\frac{1}{3}} \tag{5.37}$$

$$L_0 \gg \sqrt{\lambda L} \tag{5.38}$$

which is well obeyed. For example, we have seen that a typical value for L_0 is 15 m, whereas for $\lambda = 0.5\,\mu m$, and $L \approx 8\,km$, $\sqrt{\lambda L} \approx 60\,cm$.

The basic correctness of this hypothesis was initially checked experimentally by Tatarski. He measured the frequency f_m of maximum fluctuation noise for atmospheric paths of length L under various wind conditions and showed that the ratio $f_m \sqrt{\lambda L}/v_n$ was substantially constant. More recent work confirms his results. The hypothesis makes it unnecessary to deal with spatial and temporal statistics individually, since the one can be simply translated into the other if the velocity v_n can be estimated.

5.5.2 Frequency spectrum of fluctuations

It follows from section 5.5.1 that the frequency spectrum can be calculated from the spatial phase fluctuation spectrum. Since we know the autocorrelation function for phase fluctuations, $B_\phi(r)$, this can be directly Fourier transformed to give their power spectrum. Since we are looking at the component of the movement in the direction of the wind velocity, the Fourier transform needed is one-dimensional. Using (5.28) to express the structure function and (5.7) to relate this to the autocorrelation function, we have

$$D_\phi(x) = 6.88(x/r_0)^{\frac{5}{3}} \cdot B_\phi(x) = B_\phi(0) - \tfrac{1}{2}D_\phi(x) \tag{5.39}$$

and after the Fourier transformation we find the power spectrum in terms of spatial frequency u:

$$W_\phi(u) \sim r_0^{-\frac{5}{3}} u^{-\frac{8}{3}}. \tag{5.40}$$

Translating this by the frozen turbulence hypothesis to a frequency spectrum, we have the temporal power spectrum

$$W_\phi(f) \sim r_0^{-\frac{5}{3}}(f/v_n)^{-\frac{8}{3}}. \tag{5.41}$$

Since the power spectrum scales with wind velocity, it is convenient to define a coherence time t_0 in order to normalize values of the power spectrum under different conditions. The coherence time can be defined as the time for phase changes of order π to occur, i.e. $D_\phi(t_0) = \pi^2$, but there is no general agreement over the definition. Then t_0 is related to Fried's parameter and the wind velocity by $t_0 = 0.314 r_0/v_n$. It follows that measurements of the power spectrum presented as a function of $f t_0$ should lie on a universal curve. A typical value of t_0 is 3 msec (for $r_0 = 10$ cm and $v_n = 10\,\text{ms}^{-1}$).

Measurements of the power spectrum (Nightingale and Buscher 1991) have been made using an interferometer with a variable baseline B between its two inputs, and watching the fringe fluctuations. When $f \gg v_n/B$ the fluctuations at the two telescopes are uncorrelated, and therefore the power spectrum of the interference pattern is twice that of the individual beams, proportional to $(f/v_n)^{-\frac{8}{3}}$. When $f < v_n/B$ correlations have to be taken into account and proportionality to $(f/v_n)^{-\frac{2}{3}}$ is expected as the limit for $f \to 0$ (figure 5.6).

5.5.3 *Intensity fluctuations: twinkling*

To the unaided eye, which has an aperture much smaller than the Fried parameter r_0, the fluctuations in intensity are much more obvious than those of phase. On the other hand, for a telescope with diameter greater than r_0, the intensity fluctuations have much less effect, since they are uncorrelated in the different isophase regions and therefore are averaged by the large aperture. The problem of twinkling, or *scintillation*, is therefore of rather marginal interest to astronomical observations, which are usually made with large-aperture instruments. However, for completeness, we

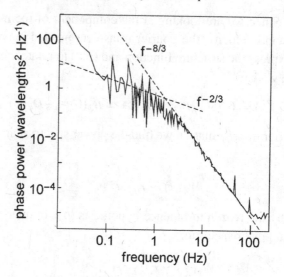

Fig. 5.6. Power spectrum for phase fluctuations, measured interferometrically using a 1 m baseline at $\lambda = 633$ nm (Nightingale and Buscher 1991). The two lines show $f^{-\frac{2}{3}}$ and $f^{-\frac{8}{3}}$ at low and high frequencies, respectively.

shall include a short analysis here. We pointed out in section 5.2 that the origin of intensity fluctuations is in the focusing or defocussing action of regions of air with nonuniform refractive index where the second spatial derivative of ϕ is important. We also saw that the intensity fluctuations develop as the light propagates away from the active region, so that high-altitude turbulence is more effective in creating twinkling than is that at ground level.

It is not easy to produce an accurate theory of scintillation. The most usual approach (Tatarski 1961) is to assume that the phase fluctuations in a layer at a given height are weak, and diffract a small amount of the incident light, which then propagates through a uniform medium to the observer. The effects of various layers can then be added by linear superposition. This approach is based on scattering theory developed for weak (i.e. dilutely dispersed) scatterers, for which it is more obviously correct because the scatterers are very localized. However, the approach appears to work (i.e. the calculations agree well with experimental results) for astronomical observations, as far as about 60° from the zenith.

To quantify scintillation, we have to measure fluctuations in intensity relative to the mean value, i.e. $\chi = \delta I/I$. This is the same as $\delta \log I$. A measure of this quantity can be obtained by integrating the power spectrum of the fluctuations. As a not-too-rigorous demonstration of this, consider a function fluctuating in time $f(t) = a[1 + bg(t)]$ during time T, where the mean is $\langle g(t) \rangle = 0$ and $\langle |g(t)|^2 \rangle = 1$. The relative fluctuation is $\chi(t) = bg(t)$ and this has variance b^2. Taking the Fourier transform of $\chi(t)$:

$$X(\omega) = bG(\omega). \tag{5.42}$$

Its power spectrum is $W(\omega) \equiv T^{-1}|X(\omega)|^2$. Integrating this between infinite limits gives

$$\int_{-\infty}^{\infty} W(\omega)d\omega = T^{-1}b^2 \int_{-\infty}^{\infty} |G(\omega)|^2 d\omega \tag{5.43}$$

$$= T^{-1}b^2 \int_{-\infty}^{\infty} |g(t)|^2 dt \tag{5.44}$$

$$= b^2 \tag{5.45}$$

from Parseval's theorem. Thus from the power spectrum of the fluctuations we can calculate the scintillation variance by integration. The same argument goes over into the spatial regime, where we relate spatial fluctuations in relative intensity to the spatial power spectrum; this will be more convenient in the framework which we have already developed, although we have to take into account that the variations in intensity now occur over the two-dimensional plane (x, y) which we designate by $\mathbf{x}$. Similarly its conjugate spatial-frequency variable will be $\mathbf{u} \equiv (u, v)$.

So again we consider a plane wave incident on a layer of air at height h with thickness δh as we did in section 5.4 where the outgoing wave was described by

$$\Psi(\mathbf{x}, h) = |\Psi(\mathbf{x}, h)| \exp[i\phi(\mathbf{x}, h)], \tag{5.46}$$

which is approximated for unit $|\Psi|$ and small ϕ by

$$\Psi(\mathbf{x}, h) = 1 + i\phi(\mathbf{x}, h). \tag{5.47}$$

The propagation down to ground level is then described by Huygens's construction by propagating each point as a spherical wave of amplitude $\Phi \sim r^{-1} \exp(-ik_0 r)$. It is also necessary to multiply by the well-known Huygens–Kirchhof prefactor $ik_0/2\pi$ to get dimensionally correct answers (section 3.1.2). This propagation is introduced by convolving (5.47) with the three-dimensional propagator Φ, which is approximated by $(ik_0/2\pi h) \exp[-ik_0(h + \mathbf{x}^2/2h)]$ when $|x| \ll h$. The resultant scattered wave at the ground $h = 0$ is thus

$$\Psi(\mathbf{x}, 0) = (ik_0/2\pi h) \exp(-ik_0 h)[1 + i\phi(\mathbf{x}, h)] \star \exp(-ik_0\mathbf{x}^2/2h). \tag{5.48}$$

It is the interference between the propagating first and second parts in the square brackets which causes the intensity fluctuations. Now the term "1" is part of the undisturbed plane wave, and the relative intensity fluctuations $\chi(\mathbf{x})$ therefore come from the imaginary part of the complex exponential divided by the unattenuated unit propagating wave:

$$\chi(\mathbf{x}) = (k_0/2\pi h)\, \phi(\mathbf{x}, h) \star \sin(k_0 \mathbf{x}^2/2h). \tag{5.49}$$

Now we can calculate the spatial power spectrum of (5.49), which is the Fourier transform $W_\chi(\mathbf{u})$ of its autocorrelation

$$B_\chi = \chi \star \chi^* \tag{5.50}$$

$$W_\chi(\mathbf{u}) = W_\phi(\mathbf{u}) \sin^2(\mathbf{u}^2 h/2k_0). \tag{5.51}$$

We have used here the two-dimensional Fourier transform of $\sin(k_0\mathbf{x}^2/2h)$ which is $(h/k_0)\sin(\mathbf{u}^2 h/2k_0)$. $W_\phi(\mathbf{u})$ is the two-dimensional transform of B_ϕ from (5.24)[†]. This can be shown to be

$$W_\phi(\mathbf{u}) = 0.03 k_0^2 u^{-\frac{11}{3}} C_n^2(h)\delta h. \tag{5.52}$$

from which

$$W_\chi(\mathbf{u}) = 0.03 k_0^2 u^{-\frac{11}{3}} C_n^2(h)\delta h \, \sin^2(\mathbf{u}^2 h/2k_0). \tag{5.53}$$

Without evaluating the expressions quantitatively, we are interested in seeing what parameters control the scintillation variance σ^2. As described above, this comes from integrating $W_\chi(\mathbf{u})$ throughout the whole $\mathbf{u}$ plane. Assuming homogeneity, so that W_χ is a function of the magnitude u only:

$$\sigma^2 = \int_0^\infty W_\chi(u) 2\pi u \, du \sim k_0^2 C_n^2(h)\delta h \int_0^\infty u^{-\frac{8}{3}} \sin^2\left(\frac{hu^2}{2k_0}\right) du. \tag{5.54}$$

By making the substitution $U \equiv hu^2/2k_0$ we then find

$$\sigma^2 \sim k_0^{\frac{7}{6}} h^{\frac{5}{6}} C_n^2(h)\delta h \int_0^\infty U^{-\frac{11}{6}} \sin^2 U dU, \tag{5.55}$$

where the definite integral can be evaluated numerically. Finally we integrate contributions from all heights (assuming the linear superposition mentioned earlier), so that

$$\sigma^2 \sim k_0^{\frac{7}{6}} \int_0^\infty z^{\frac{5}{6}} C_n^2(z) dz. \tag{5.56}$$

The $z^{\frac{5}{6}}$ within the integral indicates that the influence of higher turbulent layers is weighted progressively more strongly in the scintillation variance at ground level (figure 5.7c). We note again that we have calculated the scintillation as a variance from point to point on the ground at a certain moment in time, but this should be identical to what is observed at a given point as a function of time. The numerical

---

[†] Apart from a δ-function at the origin, the transforms of $D(\mathbf{x})$ and $B(\mathbf{x})$ just differ by a factor of 2.

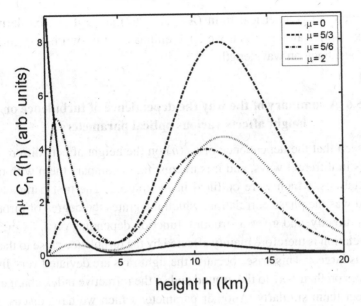

Fig. 5.7. The function $h^\mu C_n^2(h)$ indicating the relative importance of turbulence at different heights in determining (a) the phase correlations ($\mu = 0$), (b) the size of the isoplanatic patch ($\mu = \frac{5}{3}$), (c) scintillations for a small telescope ($\mu = \frac{5}{6}$), (d) scintillations averaged by a large telescope ($\mu = 2$).

value of σ^2, when all the terms are evaluated properly is (Roddier 1981):

$$\sigma^2 = 163 k_0^{\frac{7}{6}} \int_0^\infty z^{\frac{5}{6}} C_n^2(z) \mathrm{d}z. \tag{5.57}$$

Substituting typical figures (see figure 5.7) into this equation gives a value of about 0.11 at $\lambda = 0.5\,\mu$m, which is very reasonable.

It is quite easy to correct the above argument for a finite-aperture telescope. Measuring with a tiny aperture senses all spatial frequencies **u**. For larger apertures, the spatial frequencies sensed become more and more restricted, and this is introduced by integrating (5.54)–(5.55) out to a value of u corresponding to the inverse radius of the aperture, R, $u_{\max} = 2\pi/R$. For a large enough telescope, when $h u_{\max}^2/2k_0 \ll 1$, the sine in (5.54) can be replaced by its argument, and the integral becomes

$$\sigma^2 \sim k_0^2 C_n^2(h)\delta h \int_0^{2\pi/R} u^{-\frac{8}{3}} \left(\frac{h u^2}{2k_0}\right)^2 \mathrm{d}u, \tag{5.58}$$

from which (5.56) is replaced by

$$\sigma^2 \sim k_0^0 \int_0^\infty z^2 C_n^2(z)\,\mathrm{d}z. \tag{5.59}$$

This is called the geometrical limit (Reiger 1963) since it can be derived from geometrical optics alone, which we have emphasized by writing k_0^0 to show the lack of dependence on wavelength.

5.6 A summary of the way the dependence of turbulence on height affects various optical parameters

We have seen that the dependence of $C_n^2(h)$ on the height affects different optical parameters in different ways, and it is interesting to compare them with our physical expectations, which were outlined in section 5.2. The first parameter which we studied was the phase correlation, which integrates the degree of turbulence at all heights directly, and gives a structure function depending on $\int_0^\infty C_n^2(z)dz$. The phase correlation is therefore mainly affected by the turbulence close to the ground, where C_n^2 is largest. This arises because the light rays are deviated very little from straight lines on their way to the ground and so the refractive index changes at each height affect them similarly. Another parameter which we have discussed is the size of the isoplanatic region, within which phase correlations are similar. This is controlled by $\int_0^\infty C_n^2(z)z^{\frac{5}{3}}dz$; in contrast to the phase correlation, weak turbulence at great heights can make a lot of difference. This is quite expected, because the lines of sight at different angles diverge with height and therefore progressively sample more separated regions. Finally, scintillations result from focusing or defocusing of the light because of curvature of the phase, and the intensity changes develop progressively as one recedes from the region of the focusing. The result is to emphasize turbulence at large heights. In the physical optics regime, the degree of focusing is mitigated by diffraction and the height weighting is expressed by $\int_0^\infty C_n^2(z)z^{\frac{5}{6}}dz$, which indicates the growing importance of even weak turbulence at higher levels. In the geometrical optics regime, scintillation effects on a large telescope emphasize the height dependence strongly, with weighting $\int_0^\infty C_n^2(z)z^2dz$, because the mitigating effect of diffraction on the focusing is absent.

Figure 5.7 summarizes these dependencies for the turbulence structure shown in figure 5.3. This figure supports the general simplification that for some of the optical effects, at least, the turbulence can be described as being mainly in two distinct regions: one close to the ground, and one in the region of 10 km height, although one should remember that the phase correlations are almost entirely determined by the lower of these layers.

5.7 Dependence of atmospheric effects on the wavelength

The wavelength of the observing light affects the importance of atmospheric turbulence in two ways. First of all, the refractive index of air at any given pressure is a function of wavelength. Second, the phase difference introduced by a given path length is inversely dependent on the wavelength.

Because of the refractive index dispersion, if one observes a star which is at angle γ from the zenith, refraction by the atmospheric layer will cause it to appear to be at angle γ' related by Snell's law, i.e. $\gamma' = \arcsin(n^{-1}\sin\gamma)$. At $\gamma = 45°$ this small error varies from $2.79 \cdot 10^{-4}$ rad at $\lambda = 0.5\,\mu$m to $2.74 \cdot 10^{-4}$ rad at $\lambda = 1\,\mu$m. This might seem insignificant, but on the other hand the difference between these two angles, which is about 1 arcsec, is comparable with the seeing and therefore if wide-band light is being used for imaging or interferometry, dispersion has to be corrected. This can be done by using a nondeviating prism pair, made from two glasses with different dispersivities, which is constructed so that the total dispersivity can be adjusted for the angle γ (section 8.3.3). Correction is particularly relevant in speckle interferometry (section 5.9 and section 6.3), since the size of the speckles is considerably smaller than 1 arcsec, and so they appear to be smeared by the atmospheric dispersion, even if the wavelength range used is rather smaller than the $0.5\,\mu$m of the above estimate.

The wavelength dependence of phase is more pronounced. Let us consider first the dependence of the phase correlation on wavelength. Assuming that the function C_n^2 is wavelength independent, from (5.24) we see that the phase structure function $D_\phi \sim k_0^2 r^{\frac{5}{3}}$, which has a strong proportionality to λ^{-2}. As a result, phase decoherence is much weaker at longer wavelengths. Similar statements can be made about other atmospherically related parameters. In particular, Fried's parameter r_0 (5.27) depends on wavelength as $\lambda^{\frac{6}{5}}$. The isoplanatic region also has angular dimension proportional to $\lambda^{\frac{6}{5}}$ (5.35). As a result of these facts, many adaptive optics and interferometric systems have been developed in the near infrared, because not only are the phase fluctuations smaller, but larger collecting apertures can usefully be employed and the isoplanatic patch is larger.

The strength of scintillation, for a small observing aperture such as the eye, depends on $\lambda^{-\frac{7}{6}}$ (5.57), indicating that twinkling in the infrared, if we could see it, would be less obvious than in the visible. But for a large-aperture telescope, in the geometrical regime, the residual scintillation (5.59) is independent of wavelength, except to the extent that the refractive index itself is wavelength dependent.

5.8 Adaptive optics

When the aperture of a telescope D is greater than Fried's parameter r_0, light waves reaching an image from various parts of the aperture have random phase and so do not interfere constructively on average and do not contribute to making a sharper image. The resolution limit of such a telescope for exposures longer than the atmospheric coherence time is thus about λ/r_0. The only advantage arising from increasing the aperture is to make the image brighter. The idea of *adaptive optics* was proposed by H. W. Babcock in 1953 to get around this problem by correcting the relative phases of the waves from the various parts of the aperture so that they could interfere and improve the angular resolution (Babcock 1990). This would require measuring the relative phase of the incident wavefront at points separated by about r_0 and then applying a real-time correction so as to make it planar. As a result, the resolution could be improved in principle to the diffraction limit, $1.22\lambda/D$. Babcock proposed using a variant of the Foucault knife-edge test, widely used by amateur astronomers for testing the quality of their telescope mirrors by giving information on the gradient of the wavefront at each point encoded as intensity levels. He suggested the use of a continuously rotating knife-edge, but in 1958 the technology required to carry out the wavefront correction did not exist. At the time of "star wars" in the 1970s it was realized that adaptive optics could be used for improving terrestrial imaging and also for more tightly focusing intense laser beams. As a result, development of the necessary technology was funded, including in particular methods of measuring the wavefront distortions (i.e. deviation from an ideal plane wave) and also systems for deforming the wavefront in antiphase to correct the distortions. Later, in the 1990s, the developments were declassified, and all astronomers could benefit from them. For astronomy, this correction process has to be carried out using a reference star which is indeed a point source, and then the derived distortion can be used to correct the image of another, more interesting, star, provided of course that its wavefront is distorted in the same way, i.e. that it is in the same isoplanatic region as the reference star. This need for an isoplanatic reference star is in fact the main limitation of adaptive optics, and has led to some interesting ideas for artificial reference stars which will be mentioned later. A schematic layout of an adaptive optics system for a large-aperture telescope is shown in figure 5.8.

Not only is direct astronomical imaging affected by the atmosphere; interferometry is also limited by atmospheric distortion of the wavefront incident on the subapertures. If the subaperture diameters are smaller than r_0, they do not need wavefront correction within each one. The only correction necessary is to the relative phases of the subapertures, which are usually separated by more than r_0; this is called "piston correction" and corresponds to an adjustment of the path-length compensation between the interfering waves so as to stabilize the fringes. However, if the subaperture diameters are each larger than r_0, the full benefit of their diameters will contribute to improving the sensitivity and limiting magnitude of the interferometer only if they are adaptively corrected. The correction must of course be completed within a time shorter than t_0, which is of the order of 10 ms, in order to be useful. We shall not give a detailed description of adaptive optics techniques, since they have recently been described in several monographs such as Tyson (1991)

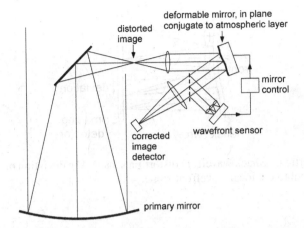

Fig. 5.8. Schematic diagram of a telescope with adaptive optical correction, operating with negative feedback.

and Hardy (1998). However, since adaptive optics have already become an integral part of several interferometers, we shall discuss the subject briefly.

5.8.1 *Measuring the wavefront distortion*

The wavefront distortion must be measured at an array of points separated by distances considerably less than r_0 along both the x and y axes in the telescope aperture. The aperture is assumed to be illuminated by a reference star which provides a plane wave incident on the atmosphere from above. There are several methods of measuring the deviation $w(x, y)$ of the received wavefront from planarity, including in particular the Hartmann–Shack technique and curvature sensing.

The Hartmann–Shack sensor

In the Hartmann-Shack detector, the incident wavefront is measured in a plane conjugate to the telescope's entrance aperture, where it is focused by means of a periodic array of small lenses, each one having aperture corresponding to a region less than r_0 in diameter (figure 5.9). From the point of view of the individual small lens, the wave incident on it is characterized completely by ∇w, which is directed along the normal to the wavefront at that point. This can be measured by locating the point to which the light is focused in its focal plane. If the incident wave were indeed an ideal plane wave, the focal plane of the array of small lenses would display an identical ordered array of focal spots. With the distorted wavefront, each spot is displaced in x and y from its ideal position; the deviation of each one indicates the local value of ∇w. Then, knowing the value of ∇w, its value can be integrated

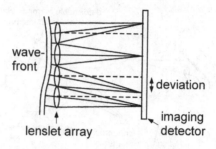

Fig. 5.9. Hartman–Shack wavefront distortion sensor. The deviation of each focus is proportional to the local wavefront slope.

to find $w(x, y)$ itself, up to a constant which is the piston error, and has to be determined otherwise. The accuracy of w achieved depends both on the accuracy with which the displacements can be measured, as well as the closeness of the spacing of measuring points; these both depend on the number of photons available for the measurement, which have to be siphoned off from the incident light and still leave enough for useful science. In many systems, correction and observation use different wavelengths. In fact, it is always necessary to determine the centroid of the focal spots to a small fraction of their actual size for the measurement to be useful. The requirement of compactness of the reference star is not as demanding as one might imagine; it has to provide coherence over each small lens, or aperture region of dimension r_0, which means only that it must not be resolvable in a long-exposure image by the *uncorrected* telescope – but no more. As a result, the star of interest may be used as its own reference provided it is bright enough. The results are then transferred to a deformable mirror and, as shown in figure 5.8, the whole is operated in a negative feedback loop which maintains the wavefront flat. This type of sensor is matched best to a deformable mirror which corrects the value of w directly.

Curvature sensing

An alternative detection and control system (Roddier 1988) measures the curvature $\nabla^2 w$ and is more appropriate to a deformable mirror which corrects curvature directly. Suppose that a certain part of the propagating wavefront has a positive value of $\nabla^2 w$. Then, as we saw in figure 5.4(d), it is converging and its intensity increases with propagating distance. Likewise, when $\nabla^2 w$ is negative, the intensity decreases. The value of $\nabla^2 w$ at each point can therefore be determined by comparing the intensity distributions $I(z_-)$ and $I(z_+)$ in two planes separated by a short distance along the axis, usually before and after a plane conjugate with the entrance aperture. More quantitatively, it can easily be shown that intensity $I(\mathbf{r})$ obeys an "intensity

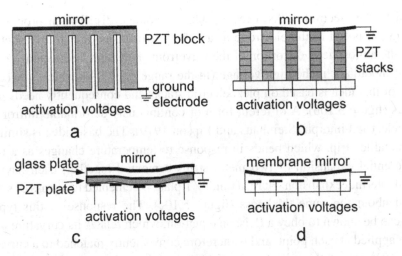

Fig. 5.10. Deformable mirrors of different types: (a) monolithic piezoelectric block, (b) discrete piezoelectric stacks, (c) bimorph mirror, (d) electrostatically deformed membrane (courtesy E. Ribak).

propagation equation" of the form

$$\frac{\mathrm{d}I(\mathbf{r})}{\mathrm{d}z} = \nabla \cdot [I(\mathbf{r})\nabla w(\mathbf{r})] = \nabla I(\mathbf{r}) \cdot \nabla w(\mathbf{r}) + I(\mathbf{r})\nabla^2 w(\mathbf{r}), \qquad (5.60)$$

in which the left-hand side is calculated from $I(z_+) - I(z_-)$ and I on the right-hand side is given by $\frac{1}{2}[I(z_+) + I(z_-)]$. The first term on the right is usually negligible, so that $\nabla^2 w$ is measured and $w(\mathbf{r})$ itself can then be obtained in principle by a double integration along x and y. However, in practice these processes are not as easy as they seem because of the effect of the boundaries, which have to be taken into account carefully. Also, the need for calculation of a difference between intensities makes it problematic for very weak light. This method of sensing is most suited to deformable mirrors which correct the wavefront curvature $\nabla^2 w$ directly.

5.8.2 Deformable mirrors

Deformable optics use piezoelectric deformation of a mirror, electrostatic bending of a membrane or can be transmission liquid crystal or micro-electro-mechanical (MEM) devices. The first practical mirror (Feinleib et al. 1974) was made from a single block of piezoelectric ceramic (electrostatically poled lead zirconium titanate, PZT) to which an array of electrodes was attached to one surface, the opposing surface being grounded. A thin mirror was cemented over the electrodes (figure 5.10a). The deformation of the mirror can then be shown to be proportional to the potential of the ceramic below it, which is determined by the voltages applied to the electrodes. As a result, when a voltage distribution proportional to $-w(x, y)$ is fed to the electrode array, the mirror takes on a form which can cancel the atmospheric distortion of the wavefront. The main disadvantage of this form of modulator is the high voltage (in the range $\pm 2000\,V$) which it needs. A variant of this idea is based on piezoelectric stacks and consequently needs lower voltages (figure 5.10b). A different form of continuously deformable mirror uses a piezoelectric bimorph (Steinhaus and Lipson 1979). The basic idea is similar to the bimetallic strip, which bends in response to temperature changes as a result of differential expansion of two metal strips cemented together, each having a different thermal expansion. Here, a thin PZT plate is cemented to a glass or silicon plate of about the same thickness (figure 5.10c). The response of this type of mirror can be shown to obey a Poisson equation, which relates its curvature to the voltage applied at each point, and is therefore conveniently matched to a curvature sensor. Voltages of the order of $\pm 400\,V$ are required to operate these devices. Membrane mirrors (figure 5.10d) use electrostatic attraction between a thin conducting membrane and an array of electrodes and once again obey a Poisson equation in which the source term is now the electrostatic force density. Since this is proportional to the square of the voltage, it is always positive and so the mirror has a built-in curvature, corresponding to the mean-square voltage, which has to be canceled optically. They are operated by voltages up to a few hundred volts.

5.8.3 Tip–tilt correction

If the wavefront aberrations in an aperture are analyzed as a polynomial series in powers of x and y, the first term is a (time-dependent) constant, the "piston" correction, and the next two are linear in x and y and correspond to tip and tilt. Their correction in each subaperture of an interferometer is very important, and essentially stabilizes the position of the star in the field of view. It is also relatively easy to carry out, since the Hartmann–Shack sensor required just has one lens corresponding to the whole aperture, and its image has to be stabilized by a mirror whose angles θ_x and θ_y can be individually controlled. This type of adaptive correction is ubiquitous on all multiple-aperture interferometers, and allows a subaperture size of about $3r_0$ to be usefully employed (Noll 1976) without causing loss of fringe contrast; larger apertures require correction to higher order. Some of the requirements for tip–tilt correction will be discussed in section 8.2.9.

5.8.4 Guide stars

In principle, an adaptive optical system is controlled by an unresolved guide star lying in the same isoplanatic patch as the star being investigated. In order to have enough intensity for the wavefront sensor to give accurate results, it should be quite bright. As pointed out above, the investigated star itself can be used as its own reference, because its structure would not be resolved by the aperture of r_0, but this is only possible when it is bright enough. In this case, a common solution is to use visible light for the wavefront sensing and infrared for observation, although since the atmospheric properties depend on wavelength, the two spectral regions must be close for the correction to be good. In general, the probability of there being a suitable guide star within the isoplanatic patch of a faint object is very small for the visible region, but becomes greater in the infrared, mainly because of the increased size of the isoplanatic patch (a few arcseconds in the visible, up to an arcminute in the far infrared).

One attractive solution is therefore to use an artificial guide star (Foy and Labeyrie 1985). A powerful laser beam (1 kw CW) is aimed along the axis of the telescope. This beam is back-scattered by two possible mechanisms; one is Rayleigh scattering in the upper atmosphere at between 15 and 30 km and the other is resonant scattering of 589 nm light by Na atoms in the atmosphere at about 93 km height. The angular position of the guide star can be adjusted to be within the isoplanatic patch around the object of interest. It is interesting to note that, since the outgoing laser beam traverses the same atmospheric layers as the light from the object, the two have the same angular deviation and so they remain in the same relative positions with no active adjustment. As a result, when using an artificial guide star, tip and tilt corrections have to be made by a different system.

5.9 Short exposure images: speckle patterns

We shall conclude this chapter where we started it by looking again at the atmospheric degradation of a stellar image. Does the K41 model of turbulence give a good picture of the point spread function of a large-aperture telescope? Although long-exposure images of point stars have long been recorded, and have been used in some ways to investigate this question, it becomes much more interesting when applied to instantaneous images which show the effect of a "frozen" statistical state of the atmosphere. This became possible with the development of image intensifiers having amplifications of the order of 10^6 and reasonable quantum efficiencies (about 10%). The resulting speckle images, of which we showed an example in figure 5.1, are the basis of the technique of speckle interferometry, initially suggested and developed by Labeyrie (1970) and discussed in detail in chapter 6. We have already discussed (section 5.5.1) the frequency spectrum of fluctuations, which determines how short an exposure is necessary to freeze an "instantaneous" image; now we shall apply the atmospheric structure analysis to modeling a speckle image.

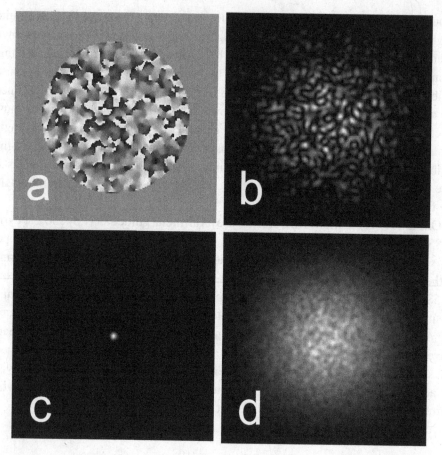

Fig. 5.11. Simulated speckle images, using the structure function (5.28), with $r_0 = 7$ units. (a) The phase field across a circular aperture, radius 64 units. Phase, modulo 2π, is indicated by gray level from white to black. (b) The point spread function corresponding to the phase field (a). (c) The ideal point spread function for the same circular aperture. (d) Long-exposure average of 50 random simulations like (b).

5.9.1 A model for a speckle image

The speckle image is that of a point source degraded by the phase distortions produced by the atmosphere, described in section 5.4 by the function $\Psi(x, h)$ (5.10). Assuming that the amplitude variations are relatively insignificant, we can describe the image as the Fraunhofer diffraction pattern (i.e. Fourier transform, $F(\mathbf{u})$) of the function $f(\mathbf{r}) = S_0(\mathbf{r}) \exp[i\phi(\mathbf{r})]$. Here $S_0(\mathbf{r})$ describes the telescope aperture, i.e. has value 1 within it and 0 outside, and would generally be $\text{circ}(r/R)$ or maybe an annular function to allow for a central obscuration (secondary mirror).

Of course, the observed image, which is the instantaneous point spread function, is $|F(\mathbf{u})|^2$.

We describe here a simple model which has the right features and illustrates the physics. The atmospheric fluctuations are modeled by representing the phase $\phi(\mathbf{r})$ as a random function which changes by amounts in the range of $\pm\pi$ from point to point on the distance scale of Fried's parameter r_0. The simulation allows any wavefront structure function to be used, although it appears that the differences between the results for the Kolmogorov structure function (5.28) and a random Gaussian approximation are very minor.

In this model, we start with the facts that the wavefront $f(\mathbf{r})$ in the aperture plane has unit modulus and unknown phase $\phi(\mathbf{r})$, but its autocorrelation function has the Kolmogorov form

$$B(\mathbf{r}) = \exp[-\tfrac{1}{2}D_\Psi(r)] = \exp[-3.44(r/r_0)^{\frac{5}{3}}]. \tag{5.61}$$

Since the *intensity* of the Fourier transform[†] $F(\mathbf{u})$ of $f(\mathbf{r})$ is the transform of its autocorrelation function, but its *phase* is arbitrary, we use the known $B(\mathbf{r})$, transform it to $b(\mathbf{u})$ and then take the most general form of the transform $F(\mathbf{u})$ to be $\sqrt{|b(\mathbf{u})|}\exp[i\Phi(\mathbf{u})]$ where $\Phi(\mathbf{u})$ is a random phase function. We then transform this back to the aperture plane, where it gives us a function $f_1(\mathbf{r})$ with random phases and amplitudes having the right correlation function. We then constrain this function by the bounding aperture S_0. Finally, we transform this back to the image plane $\mathbf{u}$ to see the speckle image or atmospherically degraded point spread function.

Figure 5.11 shows an example of results with a 512×512 pixel input field. The figure shows the phase field, in which the Fried parameter can readily be appreciated as the average size of a uniform patch, a typical speckle image, the diffraction-limited image $[\pi R^2 2J_1(uR)/(uR)]^2$ and a simulated long-exposure image obtained by superimposing 50 independent calculations.

Two further details which are of interest can easily be seen from the same simulation. First, if the range of the random ϕ is less than 2π, there is in addition a stronger spot at the center ($\mathbf{u} = 0$) into which a larger and larger fraction of the energy is concentrated as the range of ϕ gets smaller (figure 5.12a). The fraction of the wave energy concentrated in this central spot, defined by the size of the diffraction-limited spot for the complete aperture S_0, is called the *Strehl ratio*. Clearly, when the random phase range approaches zero, the Strehl ratio becomes unity[‡]. Second, the form

---

[†] We use the convention (Appendix A) that a function and its Fourier transform are indicated by capital and lower-case symbols.

[‡] This property is employed in the spatial filters commonly used with laser systems in the laboratory. The wavefront from a laser has small spatially varying deviations (in both amplitude and phase) from a plane wave. We focus the wave to as small a region as possible and insert a pinhole around the central spot. The transmitted wave is a much closer approximation to an ideal spherical wave because the components of the original wavefront with non-zero spatial frequencies have been rejected by the pinhole filter.

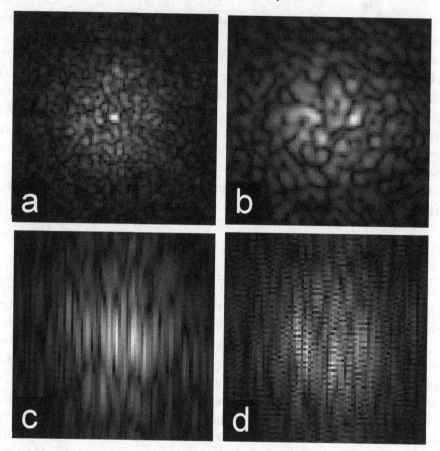

Fig. 5.12. More simulated speckle images, as in figure 5.11. (a) When the range of the phase fluctuations is less than 2π, a strong spot develops at the center. The range here is 1.95π which is close enough to 2π to allow both the speckle image and the strong spot to be seen at the same time; otherwise the image looks the same as figure 5.11(c). (b) The shape of each individual speckle is approximately a diffraction limited point spread function; in this case a small square aperture was used. (c) and (d) Single-slit and double-slit apertures. For the double-aperture telescope, each speckle is crossed by Young's fringes.

of each speckle is the point spread function of the diffraction-limited aperture; this is best seen by taking a relatively small aperture with some recognizable shape, as in figure 5.12(b), where a rectangular aperture is used, or (c), using a slit aperture. Taking this to an extreme, a pair of separated telescopes is equivalent to an aperture with two well-separated openings, and the individual speckles now show Young's fringes (figure 5.12d).

With this chapter we have completed descriptions of the various aspects of basic physics underlying astronomical interferometry. The following chapters will describe how astronomers have used this knowledge to construct various types of interferometer to improve spatial resolution and sensitivity beyond that available from a single telescope, and will provide a sample of the basic science which has been done using these instruments.

References

Babcock, H. W. (1990). *Science*, **249**, 253.

Feinleib, J., S. G. Lipson and P. Cone (1974). *Appl. Phys. Lett.*, **25**, 311.

Foy, R. and A. Labeyrie (1985). *Astron. Astrophys.*, **152**, L29.

Fried, D. L. (1966). *J. Opt. Soc. Am.*, **56**, 1372.

Hardy, J. W. (1998). *Adaptive Optics for Astronomical Telescopes*, New York: Oxford University Press.

King, J. R. (1971). *Pub. Astr. Soc. Pac.*, **83**, 199.

Kolmogorov, A. N. (1941a), (1941b). *Doklady Acad. Nauk. SSSR*, **30**, 301, **32**, 16.

Labeyrie, A. (1970). *Astron. Astrophys.*, **6**, 85.

Michelson, A. A. (1962). *Studies in Optics*, University of Chicago Press (1927), reprinted in Phoenix Science Series.

Nightingale, N. S. and D. F. Buscher (1991). *Mon. Not. R. Astron. Soc.*, **251**, 155.

Nisenson, P. and R. V. Stachnik (1978). *J. Opt. Soc. Am.*, **68**, 169.

Noll, R. J. (1976). *J. Opt. Soc. Am.*, **66**, 207.

Reiger, S. H. (1963). *Astron. J.*, **68**, 395.

Roddier, C. (1976). *J. Opt. Soc. Am.*, **66**, 478.

Roddier, F. (1981). *Progress in Optics*, ed. E. Wolf, **19**, 283, Amsterdam: North-Holland.

Roddier, F. (1988). *Appl. Opt.*, **27**, 1223.

Steinhaus, E. and S. G. Lipson (1979). *J. Opt. Soc. Am.*, **69**, 478.

Tatarski, V. I. (1961). *Wave Propagation in a Turbulent Medium*, transl. R. A. Silverman, New York: McGraw-Hill.

Tyson, R. K. (1991). *Principles of Adaptive Optics*, San Diego: Academic Press.

Weigelt, K. (1991). *Progress in Optics*, **29**, 295.

6

Single-aperture techniques

6.1 Introduction

We saw in the chapter on atmospheric turbulence that the real limitation to the resolution of a ground-based telescope is not the diameter of the telescope aperture, but the atmosphere. As a result, a telescope of any diameter will rarely give an angular resolution in visible light better than 1 arcsec[†], which is equivalent to the diffraction limit of an aperture of about 10 cm diameter (the Fried parameter, r_0, defined in section 5.4.1). This limitation has been considered so fundamental that large telescope mirrors might not even have been polished to an accuracy which could give a better resolution than this. The ideas behind the various methods of astronomical interferometry are all directed at exceeding it.

The first idea was due to Fizeau (1868) who conceived the idea of masking the aperture of a large telescope with a mask containing two apertures each having diameter less than r_0, but separated by a distance considerably greater than this. The result would be to modulate the image with Young's fringes and, from the contrast of the fringes, to glean information about the source dimensions. A few years after the publication of Fizeau's idea, Stéphan (1874) tried it out experimentally with the 1-m telescope at Marseilles and concluded (correctly) that the fixed stars were too small for their structure to be resolved by this telescope. Michelson (1891) later developed the necessary theory to make this idea quantitative and was the first to succeed in using Fizeau's technique, when he measured the diameters of the moons of Jupiter using the 12-inch Lick refractor telescope. For example, Io, the largest moon of Jupiter, subtends an average angular diameter of $\alpha = 4.4 \cdot 10^{-6}$ rad. at the Earth. The transverse coherence distance (section 3.2.4) of light from Io, at $\lambda = 0.5\,\mu\text{m}$, is thus $r_c = 1.22\lambda/\alpha = 136$ mm, which is smaller than the telescope aperture, although considerably larger than r_0. As a result, interference fringes

[†] It is reported that at Dome C on the Antarctic highlands, seeing of 0.2 arcsec is sometimes obtained, but this is truly exceptional.

crossing the image can be seen to disappear as the separation of the apertures increases through r_c. For measuring the diameter of a stellar source, the phase of the fringes (which is affected strongly by atmospheric turbulence) is of minor importance since the object can be assumed to be centrosymmetric and therefore the phase of the coherence function must be 0 within the central peak, or π in the next ring if it is observable.

Michelson's experiment was just a trial before he went to the trouble and expense of constructing his much larger stellar interferometer at Mount Wilson, which was really the father of astronomical interferometry, and is discussed in greater detail in chapter 8. However, Fizeau's experiment was not forgotten. In 1985 the reawakened interest in phase-closure techniques (section 4.2.1) in radio astronomy suggested that Fizeau's experiment could be extended to a larger number of apertures, and that the phase of the coherence function could be determined by phase closure if short-exposure images could be recorded, incorporating the fringes produced by at least three different aperture separations simultaneously. This approach was used by Haniff et al. (1987) with the Isaac Newton Telescope. The telescope aperture was imaged onto a mask with four apertures, whose separations were chosen to give simultaneously six uniformly spaced nonredundant baselines in one dimension. A short-exposure image obtained through this mask then contains a superposition of sets of fringes from each of the aperture separations which, because each set has a different period, could be isolated by Fourier analysis. After appropriate processing, which will be discussed in more detail in section 6.2, high-resolution images of double stars were obtained. More recent work using the same ideas has been reported by Tuthill et al. (2000a,b).

In practice, however, masking the telescope aperture in any way is practical only for the brightest stars. It will not be acceptable to astronomers looking at faint objects, for whom every photon detected is of immense value. To be of universal value, a technique for overcoming atmospheric turbulence degradation has to use all the light entering the telescope. Even before the advent of electronic imaging devices with sufficiently high quantum efficiency to allow images to be recorded using exposures shorter than the atmospheric stability time, visual observations showed that "instantaneous" images contained a lot of detail which gets lost in a long photographic exposure. These images are now called "speckle images" and their structure was discussed in section 5.9. Observers of binary stars with refractors as large as 1 meter had long done a primitive form of what is now called "speckle interferometry" without knowing it. One of the earliest descriptions of a speckle image is found in Jean Texereau's book *La Construction du Télescope d'amateur* in the form of a hand-drawing. A few years later, the first short-exposure photographs were obtained by Rösch, Wlerick and Boussuge using the Lallemand electronic camera at the Pic du Midi 1-m telescope. This camera used photo-electron acceleration to improve on the sensitivity of standard photographic film and, together with an additional lens, provided the high image magnification and short exposures needed to record speckles, which had been beyond the reach of emulsions even with the brightest stars. The electronic camera, although a heroic instrument with its silver emulsion exposed to photo-electrons in a high vacuum, was difficult and costly to operate. From their photographs, it was clear that individual speckles produce images with much higher resolution than the seeing limit, and Rösch et al. obtained a few usable speckle images on binary stars.

But none of these early observers had a valid theoretical model of a speckle pattern and its formation. This prevented them from anticipating that a larger aperture really improves resolution, and therefore dissuaded them from trying to use larger apertures such as those of the Mount Wilson or Palomar Telescopes. It was perhaps his background in holography, a field where speckle patterns had been extensively studied, which helped Labeyrie (1970) to better understand the atmospherically induced speckles and the possibility of exploiting them for high-resolution imaging. Following his theoretical description (Labeyrie 1970) of what became known as "stellar speckle interferometry" (section 6.3.1) he decided to verify the presence of speckles in a star image at the 200-inch telescope at Mount Palomar. To have a visual look he visited the large telescope one evening when astronomer Maarten Schmidt, famous for his discovery of the large redshift in quasars, was recording quasar spectra in the Cassegrain cage. Schmidt invited him to have a look while he went out for his midnight lunch. With the minimal equipment he had brought in his pocket, bee's wax, a small mirror and a microscope objective, it took a few minutes to steal the light beam from a bright star away from the spectrograph. As expected, the strongly magnified star image that he saw through the microscope objective had thousands of speckles, gently boiling in response to the atmospheric turbulence. This was enough to motivate him to propose a speckle interferometry program at · Mount Palomar. The proposal was first rejected, for the apparently excellent reason that the 200-inch mirror was only seeing-limited, and not diffraction-limited. Labeyrie had to insist that even so, speckle interferometry would indeed retrieve the diffraction-limited resolution, 50 times better than the seeing. He was lucky to receive the support of a Palomar astronomer, Arthur Vaughan, who had himself previously seen speckles and had considered exploiting them. The actual techniques of systematically extracting high-resolution astronomical data from the speckle patterns, to be discussed in detail in section 6.3.1, were developed by laboratory experiments and simulations, following which, in a few nights of observation with Daniel Gezari and Robert Stachnik at the 200-inch telescope, speckle interferometry proved itself by providing accurate confirmations of the early measurements achieved by Michelson, Pease and Anderson on the largest stars and the binary Capella. Later, using photon-counting television cameras, the technique also gave numerous results on hundreds of fainter sources, many of which proved to be binary stars. Some of them had measurable spectroscopic orbital elements, in which cases model-independent stellar masses could be obtained.

Speckle interferometry showed the way to use the complete aperture to get diffraction-limited information about stars, but in its original form gave a spatial autocorrelation function which could only be converted to a stellar image in the simplest cases, such as centrosymmetric sources or binary stars. This is the same information as was obtainable from intensity interferometry (chapter 7). Within a few years the first image-processing technique emerged (Knox and Thompson 1974), which allowed a true image to be created systematically from the same data. The details of this technique will be discussed in section 6.4.1. Further development of the same ideas by Weigelt (1977) produced the technique of "triple correlation" or "speckle masking," which was shown later to involve the concept of phase closure (section 4.2.1) again (Roddier 1986), and allowed more general and complicated objects to be imaged (section 6.4.2).

At the same time, an empirical approach called "shift and add" was developed based on the observation that each speckle itself approximates a diffraction-limited image of the source. By locating the centers of the brightest speckles digitally, and superimposing them, a diffraction-limited image can be built up, superimposed on a background which becomes smoother as more speckles are added, and in the long run can be subtracted. Christou et al. (1986) extended the idea to use almost all speckles in a frame, weighted by their intensity. Once an approximate image has been determined, a matched filter can be found (Ribak 1986) which enables the centers of the speckle images to be determined more accurately, even when very few photons per speckle are available. This technique works quite well if there are not too many speckles, and if the image contains a single well-defined brightest point.

6.2 Masking the aperture of a large telescope

Despite its restriction to bright objects, we shall first discuss the application of Fizeau's and Michelson's method of aperture masking, since it is a good introduction to speckle methods and has recently had some nice applications to bright stars. It is probably fair to say that, given enough light, aperture masking gives the closest approach to diffraction-limited images of compact objects attainable through the atmosphere. For more complicated objects, the method is restricted by the crowding limitation (section 4.3), and speckle masking (section 6.4.2) gives better images. The principle is as follows.

A pair of holes in the aperture plane, each one having a diameter not greater than the Fried parameter r_0, gives an image crossed by Young's fringes (figure 6.1). The envelope of the image has angular extent determined by the size of an individual

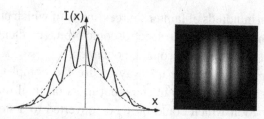

Fig. 6.1. Fringes due to two small ($< r_0$) circular holes in a mask, with an arbitrary phase difference and partial coherence ($\gamma \sim 0.3$) between them.

Fig. 6.2. Fringes due to three small circular holes in a mask, each with an arbitrary phase and each pair having a different separation: (a) mask, (b) the diffraction pattern and (c) the transform of the measured diffraction pattern (autocorrelation function).

hole, so that it has no better resolution than the long-exposure image, because that is the meaning of r_0. But as a result of the small size of the holes, the phase of illumination at each one is well-defined, and therefore the fringes have contrast (visibility, section 3.3.1) equal to the coherence function appropriate to the vector separating the apertures. The phase of the fringes is that of the coherence function, but the atmospherically induced phase difference between the two apertures is added to this, so that alone it contains no useful information. Mathematically, when the two apertures i and j are separated by r_{ij}, the fringe visibility is $|\gamma(\mathbf{r}_{ij})|$ and the phase is $\arg[\gamma(\mathbf{r}_{ij})] + \phi_{0ij}$ where ϕ_{0ij} is the atmospheric contribution. Now Michelson only used the fringe visibility in his work, and assumed a simple model for his stellar source, such as a circular disk or a binary pair, which allowed the relevant parameters to be measured.

However, suppose next that three holes are used simultaneously, each pair of the three having a different value of **r** (figure 6.2a). The image will be crossed by three different sets of fringes (figure 6.2b), corresponding to the three hole spacings. The periods and orientations of the fringes are known, because they are the Fourier transforms of the hole-pairs. The image can therefore be analyzed to extract the phases and amplitudes of the fringes. In fact this could be done by Fourier analysis of the images without knowing which fringes to expect (figure 6.2c), but in the present case, because the periods and orientations are known in advance, the extraction can be done more accurately. Thus $|\gamma(\mathbf{r}_{ij})|$ is known for three values of $\mathbf{r}$. Moreover, because the sum $\mathbf{r}_{12} + \mathbf{r}_{23} + \mathbf{r}_{31} = 0$, the sum of their phases is not affected by atmospheric fluctuations, and so the ideas of phase closure (section 4.2.1) can be used to glean information about them (but very little, when there are only three apertures). The method can be extended to larger numbers of apertures, in which case considerably larger numbers of simultaneous groups of three are defined. In general, for N apertures, there are $(N - 1)(N - 2)/2$ independent groups of three (section 4.2.1). For each one, a phase-closure relationship can be defined, so that for larger N quite a lot of information about the phase of $\gamma(\mathbf{r}_{ij})$ can be obtained from the measured fringe phases[†]. Once $\gamma(\mathbf{r}_{ij})$ has been determined at several points, images can be constructed by Fourier inversion using the Van Cittert–Zernike theorem (section 3.3.2). The detail in an image depends on the number of values of $\mathbf{r}_{ij}$ investigated and the way in which they are distributed; the resolution of the image depends on the maximum value of $|\mathbf{r}_{ij}|$, which is restricted to the aperture diameter.

In designing the mask with the N holes, it is necessary that every pair should give a different value of $\mathbf{r}$. Then the maximum number of individual values of $\gamma(\mathbf{r}_{ij})$ is obtained, and the fringes from each pair can be isolated; moreover, if two pairs were separated by the same vector, the fringes from both would superimpose to give a single fringe set with a lower contrast. A mask which satisfies this requirement is called a *nonredundant aperture mask*. It's an amusing exercise to try to design such a mask; for example in one dimension a set of holes at 1, 2, 4, 8, 16, 32 is nonredundant, but the values of r_{ij} are not very evenly spaced: 1, 2, 3, 4, 6, 7, 8, 12, 14, 15, 16, 24, 28, 30, 31; using the same number of holes in different positions 1, 3, 16, 23, 28, 32 improves the distribution while maintaining the range. The optimization of nonredundant aperture masks in two dimensions was discussed in section 4.1.3. As a recent example of their use, we cite the work of Tuthill and his group (Tuthill et al. 2000a,b) who used this method to obtain diffraction-limited resolution (better than 50 milli-arcsec) with the 10-m Keck telescope in the near infrared ($< 3\,\mu m$). The mask, figure 6.3, was situated on the secondary mirror which, although not in the aperture plane, is close enough not to affect the method significantly. With 15 and 21 hole masks, only about 0.2% of the available light was used, but this was sufficient to get good high-resolution images of some of the brightest supergiants and Mira variables (figure 6.4). To appreciate the importance of the phase information, figure 6.5 compares the results obtained by this technique assuming only zero or π phase and with the use of phase closure; the latter resolves the centrosymmetrical degeneracy of the former. This group also used

---

[†] There are, however, never enough equations to solve for the phases directly if all values of $\mathbf{r}_{ij}$ are different; see section 4.2.1.

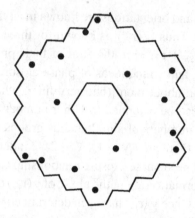

Fig. 6.3. Nonredundant aperture mask used by Tuthill et al. (2000a) on the 10-m Keck multimirror telescope.

a partially redundant approximation, consisting of a mask with an annular aperture, which transmitted about 10% of the incident light. The theory for such a mask converges on that for speckle masking, which will be discussed in section 6.4.2; however, because of the limited aperture, the amount of data to be processed is much smaller.

When using a multimirror telescope such as the Keck-I, two remarks have to be made. First of all, the holes on the aperture mask have to lie completely within the individual mirrors, even if the arrangement is not optimum. Second, although the phasing between the individual mirrors might not be exactly correct, errors can be looked at as part of the "atmosphere," even if not changing with time, and do not necessarily affect the results. Another point made by Tuthill et al. is that with nonredundant aperture imaging, the CCD camera need not satisfy the Nyquist sampling requirement of at least two sampling points per period of the fringe period. If the sampling is less than this, the data are assumed by the sampling algorithm to belong to a lower alias spatial frequency (Appendix A); but if the array is nonredundant, this alias frequency is distinguishable, and so the data are usable and do not just contribute to noise. This is an important consideration in achieving the highest resolution at short wavelengths.

6.3 Using the whole aperture: speckle interferometry

When the aperture is masked, so that many photons are wasted, diffraction-limited resolution can be obtained only for stars up to about fourth magnitude. For fainter sources, it is essential to use the whole aperture, and this leads us to the technique of speckle interferometry (Labeyrie 1970). The basic method of speckle interferometry derives the spatial autocorrelation of the object at the diffraction limit (section 6.3.1),

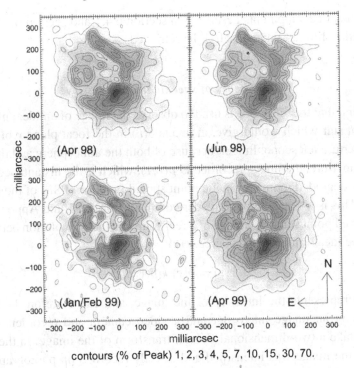

contours (% of Peak) 1, 2, 3, 4, 5, 7, 10, 15, 30, 70.

Fig. 6.4. Four high-resolution image reconstructions of IRC+10216 at 2.2 μm on different dates (Tuthill et al. 2000b).

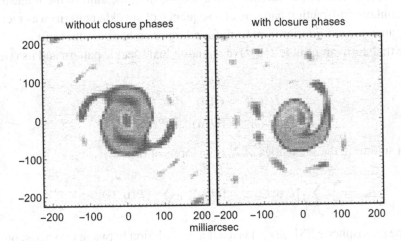

Fig. 6.5. Reconstructions of WR-104 with all phases assumed zero or π, and with phases deduced by phase closure (Monnier 2000).

but subsequent developments described in section 6.4.1 and section 6.4.2 allowed the diffraction-limited image itself to be retrieved.

6.3.1 Theory of speckle interferometry

Suppose first that the telescope is used to observe the image of an ideal monochromatic point star which would give an image $\delta(r)$ in the focal plane $\mathbf{r}$ of an ideal, infinite-aperture telescope. In fact, because of both the atmosphere and the limited telescope aperture, the image recorded is $p(\mathbf{r}; t)$, which is the instantaneous atmospherically degraded monochromatic point spread function (PSF) of the telescope at time t. This is a "speckle pattern" derived in section 5.9.1 and Appendix A for an unresolved point star. The image of a real source is the convolution between this PSF and the ideal image of the source, $o(\mathbf{r})$:

$$i(\mathbf{r}; t) = o(\mathbf{r}) \star p(\mathbf{r}; t). \tag{6.1}$$

The above represents the intensity of the image, and can therefore be recorded directly using an image intensifier and a video camera and recorder. The next stage is to take a two-dimensional Fourier transform of the image. In the original technique, the image was intensified and directly recorded on photographic film. This film was then used as a mask in a Fraunhofer diffraction experiment (section 3.1), the photographic exposure and development having been such as to provide an approximately linear relationship between the intensity of the exposure and the amplitude transmission of the film. On a second film, the sum of the intensities of the Fraunhofer diffraction patterns of a sequence of speckle patterns was recorded, giving an averaged Fourier transform for a large number of values of t.

The transform amplitude $I(\mathbf{u}; t)$ of an individual speckle pattern and its intensity are

$$I(\mathbf{u}; t) = O(\mathbf{u}) \cdot P(\mathbf{u}; t) \tag{6.2}$$

$$|I(\mathbf{u}; t)|^2 = |O(\mathbf{u})|^2 \cdot |P(\mathbf{u}; t)|^2, \tag{6.3}$$

and on summing the intensities of N transforms at times t_j:

$$\sum_{j=1}^{N} |I(\mathbf{u}; t_j)|^2 = |O(\mathbf{u})|^2 \cdot \sum_{j=1}^{N} |P(\mathbf{u}; t_j)|^2. \tag{6.4}$$

Now the atmospheric PSF $p(\mathbf{r}; t)$ is itself a convolution between two parts; one part is invariant with time and corresponds to the aperture and static aberrations of the telescope, and the second part is due to the atmosphere and is continuously varying. If the atmospheric part is envisaged as a random array of sharp points, then $p(\mathbf{r}; t)$ is the same array of telescope aperture PSFs. When its transform $P(\mathbf{u}; t)$ is taken, it is the product of the transform of the aperture PSF and that of the atmospheric

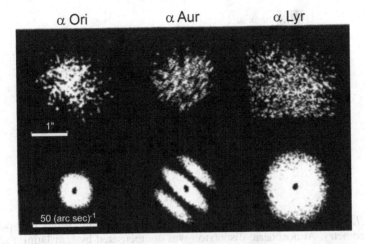

Fig. 6.6. Speckle images (above) and corresponding spatial power spectra (below). From left to right, Betelgeuse (resolved disk), Capella (resolved binary) and an unresolved reference star. The scales are r/F which are angular stellar coordinates (the bar shows 1 arcsec) and correspondingly uF which are reciprocal angular coordinates (the bar shows 50 arcsec^{-1}). The power spectra are each the sums of about 250 frames (Labeyrie 1970).

speckle pattern. The former is a smooth function[†] extending out to k_0D/F where D is the telescope aperture diameter and F its focal length. The second is also a speckle pattern, but with a sharp peak at the origin of u (Appendix A). When many values of this product are taken at different t_js, and are then squared and summed, the result is another smooth function extending out to k_0D/F, the peak at the origin remaining. It follows that (6.4) is $|O(\mathbf{u})|^2$ multiplied by a smooth function, with a strong peak at the center, extending out to the radius $u = k_0D/F$. Now, since $|O(\mathbf{u})|^2$ is known to be unity for a point star, the smooth function $\sum_{j=1}^{N} |P(\mathbf{u}; t_j)|^2$ can be determined by observation on such a star. Thus $|O(\mathbf{u})|^2$ can be determined for any other star by division, as far as $u = k_0D/F$. This function is the *spatial power spectrum* of the star image, out to the diffraction limit of the telescope and is equal to $|\gamma(\mathbf{u})|^2$ (section 3.3.3). In the photographs of the integrated transform intensities (6.4) shown in figure 6.6(d–f) the strong peak at the center has been blocked and one can see that the extent of the transform is greatest for the unresolved star.

Because the full aperture of the telescope is used, without masking, speckle interferometry has the potential to image very faint stars. An estimate of the limiting magnitude was made by Dainty (1974) for a binary star with components of equal intensity, as follows. Every speckle in the short-exposure image has two equal components, and the ideal signal would correspond to equal numbers of photons

---

[†] This function is the modulation transfer function of the telescope optics, including any aberrations (Dainty 1974).

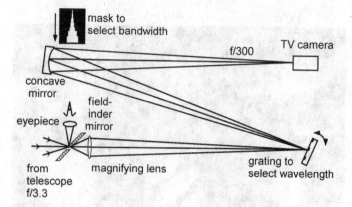

Fig. 6.7. Optics originally used by Labeyrie, Stachnik and Gezari for speckle interferometry. Atmospheric dispersion was compensated by translating the TV camera axially, the entire instrument being rotatable and oriented so that the grating dispersion was in the direction of the zenith. Analogue Fourier analysis of the recorded images used Fraunhofer diffraction.

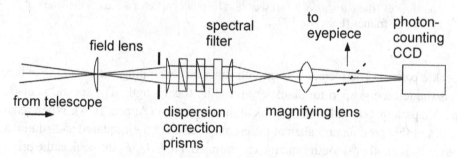

Fig. 6.8. Schematic diagram of a speckle camera with atmospheric dispersion corrector and band-limiting optical filter used at the Bernard Lyot telescope at Pic du Midi (Prieur et al. 1998). This speckle camera uses a PAPA detector.

being received in each of them. Noise corresponds to the difference between the numbers. Now the total number of photons in a speckle is related to the luminous flux from the star, the aperture of the telescope, the value of r_0 and the exposure time. The statistical distribution is assumed to be Poisson, and on this basis the expected signal-to-noise ratio can be calculated. In the case of a 4-m aperture telescope under typical atmospheric conditions, Dainty found that the limiting magnitude would be about 18 in the visible region.

6.3.2 Experimental speckle interferometry

The basic elements in a speckle camera are the same for all the various methods of analyzing the data (figures 6.7 and 6.8). An image of the star is formed at the

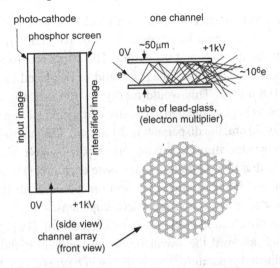

Fig. 6.9. A channel-plate image intensifier.

primary or coudé focus of a telescope with aperture D. The instantaneous image contains speckles with an angular scale of about λ/D and if this is multiplied by the focal length F it gives a size $\lambda \times f\#$; for $f\# = 8$ this means about $4\,\mu$m, requiring further magnification to ensure correct sampling by the camera. The individual speckles are too weak to be recorded directly, so that an image-intensifier had to be employed; it was the development of these devices which made speckle interferometry practical (figure 6.9). A typical microchannel plate intensifier has a hexagonal array of amplifier tubes which are about $50\,\mu$m in diameter, so that the speckle image has to be magnified considerably onto the intensifier, by a factor of at least 30 to allow each speckle to be sampled by a few microchannels. The output of the image intensifier is of order 10^6 brighter than the input and is coupled, either using a lens or by direct contact, to an electronic camera which can be gated to record individual images with exposures in the range 2 to 50 msec. The recent development of "electron-multiplier CCDs" has made the use of separate image intensifiers obsolete.

The speckle pattern itself is a function of wavelength, so that off-axis speckles are elongated radially. This means that the light has to be filtered, usually by an interference filter with a bandwidth between 10 and 50 nm[†]. In addition the atmosphere acts as a weak prism and causes the images at different wavelengths to be displaced relative to one another. The order of magnitude of the atmospheric dispersion is small enough not to be of importance for long-exposure imaging, but affects the speckle structure. The refractive index of air has dispersion $dn_{air}/d\lambda \approx 10^{-5}\,\mu$m^{-1}

[†] A rule of thumb suggested by Nisenson is that the bandwidth should be 1000 Å/$(S \cdot D)$ where S is the seeing in arcsec and D is the diameter of the telescope in m.

for green light at atmospheric pressure. As a result, the variation of the apparent angular positions of a star with wavelength is about $10^{-5} \tan \gamma \, \mu \mathrm{m}^{-1}$, where γ is the angle between the star axis and the zenith. The effect is zero at the zenith, but at $45°$ declination, the red and blue images would be displaced vertically by about $0.4 \cdot 10^{-5}$ rad ≈ 0.8 arcsec. This would hardly be noticeable in a long-exposure image, but of course is large compared to the size of a speckle. Even within the bandwidth of a filter, say 50 nm, the dispersion is 0.1 arcsec. This has to be compensated by the use of two nondeviating dispersing prism pairs whose relative orientation can be changed (so that they partially compensate each other) in accordance with the declination of star being studied (see section 8.2.3). Both the filter and the dispersion corrector are placed after the microscope lens.

The first speckle observations were made by Gezari et al. (1972) using a grating filter system, which allowed the wavelength, the bandwidth and the dispersion correction to be adjusted separately. This is shown in figure 6.7(a). A typical system in use today is shown in figure 6.8 (Prieur et al. 1998).

Processing the data is today done digitally. The first experiments, however, used analog processing, since the operation was a Fourier transform which could conveniently be carried out by Fraunhofer diffraction. The output of the image intensifier was photographed on film as a long series of speckle images. These images were then used as masks in a Fraunhofer experiment (figure 6.7b); in the Fourier plane one has directly the intensity of the transform, and these could be summed directly on film by multiple exposure, as the speckle film was run through the mask plane using a ciné projector. It is worth reminding the reader here that exact positioning of the frames in the mask plane is immaterial, because the position does not affect the intensity of the diffraction pattern (only its phase, which is not recorded). However, the optical quality of the recording film, which acts as an additional phase mask superimposed on the speckle image, was not too good, and resulted in some degradation of the resolution obtained.

The combination of an image-intensifier and an electronic camera can be replaced by a low-light-level intensified camera. A camera specially devised for this purpose is the PAPA (Precision Analog Photon Address) camera (Papaliolios et al. 1985), a single-photon detector giving a direct position-sensitive binary output (figure 6.10). In this device, multiple images of the intensifier output, created by a lens array, fall on a series of carefully aligned binary masks. The light transmitted by each mask is recorded by a small photomultiplier. One mask is clear, and is used to signal the detection of a photon and to activate all the other channels. Then signals 1 or 0 transmitted by each mask give a binary-coded output of the input photon position in the field[†]. This allows speckle interferometry to be carried out at extremely low light

---

[†] The masks used are Gray code, rather than binary, which ensures that the stripe edges of the various masks do not coincide, so that any ambiguity arising from a photon falling on an edge is limited to one bit only in each of the x and y directions.

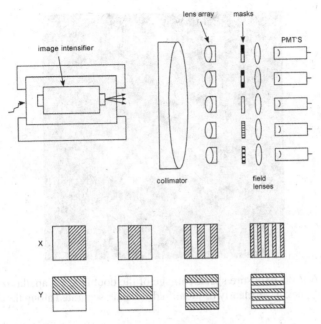

Fig. 6.10. The PAPA camera.

levels; there must not be more than one photon detected within the decay time of the intensifier phosphor (of the order of 200 nsec). Since the address of each photon is recorded individually, there are no problems with nonlinear responses, and the integration time (which depends on atmospheric conditions) can be optimized *post factum* by analyzing the response to a reference point star.

6.3.3 Some early results of speckle interferometry

In a few nights of observation with the Palomar 200-inch telescope, speckle interferometry initially provided accurate confirmations of the early measurements achieved by Michelson, Pease and Anderson on the largest stars and the binary star Capella. The two-dimensional autocorrelation patterns obtained were seen to be nearly axially symmetrical, rather than oblate, thus showing that the stars themselves had little oblateness. Results were obtained in several colors, from blue to the near infrared at 1 micron, using the tunable filtering device shown in figure 6.7, and just tuning the color while looking at the image led to the discovery of a sharp size variation with wavelength in both o Ceti and R Leo, which are members of the same class of long-period variable stars. The speckle size was unexpectedly found to double at the wavelength where the spectrum has a broad absorption band of TiO. This proved that both stars have a TiO atmosphere extending well beyond its photosphere.

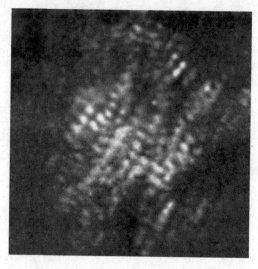

Fig. 6.11. A short-exposure speckle image of the double star Capella (α-Aur), in which each speckle can clearly be identified as a pair, separated along the diagonal.

6.4 Speckle imaging: getting round the limitations of the spatial autocorrelation function

From the spatial power spectrum function $|O(\mathbf{u})|^2$ one can calculate by Fourier transform the spatial autocorrelation, but not the actual star image[†]. Thus, since the spatial autocorrelation function cannot uniquely be translated to an image, the application of speckle interferometry seems to be limited even though it does provide a wealth of information about simple stellar objects. For example, a binary pair can easily be recognized, and the separation of the elements, their relative intensities and their orientation deduced; but one cannot distinguish between two centrosymmetric possibilities for the image (e.g. brighter element on the left or on the right). Again, limb darkening or other intensity variations as a function of radius of the star determined this way must also be assumed to be centrosymmetric. On the other hand, if you look at the speckle image of a binary star (figure 6.11), for example, you can see immediately which component is on the left and which on the right, because the individual speckles are like diffraction-limited images. This makes it clear that the technique should not really be limited by the spatial autocorrelation function; and after publication of the first speckle results it quickly became a challenge to find algorithmic methods to extract image information directly.

[†] In the original manuscript of his initial article on speckle interferometry, Labeyrie mentioned without proof that it should eventually be possible to reconstruct a true image of the object and not just the autocorrelation provided by his basic method. But an unknown referee considered this impossible and asked that the statement be deleted; this did not prevent later workers finding a solution to the problem.

6.4.1 The Knox–Thompson algorithm

The first successful method of deducing the phases of the Fourier components was due to Knox and Thompson (1974). We recall from (6.1) that the instantaneous image at time t is $i(\mathbf{r}; t) = o(\mathbf{r}) \star p(\mathbf{r}; t)$. Since the algorithmic methods of speckle imaging cannot easily be carried out by analog methods, we assume this to be transformed digitally to $I(\mathbf{u}; t) = O(\mathbf{u}) \cdot P(\mathbf{u}; t)$. Since this is a digital process, the complete complex transform is known; compare this to photographing the Fraunhofer diffraction pattern, which only records the modulus of $I(\mathbf{u}; t)$. A small shift vector $\delta\mathbf{u}$ is now chosen, with some discretion as described below. We can then calculate an overlap function

$$I(\mathbf{u}; t) \cdot I^*(\mathbf{u} + \delta\mathbf{u}; t) = O(\mathbf{u}) \cdot\cdot O^*(\mathbf{u} + \delta\mathbf{u}) \cdot P(\mathbf{u}; t) \cdot P^*(\mathbf{u} + \delta\mathbf{u}; t). \quad (6.5)$$

Introducing the phases explicitly in the form $S(\mathbf{u}) = |S(\mathbf{u})| \exp[i\phi_S(\mathbf{u})]$, where S can denote $I(t)$, O or $P(t)$, and denoting $\phi_S(\mathbf{u} + \delta\mathbf{u}) - \phi_S(\mathbf{u})$ by $\delta\phi_S$, we have

$$|I(\mathbf{u}; t)| \cdot |I(\mathbf{u} + \delta\mathbf{u}; t)| \exp[i\delta\phi_I(\mathbf{u}; t)]$$
$$= |O(\mathbf{u})| \cdot |O(\mathbf{u} + \delta\mathbf{u})| \exp[i\delta\phi_O(\mathbf{u})]$$
$$\cdot |P(\mathbf{u}; t)| \cdot |P(\mathbf{u} + \delta\mathbf{u}; t)| \exp[i\delta\phi_P(\mathbf{u}; t)]. \quad (6.6)$$

Now (6.6) is summed over a large number of frames at times t_j between which the atmospheric contribution changes randomly. The important point is that the phases $\phi_P(\mathbf{u}; t)$ due to the atmosphere change randomly, so that $\langle\delta\phi_P\rangle = 0$. It follows that

$$\sum_j I(\mathbf{u}; t_j) \cdot I^*(\mathbf{u} + \delta\mathbf{u}; t_j)$$
$$= |O(\mathbf{u})| \cdot |O(\mathbf{u} + \delta\mathbf{u})| \exp[i\delta\phi_O(\mathbf{u})]$$
$$\cdot \sum_j |P(\mathbf{u}; t_j)| \cdot |P(\mathbf{u} + \delta\mathbf{u}; t_j)|. \quad (6.7)$$

To use this result, we require that the chosen $\delta\mathbf{u}$ be small enough that $|O(\mathbf{u})|$ and $|O(\mathbf{u} + \delta\mathbf{u})|$ can be taken as equal; this depends on the outside angular dimensions of the source (often called the *support*). More importantly, this will result in $\delta\phi_O$ being much smaller than π so that there is no ambiguity in its value when calculated from $\exp(i\delta\phi_O)$. The value of $\sum_j |P(\mathbf{u}; t_j)| \cdot |P(\mathbf{u} + \delta\mathbf{u}; t_j)| \approx \sum_j |P(\mathbf{u}; t_j)|^2$ can then be determined from the data for an unresolved star, as in section 6.3.1 and is a generally smooth function out to $k_0 D / F$, except for the peak at the origin. Following this, both $\exp[i\delta\phi_O(\mathbf{u})]$ and $|O(\mathbf{u})|$ are known. By carrying out this process for two orthogonal values of $\delta\mathbf{u}$, say $(\Delta u, 0)$ and $(0, \Delta v)$, the values of ϕ_O can be calculated at points on a grid $(u, v) = (m\Delta u, n\Delta v)$ by summing the $\delta\phi$s out from a reference point (e.g. $\mathbf{u} = 0$)[†]. This can usually be done by several different routes in the (u, v)

[†] Choice of a different reference point for zero phase would simply move the star to a different point in the sky.

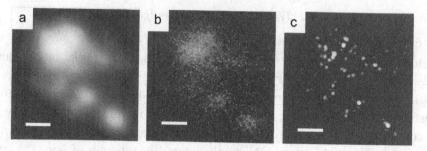

Fig. 6.12. A diffraction-limited image retrieved by triple-correlation, courtesy of
G. Weigelt: (a) shows the long-exposure image of R136 in the 30 Doraldus nebula;
(b) a single short-exposure image; and (c) the reconstructed image of the source.
The scale bars correspond to 1 arcsec. (Pehlemann et al. 1992).

plane, and the results averaged to improve the accuracy. The averaging is best done
by calculating the product $\prod_j |O(\mathbf{u_j})| \exp[i\delta\phi_O(\mathbf{u_j})]$ along each route, and taking
the weighted average of the various values using $\prod_j |O(\mathbf{u_j})|$ as the weight. Once the
phases are known, then we have the complete function $O(\mathbf{u}) = |O(\mathbf{u})| \exp[i\phi_O(\mathbf{u})]$
from which the true image $o(x, y)$ can be reconstructed. Note that the grid can only
be extended as long as $|P(\mathbf{u})|$ has non-zero value (i.e. out to $k_0 D/F$) which shows
that the resolution of the image is indeed the diffraction limit of the full telescope
aperture.

6.4.2 Speckle masking, or triple correlation

Another method of determining the phase of the transform $S(\mathbf{u})$ is to use the ideas
behind phase closure (section 4.2.1). The idea was proposed by Weigelt (1977) and
has the advantage that larger discrete steps in $\mathbf{u}$ can be used, so that it is less error-
prone than the Knox–Thompson algorithm. Despite the large computation times
involved in processing speckle sequences by this method, it has produced excellent
quality images of quite complicated scenes. The image shown in figure 6.12 was
subsequently confirmed by the Hubble Space Telescope, which is the best evidence
of the reliability of the method. In fact, by showing that what was considered to be
a supermassive star was in fact a cluster, it cast suspicion on the very existence of
supermassive stars.

Before developing a mathematical analysis of the triple-correlation method, we
can demonstrate it graphically for the case of the speckle image of a binary star
with two components of different strength. Consider first the hypothetical situation in which we had, within the same isoplanatic patch, an unresolved star and
the binary, and we recorded both their speckle patterns simultaneously. Then, the

cross-correlation between the two speckle images would clearly contain an over-whelming component which is the convolution of each binary speckle with the corresponding speckle of the point star, which would be a true image of the binary. There would also be a noisy background, from each binary speckle convolved with a different point-star speckle, but with enough frames, this should average to a smooth and weak function. This is called "speckle holography"[†]. In triple correlation we provide an approximation to the point-star speckle from the image itself. This is easy to do in the case of a binary star. Using the classical speckle interferometry method (section 6.3.1), we determine the separation vector of the binary. Then we take the product of the speckle image with itself shifted by that vector; this superimposes one of the binary components in one image on the other in the other image and therefore provides one spot per speckle, which is just the point-star speckle image. This product is then correlated with the speckle image again, to provide the result of speckle holography. The process is illustrated diagrammatically in figure 6.13 for a single frame.

The genius of speckle masking was to realize that when the above process is described mathematically, it can work for objects much more complicated than a binary star. First, we recall that the autocorrelation function $c(x_1)$ of $f(x)$ in one dimension compares pairs of values of f separated by distance x_1 (section 4.1.3). Now we define a *triple correlation* c_3 depending on two distances x_1 and x_2 in a similar manner:

$$c_3(x_1, x_2) = \int_{-\infty}^{\infty} f(x)f(x - x_1)f(x - x_2)\, dx. \tag{6.8}$$

Its Fourier transform depends on *two* reciprocal space parameters u_1 and u_2 and is easily expressed in terms of $F(u)$ (the proof follows the same lines as that of the convolution theorem):

$$
\begin{aligned}
C_3(u_1, u_2) &= \int c_3(x_1, x_2)\, \exp[-i(u_1 x_1 + u_2 x_2)]\, dx_1\, dx_2 \\
&= F(-u_1 - u_2) \cdot F(u_1) \cdot F(u_2) \\
&= F^*(u_1 + u_2) \cdot F(u_1) \cdot F(u_2),
\end{aligned}
\tag{6.9}
$$

since $f(x)$ is real. The same development can be made for a two-dimensional $f(x, y)$ in which case C_3 is a function of the four-dimensional vector[‡] $(\mathbf{u}_1, \mathbf{u}_2)$. We can now use $i = o * p$ to replace f as in (6.1), which can be calculated for every individual speckle frame. The values of $O(\mathbf{u})$ at $\mathbf{u}_1, \mathbf{u}_2$ and $\mathbf{u}_1 + \mathbf{u}_2$ are then related

---

[†] It has found little use in practice because of the very restrictive condition of needing a reference star of comparable magnitude to the star of interest in the same isoplanatic patch.

[‡] This four-vector has only mathematical significance; it doesn't correspond to a physical entity.

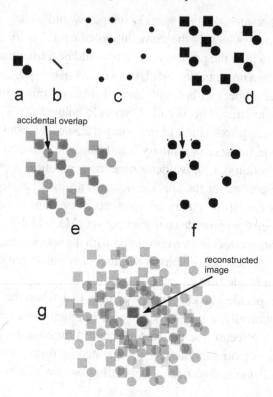

Fig. 6.13. The idea behind triple correlation, illustrated for a binary with unequal components. (a) shows the true image of the binary star and (b) the vector separating the two elements, as determined by speckle interferometry. (c) shows the atmospheric point spread function, i.e. the image of a point star. (d) is the convolution of (a) and (c), i.e. the speckle image observed. (e) shows the overlap of (d) with itself shifted by the vector (b), the product (f) being the retrieved speckle image of a point star, which should be compared with (c). (g) shows the correlation of (d) with (f), created by rotating (b) by 180° and centering it on each of the speckles of (f) successively. At its center, one image of (a) stands out above the noisy background.

to $C_3(\mathbf{u}_1, \mathbf{u}_2)$ by

$$C_3(\mathbf{u}_1, \mathbf{u}_2)$$
$$= O^*(\mathbf{u}_1 + \mathbf{u}_2) \cdot O(\mathbf{u}_1) \cdot O(\mathbf{u}_2) \cdot P^*(\mathbf{u}_1 + \mathbf{u}_2; t) \cdot P(\mathbf{u}_1; t) \cdot P(\mathbf{u}_2; t). \quad (6.10)$$

Once again, on averaging the time-dependent terms by summing many frames, the product of the P terms is a smooth function with zero phase. Thus, equating the phases of the two sides of (6.10):

$$\text{phase}[C_3(\mathbf{u}_1, \mathbf{u}_2)] = \phi_O(\mathbf{u}_1) + \phi_O(\mathbf{u}_2) - \phi_O(\mathbf{u}_1 + \mathbf{u}_2). \quad (6.11)$$

This should remind you of the result obtained for phase closure in section 4.2.1, and the rest follows the same theme. Of course, the phases of (6.10) can be equated accurately (in the presence of noise) only if both the products $|O(\mathbf{u}_1 + \mathbf{u}_2)| \cdot |O(\mathbf{u}_1)| \cdot |O(\mathbf{u}_2)|$ and $|P(\mathbf{u}_1 + \mathbf{u}_2)| \cdot |P(\mathbf{u}_1)| \cdot |P(\mathbf{u}_2)|$ are well above the noise level; to ensure this, a preliminary inspection of $|O(\mathbf{u})|^2$ as determined by speckle interferometry (section 6.3.1) has to be made. Defining the phases as zero at two noncollinear values of $\mathbf{u}$ (which fixes the position of the center of gravity of the image), one then steps around the Fourier plane from point to point on a square lattice determining the phases at each point from measurements of phase $[C_3(\mathbf{u}_1, \mathbf{u}_2)]$, until the resolution limit is reached at $k_0 D / F$ when the amplitude terms in (6.10) become too small for the equation to be reliable. In this case, the array used is redundant, and so phases can be calculated absolutely.

It should be appreciated that the amount of computation involved in this technique is very large, because of the four-dimensional net $(\mathbf{u}_1, \mathbf{u}_2)$. Some simplifications can shorten this process. For example, one can work with one-dimensional projections in several directions, in the spirit of (6.9), and then reconstruct the complete image by tomography.

6.4.3 Spectral speckle masking

A serious drawback to non-real-time imaging methods is the potential loss of spectral information. In section 6.3.3 we mentioned how the stellar diameter was in some cases found to be a strong function of the observation wavelength. The question is whether spectral information can also be coded into speckle images. A method suggested by Weigelt (1991) uses a one-dimensional projection of the speckle image, produced physically by the use of a cylindrical lens, which is then dispersed in the orthogonal direction by a prism. The one-dimensional image can then be used to get high angular resolution in the direction parallel to its length, with appropriate spectral resolution provided by the prism dispersion.

References

Christou, J. C., E. K. Hege, J. D. Freeman *et al.* (1986). *J. Opt. Soc. Am.* **A 3**, 204.
Dainty, J. C. (1974). *Mon. Not. R. Astr.* Soc., **169**, 631.
Fizeau, H., C. R. Acad. (1868). *Sc. (Paris)*, **66**, 932.
Gezari, D., A. Labeyrie and R. V. Stachnik (1972). *Astrophys. J. Lett.*, **173**, L1.
Haniff, C. A., C. D. McKay, D. J. Titterington *et al.* (1987). *Nature*, **328**, 694.
Knox, K. and B. J. Thompson (1974). *Astrophys. Jour.*, **193**, L45.
Labeyrie, A. (1970). *Astron. Astrophys.*, **6**, 85.
Michelson, A. A. (1891). *Nature*, **45**, 160.
Monnier, J. D. (2000). p. 224 in *Principles of Long Baseline Interferometry* ed.
    P. R. Lawson, NASA-JPL.

Papaliolios, C., P. Nisenson and S. Ebstein (1985). *Appl. Opt.*, **24**, 285.
Pehlemann, E., K.-H. Hoffman and G. Weigelt (1992). *Astron. Astrophys.*, **256**, 701.
Prieur, J.-L.. L. Koechlin, C. André, G. Gallon and C. Lucuix (1998). *Experim. Astron.* **8**, 297.
Ribak, E. (1986). *J. Opt. Soc. Am.*, **A 3**, 2069.
Roddier, F. (1986). *Opt. Comm.*, **60**, 145.
Stéphan, H. (1874). *C. R. Acad. Sc. (Paris)*, **78**, 1008.
Tuthill, P. G., J. D. Monnier, W. C. Danchi *et al.* (2000a). *Proc. Astron. Soc. Pacific*, **112**, 555.
Tuthill, P. G. *et al.* (2000b). *Astrophys. J.*, **543**, 284.
Weigelt, G. (1977). *Opt. Comm.*, **21**, 55.
Weigelt, G. (1991). *Progress in Optics*, Ed. E. Wolf, **29**, 295.

7

Intensity interferometry

7.1 Introduction

The idea of using measurements of the correlation between temporal fluctuations in light intensity at different field points was proposed by R. Hanbury Brown as an alternative to interferometry for measuring the spatial coherence function and therefore obtaining stellar data with high resolution. He called it *intensity interferometry*. Basically, in terms which should by now be familiar to readers of this book, an extended body of angular diameter α, consisting of many incoherently emitting sources, produces a speckled wavefront at the observer in which the speckles have typical size λ/α and typical lifetime τ_c. A pair of observers separated by a distance considerably less than λ/α are in the same speckle and therefore see the same intensity fluctuations. Observers separated by larger distances are likely to be in different speckles and see fluctuations with lesser correlation. The method was originally used for radio astronomy, in order to overcome the problem of providing identical phase references at two receivers separated by a very long distance (Hanbury Brown et al. 1952). It was then noticed that the measured correlations were immune to severe fluctuations produced by ionospheric instabilities, since these were in a frequency range very different from those of the intensity fluctuations being correlated. This provided the incentive to extend the method to the optical region. One should remember that at that time, the Michelson stellar interferometer was the only interferometric instrument which had provided resolution exceeding the atmospherically limited seeing, having successfully measured the diameters of six stars, and Pease's attempts to extend the baseline from 6 to 15 meters had proved impractical because of problems of atmospheric turbulence and mechanical stability. Hanbury Brown's method promised to overcome both of these problems, and indeed the intensity interferometer he built at Narrabri provided accurate measurements of stellar diameters of a further 32 bright stars with magnitude less than 2.5, as well as identifying several binary pairs. Unfortunately, Twiss

(1969) showed that the interferometer, as built, would be limited by signal-to-noise considerations to stars brighter than about second magnitude, although he made several suggestions for improvements which together might have improved the sensitivity to 7.5 magnitude. This estimate indeed proved to be correct, so that now that all the possible measurements (in the southern hemisphere) have been made, research using this technique has been discontinued in the optical waveband. It has recently been revived for measurements of X-ray sources. This chapter is therefore historical, in the sense that it is unlikely that a bigger intensity interferometer than that at Narrabri will ever be built due to the more recent success and prospects of amplitude interferometry – unless, of course, some brilliant new discovery in the field of higher order correlations comes about[†]. A fascinating account of the history, physics, engineering and achievements of this technique can be found in R. Hanbury Brown's book *The Intensity Interferometer* (1974), from which much of the material of this chapter was gleaned.

7.2 Intensity fluctuations and the second-order coherence function

7.2.1 The classical wave interpretation

Intensity interferometry is most clearly understood by looking at a typical waveform of light emitted by a quasimonochromatic source. The simulation presented in section 3.4 (figure 3.22) is an example of this, and is repeated in figure 7.1. It was performed by adding together the sinusoidal waves emitted by a large assembly of independent oscillators with frequencies ω distributed randomly in the given frequency band $\omega_0 \pm \delta\omega$ of the source. You can see that the wave looks like a random succession of wave groups, each one having about $\omega_0/\delta\omega$ periods within it. The total intensity and the phase fluctuate in a rather random fashion; you can see that within a wave group, when the amplitude is large, the phase is fairly stable, but it changes rapidly between groups, when the amplitude is small. The length of a wave group is about $\delta\omega^{-1}$, which is the coherence time τ_c, and this clearly corresponds to the definition in section 3.2.5 as the time for which the phase is approximately stable. This classical approach is clearly appropriate to radio waves, for which quantization obviously could be ignored because of the negligible energy of a photon. However, when the technique of intensity fluctuation correlation was first proposed in the optical region, it was thought that these wave groups might correspond to individual photons; but since the number of wave groups per unit time does not depend on the mean intensity, this interpretation was eventually dispelled.

[†] One wonders whether intensity interferometry could be useful in space, where the advantage of not needing high stability in the receiver separation might be important.

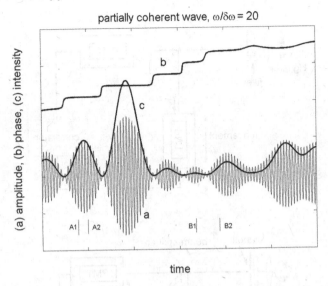

Fig. 7.1. A partially coherent wave simulated by superposing waves with random frequencies in a band of width 0.05 times the center frequency. (a) shows the wave amplitude, (b) the phase (compared with a pure sine wave at the center frequency) and (c) the fluctuating intensity of the wave.

Now let's see what we can learn from the measurement of correlation between intensity fluctuations, i.e. the *deviations* of the envelope from its mean value. Consider the train of waves of figure 7.1. Suppose we take two samples of the intensity (c) which follow one another closely in time, such as those at A1 and A2 separated by a time much shorter than the average length of a wave group. They are both likely to have about the same magnitude, since they probably belong to the same wave group, or are both close to a minimum, and therefore their deviations from the mean intensity will have the same sign. Therefore their correlation is strong. However, as the interval between them increases, the likelihood of the samples belonging to the same wave group decreases, and so does the correlation. When the interval is large, and two samples such as B1 and B2 are separated by a time greater than the typical length of a wave group, the probabilities of the deviations having the same sign or opposite sign become equal, and the correlation becomes negligible. As can be seen from the figure, the phase of the wave varies from wave group to wave group. It follows that the correlations between phases and the correlation between intensity fluctuations should give the same information. Now, from the Van Cittert–Zernike theorem, we know that phase correlations, measured as the coherence function $\gamma(\mathbf{r}, t)$, have an analytical relationship to the source properties, and so correlation between intensity fluctuations can also be used for astronomy. To prove their point, Hanbury Brown and Twiss (1956a) first showed that fluctuation correlations in time

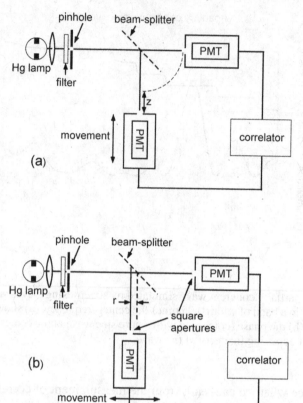

Fig. 7.2. Hanbury Brown and Twiss's experiments to show correlation between intensity fluctuations of two waves from the same source: (a) temporal correlation, as a function of the time delay z/c; (b) spatial correlation, as a function of the lateral displacement r. PMT indicates a photomultiplier tube.

could be related to the spectrum of the light, analogously to Fourier spectroscopy (although that term was coined several years later). Their experiment consisted of splitting the beam from a small source of light (a mercury arc) into two beams and sampling each one with a fast photomultiplier, with a variable time interval between the samples (figure 7.2a). The set-up clearly reminds us of a Michelson interferometer, but there is no actual interference between the waves because they are not returned by mirrors to overlap again after the beam-splitter. The electric signals from the two photodetectors were correlated (it is easy to state this, but in practice the electronics were a tour de force – one should remember that this was in 1956!) and the expected correlations were found within an interval $z_c = c/\delta\omega$. This experiment paved the way for a second experiment in which the photomultipliers, whose sensitive regions were masked by apertures, were optically equidistant

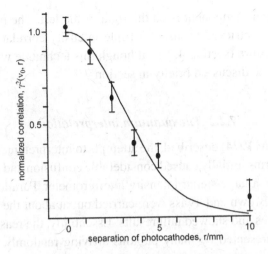

Fig. 7.3. Results of Hanbury Brown and Twiss's second experiment (figure 7.2b) showing spatial correlation between intensity fluctuations in waves from a pinhole 0.19 mm diameter in Hg light $\lambda = 435.8$ nm. The curve shows the theoretical result (Hanbury Brown and Twiss 1956b).

from the source, but displaced laterally (figure 7.2b). In this experiment, when the angular dimension of the source as seen by the photomultipliers was α, correlations were found within the coherence distance $r_c = \lambda/\alpha$ (figure 7.3).

How is the intensity correlation related to the coherence function? For intensity fluctuation studies we use the second-order coherence function $\gamma^{(2)}(\mathbf{r}, \tau)$ (3.50) defined in an analogous way to (3.19):

$$\gamma^{(2)}(\mathbf{r}, \tau) \equiv \frac{\langle I(\mathbf{r}_1, t)I(\mathbf{r}_1 + \mathbf{r}, t + \tau)\rangle}{\langle I(\mathbf{r}_1, t)\rangle \langle I(\mathbf{r}_1 + \mathbf{r}, t + \tau)\rangle}, \tag{7.1}$$

in which the averages $\langle \ldots \rangle$ are taken over t and in principle over $\mathbf{r}_1$ (but not always). Now the same technique as we used to demonstrate the fluctuation properties of the quasimonochromatic wave (section 3.4) can be used to calculate this quantity, and we find the simple result:

$$\gamma^{(2)}(\mathbf{r}, \tau) = 1 + |\gamma(\mathbf{r}, \tau)|^2. \tag{7.2}$$

It follows that the visibility amplitude can be measured this way, but phase information is lost. This loss is not surprising, since no actual wave interference has been employed, but of course it is also the price paid for immunity to atmospheric and mechanical disturbances. However, as we saw from the Michelson stellar interferometer, stellar angular diameters, ellipticities and centrosymmetric details such as limb darkening can be measured without knowing the phase of γ. Moreover, binary pairs can be identified and oriented and measured even if one cannot decide whether the stronger component is on the right or the left. The phase information can in principle be retrieved by using a triple intensity correlator (Gamo 1963), based on phase closure (section 4.2.1) although this technique was never used for astronomy. It will be discussed briefly in section 7.7.

7.2.2 The quantum interpretation

As Hanbury Brown (1974) described, the attempts to interpret intensity interferometry in quantum terms initially caused considerable confusion and almost grounded the project to build a large stellar intensity interferometer. Parallel experiments to those of Hanbury Brown and Twiss were carried out at about the same time using photon coincidences and showed null results. Essentially, the reason was that each wave group is represented by many photons arriving randomly with probability proportional to the instantaneous intensity. Coincidence measurements would represent correlations between the photon events themselves, and not between their average rates of arrival (see figure 7.4).

It is interesting, however, that an assembly of photons does have fluctuation characteristics similar to those of the classical superposition of random waves, although apparently for quite different reasons. The number of photons n emitted in a time T by a source of power $P(t)$ is $PT/\hbar\omega$ and because the emission process is Poissonian, the fluctuation in this number is equal to its square root. However, $P(t)$ itself is also fluctuating macroscopically, with amplitude about $P(t)$ on a time-scale of τ_c. The number of wave groups in time T is thus T/τ_c and this number too fluctuates with Poisson statistics provided that $T \gg \tau_c$. It follows that the total fluctuation Δn in the number of photons in δt is given by

$$(\Delta n)^2 = n + n^2\tau_c/T \qquad (T \gg \tau_c), \qquad (7.3)$$

in which the first term corresponds to the random photon noise and the second to the wave fluctuations which are to be correlated. A photon stream with this property is called "super-Poisson" because during short times the fluctuations are greater than those of a Poisson distribution. In the quantum interpretation, they correspond to the fluctuations of a Bose–Einstein condensate. Figure 7.4 shows super-Poisson distributions and also explains why photon correlations are negligible. The fluctuating wave shown in figure 7.1 is shown again in (a). A stream of photons arriving randomly during the same period is shown in (d); this has a Poisson distribution. In (b) and (c) we see two streams of photons arriving randomly with probability at any time proportional to the intensity of the wave in (a) at that time. These have super-Poisson distributions and one can see that the fluctuations in arrival rate in (b) and (c) are noticeably larger than those in the Poisson distribution, (d). The streams (b) and (c), which have the same total number of events as (d), simulate the photo-electrons received from the source at two detectors during the same period

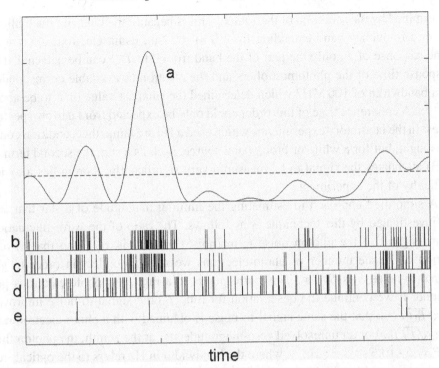

Fig. 7.4. Correlation between intensity fluctuations and individual photon events. (a) The intensity of the wave shown in figure 7.1. The mean intensity is shown by the broken line. (b) and (c) Two independent streams of photons generated randomly with probability at each time proportional to the intensity of (a) at that time. These have "super-Poisson" distributions. (d) A stream of photons generated randomly with probability proportional to the mean intensity of (a), showing a Poisson distribution. The three sequences (b)–(d) total the same number of events. (e) Coincidences between the photon events in (b) and (c) using time-slots narrower than the average interval between the photons in (d). The coincidences are almost nonexistent, which is why photon coincidence experiments failed to confirm the original intensity-correlation experiments.

of time. In a coincidence-counting experiment, one divides time into slots considerably shorter than the average time between photons and asks how often photons are seen in both (b) and (c) in the same slot. Even though the two correspond to the same intensity fluctuations, the coincidences (e) found between the individual photon arrival times in (b) and (c) is very small. This is the reason that the photon correlation experiments initially failed to confirm Hanbury Brown and Twiss's intensity correlation measurements (Purcell 1956).

7.3 Estimating the sensitivity of fluctuation correlations

As we saw in (7.3) above, the ratio between wave fluctuations and photon-counting fluctuations is greatest if T is of the order of the coherence time, τ_c, which is determined by the spectrum of the source. This is because the intensity fluctuations cover a frequency band extending from 0 to τ_c^{-1} but, using electronics having a time response of T, only the part of the band from 0 to T^{-1} can be detected. The response time of the photomultipliers and the electronics available corresponded to a bandwidth of 100 MHz which determined the smallest value of T to be about 10^{-8} s. A coherence time of this order could only be expected from narrow spectral lines. In the laboratory experiments, which used a Hg arc lamp, this condition could be sought, but for a white or broad-band source, such as a star, the second term in (7.3) attenuates the signal to be measured very significantly; therein lies a major difficulty of the experiment.

A simplified approach to estimating the limiting magnitude of a star that can be investigated by this technique is as follows. The ratio of the wave-fluctuation term (signal) to the photon noise term (noise) in (7.3) is τ_c/T, so that in one period of T, the number of photo-electrons we need to collect in order to get signal-to-noise S is ST/τ_c. This corresponds to a rate S/τ_c photo-electrons per second. If we continue the observation for time T_0, the signal-to-noise improves as $\sqrt{T_0/T}$, so that the same signal-to-noise is obtained with a photo-electron rate $S/\tau_c\sqrt{T/T_0}$. For an unresolved zero-magnitude star at the zenith, the photon flux is $P = 5 \cdot 10^{-5} \text{s}^{-1}\,\text{m}^{-2}\,\text{Hz}^{-1}$, where the bandwidth in Hz refers to the optical frequency bandwidth, which is equal to τ_c^{-1} (section 3.2.5). The *magnitude* m_λ of a star is defined such that the photon flux at wavelength λ is $2.5^{-m_\lambda} P$. Thus a collector of area A results in a photo-electron current of $2.5^{-m_\lambda} P A \eta/\tau_c$, where η is the conversion efficiency, which includes the quantum efficiency of the detector, reflectivity of the mirror and transmittance of the optics and the atmosphere (compared to that at the zenith). Equating this photo-electron current to the required value gives:

$$2.5^{-m_\lambda} P A \eta \tau_c^{-1} = S \tau_c^{-1}\sqrt{T/T_0}. \tag{7.4}$$

It is noteworthy that the optical bandwidth τ_c^{-1} in this equation cancels. In the astronomical scenario, the longest integration time T_0 cannot be expected to be much more than one hour. The biggest practical size of mirror turned out to have area $A = 30\,\text{m}^2$; its area cannot even theoretically be increased *ad infinitum*, since above a certain diameter, equal to the spatial coherence distance r_c and thus determined by the angular size of the source, different regions would start to contribute uncorrelated fluctuations which would cancel the effect being measured. The conversion efficiency cannot be assumed better than 5% and to measure correlation as a function of receiver spacing in order to determine stellar parameters, the basic signal-to-noise S, which corresponds to zero separation, must be at least 5. Inserting these numbers into (7.4) gives a limiting magnitude $m_\lambda = 2.5$.

Preliminary experiments (Hanbury Brown and Twiss 1956b) to prove the feasibility of intensity interferometry were carried out at Jodrell Bank, near Manchester

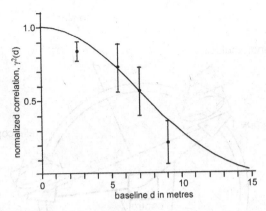

Fig. 7.5. Correlation measured for Sirius with baselines up to 9 m in 1956 (Hanbury Brown 1974). This can be compared with the later data in figure 7.9.

in England. It was appreciated that the receiver mirrors need not have good optical quality, since their purpose was to collect enough energy to get good signals from the photomultipliers. In fact, path-length errors up to a fraction of $cT \sim 0.3$ m could be tolerated, and military searchlight mirrors 1.5 m in diameter were used. Similarly, the signals from the two photomultipliers had to be synchronized to an accuracy of a fraction of T, which was not difficult to achieve by the use of varying cable lengths. The bright star Sirius ($m_{500nm} = -1.47$) was observed with baselines up to 9 m and after 18 hours of data collection during five months, a satisfactory result of $(7.1 \pm 0.5) \times 10^{-3}$ arcsec was obtained for its diameter (figure 7.5). In particular, because Sirius never rose more than 20° above the horizon during the experiments, the immunity of intensity interferometry to atmospheric disturbances was definitely established.

After this successful demonstration, and staving off the criticism leveled at his ideas from all directions (Purcell 1956), Hanbury Brown continued by constructing the first and only intensity interferometry observatory at Narrabri in NSW, Australia.

7.4 The Narrabri intensity interferometer

Hanbury Brown's book (1974) gives a colorful description of the funding, engineering and other practical aspects of the Narrabri interferometer. Its layout is shown schematically in figure 7.6. Basically, two 6.5-m diameter mosaic reflectors could be moved to any positions along a circular railway track with a diameter 166 m, which therefore represents the maximum baseline. Each reflector had a box at its focus containing a 45-mm diameter photomultiplier and an interference filter transmitting (443 ± 4) nm. The filter was sandwiched between positive and negative

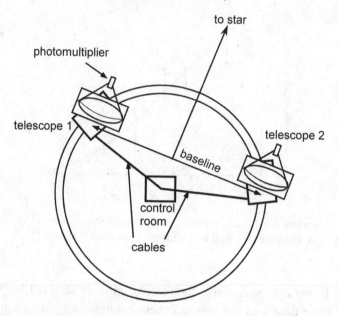

Fig. 7.6. Layout of the Narrabri intensity interferometer. Notice that the baseline is always normal to the direction of the star, so that with equal-length cables, the signals arrive simultaneously at the correlator.

collimating lenses, needed because the wavelengths of the transmission band depend on the angle of incidence. As shown above (see also Hanbury Brown 1974, p. 50) the signal-to-noise is independent of the optical bandwidth, but the narrow band-pass was useful to define the effective wavelength of the observations, an important parameter in determining the angular diameters from correlation measurements. Broader band-passes would have required a more complex analysis with effective wavelength becoming a function of spectral type. An additional practical consideration was the fact that, in the case of bright stars, a broad band-pass would lead to saturation of the photomultipliers.

Signals were transmitted to and from a control station at the center of the track by overhead cables. Each reflector could be controlled in elevation and azimuth, but in order to maintain synchronism between the detectors, the mirrors tracked the star in azimuth by traveling around the railway track so that the baseline was always normal to the direction of the star. A second photomultiplier receiving light from one element of the mosaic only was used as an autoguider to correct small orientation errors resulting from irregularities in the track.

7.4.1 The electronic correlator

The correlator measured directly the correlation between the fluctuations around the mean intensity at each receiver. This is easily shown to be $\gamma^{(2)}(\mathbf{r}) - 1 = |\gamma(\mathbf{r})|^2$.

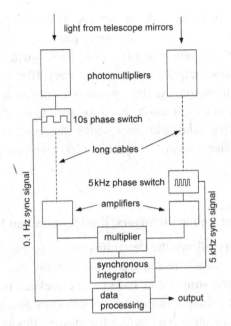

Fig. 7.7. Schematic diagram of the correlator and integrator system (after Hanbury Brown 1974).

Most of the experimental results were presented this way (e.g. figures 7.3, 7.5, 7.8). The principle of the correlator was to accept signals from the two photomultipliers and to integrate the product of the fluctuations in these two signals about their average values. There were two main parts: the multiplier and the integrator, and several synchronous switching devices (figure 7.7). The multiplying circuit was basically an amplifier in which the degree of amplification of one signal was controlled linearly by the other signal. This was done by applying one signal to the emitter of a transistor, which controlled the degree of amplification, and the second as input to the base. The output then contained a term which was proportional to the product of the two. The design, which we will not go into, was such that for an acceptable range of signal magnitudes this term is dominant over other terms which might contain products of higher powers of the signals. The output from the multiplier then entered an integrator, which output a value of the correlation integrated over a stated time (10 s). Several additional techniques to reduce noise were based on extensive experience in radio astronomy. Since integration is very sensitive to drift in the baseline of the input, resulting from the amplifiers of the multiplier, the phase of one of the photomultiplier signals was periodically switched by π at rate (5 kHz) slow compared with the fluctuations but fast compared with the integration time, and the integrator input switched in sign simultaneously. As a result the fluctuation correlations received by the integrator had a consistent sign, but amplifier baseline drifts, which occur between the two switches, appeared as a 5 kHz signal and could be integrated out. The principle is somewhat similar to that used in the ubiquitous laboratory phase-sensitive ("lock-in") amplifier. A second very low frequency (0.1 Hz) input phase-switching stage resulted in the integrated correlation output changing sign every 10 s. Alternate values were stored in separate accumulators. At 100 s intervals the difference between the stored values was computed as a final result. It can be seen that, with appropriately fast electronics, the effective bandwidth over which the correlation was evaluated is from the photomultiplier response frequency (100 MHz) to the 5 kHz switching frequency.

7.5 Data analysis: stellar diameters, double stars and limb darkening

Intensity interferometry allows the determination of $|\gamma(\mathbf{r})|^2$ for a stellar object, but not the phase of γ. As a result, the measurements that can be made either assume centrosymmetric structure or have to be correlated to a supposed model, or both. The simplest example is the determination of a stellar diameter which is obtained by fitting the data to a circular (or maybe elliptical) uniformly bright disk. A variation of this is to introduce to the model a monotonically decreasing radial dependence of the brightness, which is called "limb darkening" (section 3.3.3). The other feature which can be usefully investigated using $|\gamma|$ alone is double, or possibly multiple, component structure, although in this case there will be centrosymmetrically related solutions which cannot be distinguished. The data obtained by intensity interferometry is not accurate enough at large baselines for more complicated structures to be determined, since it is the square of γ which is measured. Accuracies of γ^2 better than about 1% were never obtained (and that only with Sirius) and so the smallest value of γ measurable was about 0.1.

7.5.1 Double stars

Although the main use of intensity interferometry has been to determine the angular diameters of single stars, it will be convenient to discuss first the determination of the separation and relative intensity of the components of double stars. The second-order coherence function for a pair of point stellar sources with relative intensities β separated by an angle whose projection onto a line x parallel to the baseline is Δ, can be written[†]

$$|\gamma|^2 = \frac{1}{(1+\beta)^2}[1 + \beta^2 + 2\beta \cos(k_0 x \Delta)]. \qquad (7.5)$$

[†] It is easy enough to rewrite this for the case of disk-like stars.

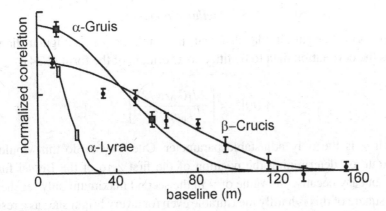

Fig. 7.8. Correlation data measured for three stars, showing the dependence on their angular diameters (after Hanbury Brown 1974)

This function is a cosine function plus a constant, and determination of Δ and β require finding the period of the cosine and the value of the constant. Then by observing the change in Δ during the night, as the baseline rotates, the angle of the line joining the two components can be found. This can be done, in principle, if Δ lies between two limits. The smallest value is determined by the maximum value of the baseline, which has to correspond to at least one period of the cosine. For 166-m baseline and $\lambda = 400\,\text{nm}$, this is $\Delta_{\text{min}} = 2.5 \cdot 10^{-9}$ rad $= 0.5$ mas. On the other hand, if Δ is too large, and a 6.5-m diameter reflector could in principle (i.e. if it were diffraction-limited) resolve the components, then the fluctuations in different regions of the reflector become uncorrelated and this reduces the measured correlation even for zero baseline. This limit is given when the argument of the cosine is greater than about $\pi/2$ at a baseline equal to the diameter of the reflector which is $\Delta_{\text{max}} = 16 \cdot 10^{-9}$ rad $= 3.3$ mas.

A check that this was not the case was always made before interpreting the data. A "zero-baseline correction" was determined by extrapolating several measured correlations to zero baseline, which would give the correlation expected for point stars. This could be compared to the value actually obtained for a known unresolvable source. If the value of this figure was greater than 0.75, any companion star must be less than two magnitudes weaker than the major one and the data could be interpreted directly. However, even in the cases where the star was determined to be multiple, the data could be used to determine the angular diameter of the major component (e.g. β-Crucis in figure 7.8). The rather limited region of angular separations which could be usefully measured resulted in only one double star, the spectroscopic binary Spica (α Vir), being studied in detail.

7.5.2 Stellar diameters

For a uniformly bright circular disk star, measurement of its angular diameter α requires the correlation data to be fitted to a formula of the form (3.34):

$$|\gamma(r)|^2 = \left[\frac{2J_1(k_0 r\alpha/2)}{k_0 r\alpha/2}\right]^2, \tag{7.6}$$

in which α is the only adjustable parameter. One way to do this would be to concentrate on determining the position of the first zero of the Bessel function. This is not easy because the value of γ in the second maximum only reaches -0.2 and the square of this is hardly measurable even for a very bright star; as a result, the value of baseline at which γ changes sign is not well-determined. So the numerical values of γ as determined within the first peak must be used. Empirically, it was found that the best way to make the fit used the baseline at which the correlation was about 0.3, assuming that the extrapolated value of $\gamma(0)$ fitted the criterion discussed in section 7.5.1. In one case, Altair (α Aql), variation of the diameter as measured along several axes suggested that the star was elliptical in shape; this observation was subsequently verified in 2000. Some examples of measured data are shown in figure 7.8.

7.5.3 Limb darkening

The expression (7.6) represents the coherence function expected for a uniformly bright disk. This can be re-expressed for a theoretical model with one parameter which allows for absorption in a uniformly thick stellar atmosphere, whose optical thickness therefore varies as the secant of the angle of emission. In principle, the results obtained for $|\gamma(r)|$ could be used to determine the value of this parameter and a corrected diameter calculated, but as explained in section 3.3.3 this really requires reliable data in the region of the second maximum of (7.6). Limb darkening could only be estimated with intensity interferometry for Sirius (α CMa), for which the data are shown in figure 7.9.

7.6 Astronomical results

Hanbury Brown et al. (1974) reported all the measurements of single stars made with the Narrabri interferometer, and these are summarized in Hanbury Brown's book (1974), so it would be superfluous to repeat them here. In all, measurements on 32 stars were made, their diameters lying between 6.6 mas and 0.4 mas. The results could be corrected for an accepted theoretical model for limb-darkening. Many of these measurements have now been confirmed by amplitude interferometry. Fifteen

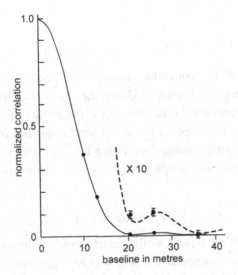

Fig. 7.9. Correlation data measured at Narrabri for Sirius, showing in particular the second peak, whose height is critical in determining details of limb darkening (Hanbury Brown 1974).

of the stars have large enough parallax for their distances to be determined, so that angular diameters could be translated to absolute radii.

Observation of a zero-baseline correlation significantly below the theoretical value allowed nine stars to be determined as multiplets, with separations too large to be determined directly by intensity interferometry. The only double star system with separation in the accessible range which was investigated in detail was the spectroscopic binary Spica (α Vir) for which considerable detail of the orbit and the components was obtained.

7.7 Retrieving the phase of the coherence function from intensity correlations

Since the second-order coherence function is $\gamma^{(2)} = 1 + |\gamma|^2$ it is clear that by measuring correlations with two detectors no information on the phase of γ can be acquired. However, just as in speckle masking (section 6.4.2), information on the phase can be found by using correlations between more than two detectors. This technique was suggested by Gamo (1963) but it was never used for astronomy. Gamo's basic idea was applied to spectral analysis, i.e. using the time domain.

The triple correlation between the photocurrent fluctuations ΔI at three detectors receiving light at times t, $t + t_1$ and $t + t_2$ is, using normalized values,

$$\gamma^{(3)}(t_1, t_2) = \langle \Delta I(t)\Delta I(t + t_1)\Delta I(t + t_2)\rangle_t$$
$$= K|\gamma(t_1)||\gamma(t_2)||\gamma(t_2 - t_1)|\cos(Q), \qquad (7.7)$$

where

$$Q = \phi(t_1) + \phi(t_2) - \phi(t_2 - t_1). \qquad (7.8)$$

and ϕ is the phase of the coherence function γ. The constant K describes the photo-electric conversion efficiency. Using the same procedures as described in section 4.2.1 the phases can be determined if the array t_j used is nonredundant. Sato et al. (1978) used this technique in the spatial domain for imaging nonsymmetrical objects. Recently, Holmes and Belen'kii (2004) have suggested the use of phase retrieval for acquiring the phases algorithmically from the moduli of γ.

7.8 Conclusion

After the demise of Michelson's stellar interferometer, the technique of intensity interferometry brilliantly provided a new way to determine stellar diameters and other parameters for bright enough stars. It was exploited fully by Hanbury Brown and his group at Narrabri, who apparently observed all the relevant stars in their section of the sky, and although Hanbury Brown did suggest in about 1971 how an improved model could be constructed, it was never built. Regarding this project, John Davis has written: "In hindsight, if we had gone ahead with the more sensitive intensity interferometer we proposed around 1971 we could have built it and made it work without difficulty, based on our experience with the Narrabri instrument; there were no unknowns. Nobody predicted – least of all ourselves – how long it would take to solve the problems with amplitude interferometry and to reach the performance we would have had with a new intensity interferometer. We would have had some 20 years more of pioneering work before amplitude interferometers competed in sensitivity – and even now they do not compete with the resolution we would have had which is needed for the hot stars!" (Davis, personal communication, 2005).

In fact, Davis and Tango (1986) published a direct experimental comparison between measurements of the diameter of Sirius at the same wavelength made by intensity interferometry and by a new relatively unsophisticated amplitude interferometer with two 150-mm apertures and an 11.4-m baseline (the forerunner of SUSI, section 8.4.5), and showed that equally accurate results could be obtained by the latter instrument in 1/40 of the time. So intensity interferometry was discontinued in the optical region, although space interferometers might one day bring it back again.

But the guiding principle, that of spatial correlations between fluctuations in the intensity of partially coherent waves, has not been forgotten. Hanbury Brown–Twiss correlations have recently been used in elementary particle physics to investigate the size of the source of pion emitters, by observing correlations between arrivals at a pair of detectors as a function of their separation, just as in the astronomical experiments (Baym 1998; Alexander 2003). The pions have spin 1, and therefore are bosons, like photons; the analysis is therefore identical to the astronomical case and the results show good correlation with the expected nuclear radius, which is proportional to the cube root of the atomic mass. In the elementary particle case, a new scenario is possible: if the emitted particles were fermions, there would be anti-bunching, and the stream of particles would be smoother than Poisson ... but that will take us too far astray from the theme of this book. Other recent applications of intensity interferometry are to correlations in atomic beams (Yasuda and Shimizu 1996) and to measuring the properties of ultrashort optical pulses.

References

Alexander, G. (2003). *Rep. Prog. Phys.*, **66**, 481.

Baym, G. (1998). *Acta Phys. Polonica*, B **29**, 1839.

Davis, J. and W. J. Tango (1986). *Nature*, **323**, 234.

Gamo, H. (1963). *J. Appl. Phys.*, **34**, 875.

Hanbury Brown, R., R. C. Jennison and M. K. Das Gupta (1952). *Nature* **170**, 1061.

Hanbury Brown, R. and R. Q. Twiss (1954). *Phil. Mag. ser 7*, **45**, 663.

Hanbury Brown, R. and R. Q. Twiss (1956a). *Nature*, **177**, 27.

Hanbury Brown, R. and R. Q. Twiss (1956b). *Nature*, **178**, 1046.

Hanbury Brown, R. (1974). *The Intensity interferometer*, London: Taylor & Francis.

Hanbury Brown, R., J. Davis and L. R. Allen (1974). *Mon. Not. R. Astron. Soc.*, **167**, 121–36.

Holmes, R. B. and M. Belen'kii (2004). *J. Opt. Soc. Am.*, **21**, 697.

Purcell, E. (1956). *Nature*, **178**, 1449.

Sato, T., S. Wakada, J. Yamamoto *et al.* (1978). *Appl. Opt.* **17**, 2047.

Twiss, R. Q. (1969). *Optica Acta*, **16**, 423.

Yasuda, M. and F. Shimizu (1996). *Phys. Rev. Lett.*, **77**, 3090.

8

Amplitude interferometry: techniques and instruments

8.1 Introduction

This chapter is about the practical side of interferometric astronomy. There are many instruments which have been built to use coherence measurements in order to gather information about stellar and other cosmic structures, all based on the principles which have already been discussed theoretically. The plan of the chapter is to discuss first the building blocks and techniques which are common to many interferometers, with some mention of advantages and disadvantages of different approaches. Then we shall describe the way in which various interferometers use these building blocks, with one or two examples of their results. Most of the material of this chapter is based on published material, and some valuable references which have been used intensively are the chapters in the course notes from the 1999 Michelson Summer School (Lawson 2000) and the Proceedings of the SPIE meetings on Stellar Interferometry up to 2004 (SPIE 1994–2004). The latter are unfortunately not available freely to the public; we have therefore made an effort to cite work published in the open literature as far as possible.

Every interferometer is built from several definable blocks, or subsystems, of which the maximal set is illustrated schematically in figure 8.1. First, the global "aperture" of the interferometer, which defines its angular resolution limit, is the bounding region within which there are several subapertures, separated by the vectors $\mathbf{r}$ which will be the arguments of the measured values of the coherence function. Each subaperture may simply be a defined opening, or it may be a complete telescope in its own right, or the various subapertures may be parts of a single telescope defined by a mask. The light beams from the several apertures are then transported by beam lines to the beam-combining system. On the way there, the optical paths from the star to the beam-combiner have to be equalized, to a certain degree of accuracy, which depends on the optical bandwidth being used. After the beam-combiner, the interference outputs have to be measured, using either imaging or single-pixel detectors and the associated data-processing equipment. At the

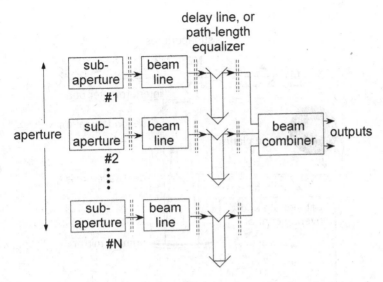

Fig. 8.1. The blocks, or subsystems, from which a stellar interferometer is composed. Extra optics for focusing, filtering, etc. may be inserted at any of the positions indicated by vertical double broken lines.

beam-combining system, some of the light may be extracted to be used for adaptive correction of the subapertures.

The figure shows these building blocks. The vertical broken double lines indicate places where additional optical components may be placed for beam-compression, tip–tilt correction, filtering, etc. Now all the components in the interferometer are individually designed by the experimenters in accordance with their scientific aims, available and proposed techniques and of course budget considerations. As a result, every interferometer is different. First, let us look at the way that these blocks were relevant to the techniques already described in previous chapters.

8.1.1 The Michelson stellar interferometer

The Michelson stellar interferometer (MSI) was the progenitor of all the modern interferometers, and in that instrument the components can be identified clearly. The subapertures were two plane mirrors with no additional optics. The beam lines were the periscope-like structures parallel to the mechanical beam, and were equalized in optical length by careful positioning of the components, although an additional small path-length correction was provided by a double glass wedge in one beam and a glass plate in the other (figure 8.2). By using the telescope mount to follow the star diurnally, no dynamic path-length equalization was necessary in order to see fringes.

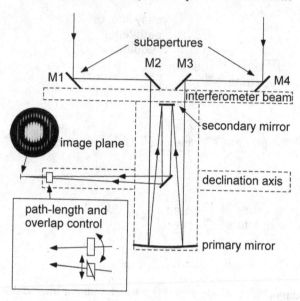

Fig. 8.2. Michelson stellar interferometer, showing the path-length corrector and the tilt plate used to ensure overlap of the two images.

The beams were combined as they interfered on the detector surface (the Fizeau configuration, or image-plane combination). Since the subapertures were smaller than r_0, Fried's atmospheric parameter (section 5.4.1), the image observed would be diffraction-limited by these subapertures and the appearance of the image was their diffraction pattern crossed by fringes, with spacing determined by the angle of inclination between the two beams. The data processing system was Michelson's eye and brain.

8.1.2 *The Narrabri Intensity Interferometer*

In the intensity interferometer, discussed in chapter 7, there were two subapertures in the form of reflecting telescopes. The light went directly through filters to the detectors (photomultipliers) and the data processing system; the layout with two telescopes traveling around a circular track played the part of the path equalizer, and a second electronic fine-tuning path-length adjustment stage was incorporated in the data processing. What is unique to this experiment is that there was no beam-combiner, since the correlations were deduced electronically from the photocurrents.

8.1.3 Aperture masking

As discussed in chapter 1, aperture masking using two subapertures with variable separation was the first form of amplitude interferometry to be used, as suggested by Fizeau in 1868 and carried out by Stéfan in 1871. In the recent work on aperture masking, described in chapter 6, there are several subapertures in a mask covering the aperture of the Keck I telescope, the focusing mechanism of which brings all the beams onto the imaging detector. As in the Michelson stellar interferometer, the fact that the whole telescope can be steered obviates the necessity for path equalizers and the beams interfere on the detector itself (Fizeau configuration).

8.2 What do we demand of an interferometer?

Astronomical interferometers are used in general for two different purposes, imaging and accurate astronomical mapping, called *astrometry*. Imaging includes measuring the angular sizes of stars and their surface characteristics, visualization of multiple stars and galaxies and, hopefully, observing a planetary system like ours around a nearby star. For example, measurements of the time-dependent angular diameter of a pulsating star can be compared with the time-varying Doppler shift of its spectral emission lines to determine the distance of the star absolutely (section 11.2.3). Astrometry consists of the determination of relative stellar positions on the celestial sphere with high precision. It facilitates the absolute determination of stellar distances via parallax measurements and also measuring the orbital characteristics of multiple stars, thereby contributing to the determination the component masses. Hopefully, high-resolution astrometry will lead to the discovery of exoplanets by detecting the induced periodic displacement of the parent star; invisible brown dwarf companions have already been discovered this way. The displacement of the Sun caused by Jupiter's motion is about 1 mas when observed from a distance of 10 parsec, whereas that caused by the earth at 3 parsec is about 8 μas, which both present a significant challenge to present techniques.

The requirements of an interferometer for these disciplines are rather different. For imaging, very high angular resolution is required, and also a sharp point spread function if more than just model-fitting is demanded. This means, in turn, the use of long baselines between several strategically placed subapertures in order to get a uniform distribution in the (u, v) plane (see section 4.1 and figure 8.42). For astrometry, the goal is to measure the angular position of the center of light of a star, relative to certain standard fiducial markers. The present-day accuracy of the Hipparcos star catalog is about 0.2 mas, and an improvement of more than an order of magnitude to 10 μas is expected. However, for astrometry there is no need to resolve the star's diameter, and so long baselines are not needed. In fact, if the baseline is long enough to resolve the star, the interference fringes will necessarily have poor visibility, and this is counter-productive. What is more appropriate is a relatively short baseline, whose length is known very accurately. Then, given an equally accurate determination of the path-length difference, the angle between star and baseline can be calculated as arcsin(delay/baseline). One might argue that if the baseline is constant, even if it is not known accurately, relative positions of two stars can be determined from the path-length differences alone, but this is only true if the two are in the same field of view and can be observed simultaneously. Otherwise small changes in the baseline for geological, meteorological or mechanical reasons can introduce systematic errors into sequentially determined stellar positions.

For example, consider a single measurement of declination accurate to 1 mas $= 5 \cdot 10^{-9}$ rad. Using a baseline of 10 m (recall that Michelson's 6 m beam interferometer could resolve the diameters of the largest stars, so that this figure represents a maximum useful value), this means that the baseline has to be measured to an accuracy of $5 \cdot 10^{-8}$ m, or about 0.1 wavelengths in the visible. So very accurate laser-based metrology systems must be added to the requirements of the interferometer. However, in small-angle astrometry, where two stars separated by θ (of the order of arcseconds, 10^{-5} rad, so as to be in the same isoplanatic patch) are measured simultaneously, the change in path-length to be measured for baseline B is $\approx \theta B$. If a change in θ is to be determined, we have $\delta(\theta B) = \theta \, \delta B + B \, \delta\theta \approx B \, \delta\theta$ if θ is very small. For example, if $\delta\theta$ is to be of order $10 \, \mu$as and the accuracy of measurement of the path-length delay is 50 nm, B must be 1000 m but its uncertainty can be as large as 1 mm before the measurement accuracy is compromised. This is a promising method, developed particularly at PTI (section 8.4.8), for observing the periodic wobble in a stellar position, using a distant marker star as reference.

This argument is essentially two-dimensional. For three-dimensional determinations, a nonlinear array of subapertures is necessary. It turns out that, with one or two accurate baseline measurements, the rest can be determined from the data itself by requiring internal consistency, provided that enough stars are measured using the same set of aperture positions. But it would seem that doing this in order to save baseline metrology can only compromise the accuracy of the deduced stellar positions.

8.3 The components of modern amplitude interferometers

In order to understand the differences between the various interferometer concepts, we shall now discuss their various components, with illustration from some of the systems in use or being built today, and which are described individually later in the chapter.

8.3.1 Subapertures and telescopes

Starlight is collected by a number of subapertures, each of which defines a collecting area and sends its output to a beam-processing laboratory. A subaperture may be simply a hole or mirror, usually circular, or it may be a complete astronomical telescope. Pointing to the star can be accomplished either by rotating the telescope or by the use of a siderostat, which is a plane mirror tilted so as to direct the starlight in a direction which does not change during the day; the siderostat may feed a plain subaperture (Mark-III, SUSI), or the entrance to a fixed telescope (COAST, ISI). If the complete telescope is pointed, then an additional tilt mechanism is necessary to bring the output light to a constant direction (for example I2T, CHARA, GI2T). This is the same as the well-known coudé system, where stellar light is brought to a fixed spectroscopic laboratory. Although older telescopes used equatorial mounts, whereby the pointing is achieved by the use of a single constant-speed motor rotating the telescope once a day about an axis parallel to the Earth's axis, it is more usual today to use an "alt–az" mounting where computer-controlled rotation about one vertical and one horizontal axis achieves the same end, although the image-plane rotation then has to be compensated. By using a horizontal axis of rotation, pointing in the direction of the interferometric optics, we have an "alt–alt" mounting in which the starlight can be projected in the required direction by the aid of a coudé mirror which rotates at half the rate of the parallel alt axis. No beam-reducing telescope is needed if the subapertures are small, and the siderostat can be used alone, which makes for a much simpler mechanical system. In addition, the center of rotation of a siderostat can be defined precisely, which is important in astrometric work. In systems such as the ISI and COAST, a large siderostat is installed before the beam-reducing telescope. The size of the siderostat mirror must be considerably larger than that of the sub-aperture to allow for the fact that it is always at an angle to the axis of the telescope, sometimes a large angle. For example at ISI 2-m siderostats feed 1.65-m diameter telescopes. Since large plane mirrors are more difficult to test during production than concave ones, but on the other hand the mechanical operation of a siderostat is simple, the choice is not obvious.

Because of atmospheric turbulence, the size of subaperture chosen depends on the wavelength of operation and the decision on whether to add adaptive components. If the subaperture is smaller than r_0 (function of wavelength), the main diffraction lobe of the subaperture is larger than the seeing variations, and so no adaptive optics are needed to correct the wavefront errors, although piston errors will usually be compensated later in the system (section 8.3.9). This approach is exemplified by the MSI and by SUSI. A telescope at the subaperture reduces the beam diameter and thus increases its angular spread, but this can be tolerable if it keeps the size of the collecting optics manageable. For example, the main diffraction lobe from a subaperture $D = 15$ cm diameter (about r_0 in the visible) at $\lambda = 1\ \mu$m has angular width $\lambda/D \approx 6 \cdot 10^{-6}$ rad. Even at the end of a baseline of 100 m, the diffractive spread of the beam is only $6 \cdot 10^{-4}$ m or 0.6 mm. Using a telescope to reduce the beam width to 3 cm, for example, to allow smaller optics to be used increases this spread by a factor of five, to 3 mm. But if the telescope focuses the star image at the entrance of the beam-combiner, the beam size is in fact much reduced and smaller collecting optics can be used.[†] If the subaperture is larger than r_0, some adaptive correction needs to be made if most of the collected light is to be used. Otherwise, in pupil-plane interferometry essentially only one speckle of the ensuing interferogram is used for analysis, and in image-plane detection, it is difficult to use the resulting wrinkled fringes efficiently. The first adaptive correction that is done is to correct tip–tilt by stabilizing the stellar image from each subaperture independently. The optimum subaperture, at which the main diffraction lobe has angle equal to the stabilized seeing, is now larger than r_0 (Noll 1976)(COAST and MIRA-I2).

The telescope optics used are almost invariably Cassegrain, with output beams having very large f-number, even infinite (figure 8.3). The mirrors are paraboloidal; because the field of view is very small, more elaborate optical designs such as Ritchey–Chretien, Schmidt, etc. would give no advantage here. The idea is to reduce the diameter of the collecting optics at the beam-combiner to the size of conventional optical components. Adaptive correction, either tip–tilt or higher order and interaperture piston, is applied either on the secondary mirror or after it so that the physical size of the adaptive corrector is minimal; the wavefront sensor is preferably located inside the beam combiner so that "seeing" generated by turbulence within the coudé beam is also corrected. In some interferometers, an important general consideration in the optical design has been the preservation of polarization. Mirrors and their coatings have birefringent properties when the angle of incidence is not normal, which introduce phase differences between the principal polarizations. These become a problem when two beams with different (generally elliptic) polarizations are asked to interfere. The way around this problem is to ensure that the optical trains from each telescope to the beam-combiner contain the same number of mirrors working with the same angles of incidence; then, one hopes, the

---

[†] This issue will be of major importance with the extremely long baselines, $10^5 - 10^6$ km, of the proposed Neutron Star Imager in space (section 12.4). Subapertures as large as 8 m, with a flat or slightly concave mirror, and collecting optics of similar size, are then needed at visible wavelengths to avoid diffractive losses in the intermediate beam.

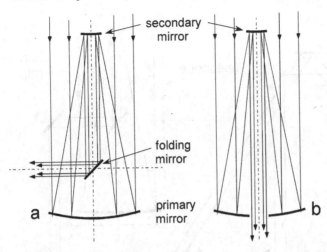

Fig. 8.3. Cassegrain optics (a) as a telescope, (b) as a beam-compressor. In (a), the flat folding mirror could equivalently, although not in terms of cost, be a large mirror before the telescope, in which case the telescope is fixed in orientation. Otherwise, the telescope is pointed towards the star, and the small flat mirror is best located at the mechanical node where both axes of rotation intersect. The vertical axis of rotation does not coincide with the optical axis of the telescope, but intersects the horizontal one on the folding mirror. See also figure 8.4.

elliptical polarizations at the beam-combiner will be identical and the phase differences between the principal components from the two telescopes will be the same. As examples of these, see the details in the layouts of COAST, NPOI and CHARA and figure 8.4. Otherwise, it is necessary to introduce a linear polarizer before the detector, which inevitably absorbs or reflects some light, to detect simultaneously both polarized interferograms or to correct the polarization state adaptively.

8.3.2 Beam lines and their dispersion correction

In several interferometers, the beam lines from the subapertures to the beam-combiner are within evacuated tubes. There are two reasons for this. First, there is the problem of atmospheric turbulence. The beams obviously travel from subaperture to beam-combiner close to the ground, where the fluctuations are greatest, particularly out of doors. The turbulence is uncorrelated in the various beam regions, so that no natural compensation occurs. Second, there is a problem of atmospheric dispersion, which means that the different wavelengths have different optical paths and it will be impossible to equalize all components together. Indeed, if all telescopes are in a common horizontal plane, the optical path lengths through the atmosphere, from its horizontal upper layers down to each subaperture, are equal at

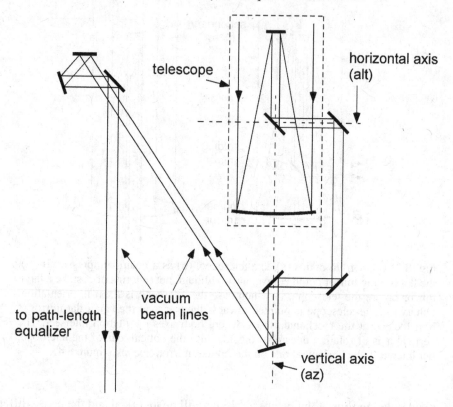

Fig. 8.4. Example of the sequence of mirrors in one beam line at CHARA, designed in order to control polarization effects. Each beam line has the same number of mirrors reflecting at the same angles.

a given wavelength. But the unequal horizontal distances between the subapertures and the beam combiner, required to compensate the optical path difference, introduce dispersion in the fringes unless the beams propagate in vacuum. Neither of these reasons is a great obstacle for short baselines, so that the MSI (baseline 6.5 m.) and I2T (baseline 13.8 m) worked without them, both using glass prism pairs with variable thickness or interchangeable glass plates to compensate the atmospheric dispersion. Of course, even vacuum beam lines do not solve the dispersion problem completely; only if the path-length equalizers are also in vacuum does it disappear.

There is one interesting counter-example to the above. ISI works with heterodyne detection (section 4.4.1) and the beams which travel down the beam lines are the local oscillator laser beams and not the starlight. The method used here is to send these laser beams back again to their source down the beam lines and then to measure interferometrically the path-length fluctuations, which can then be compensated.

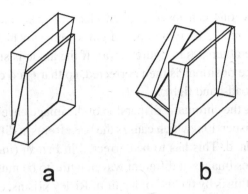

Fig. 8.5. Dispersion correctors: (a) path-length and dispersion; (b) angular, using two Risley prism pairs.

Of course, with the very narrow bandwidth implied by heterodyne detection, no dispersion correction is needed.

The beam pipes do not need to have very high vacuum. In fact, even pipes full of static air do a lot to prevent turbulence. But in order to avoid appreciable refractive dispersion, we can calculate the necessary vacuum. At standard pressure and temperature, the refractive index of air is $n = 1.000\,29$. This means that in $L = 100\,\text{m}$ of air, for example, the accumulated phase difference compared to vacuum is $2.9 \cdot 10^{-4} \times 2L\pi/\lambda \approx 3 \cdot 10^5\,\text{rad}$. The variation in this with wavelength over the visible spectrum is about 1%, i.e. $3 \cdot 10^3$ rad. Since $n-1$ is proportional to pressure, this can be reduced to 0.1 rad by using air pressure $5 \cdot 10^{-5}\,\text{atm}$, or about $0.05\,\text{torr}$.

8.3.3 Correction of angular dispersion

Interferometers using two or more subapertures are affected by the chromatic dispersion of the atmosphere in the same way as the unmasked aperture of which they are parts. Except when the observed star is at the zenith, the atmosphere acts like a prism, and slightly deviates the light rays, increasingly so at shorter wavelengths and at larger angles from the zenith. As a result, the image from the unmasked aperture is both displaced and spectrally dispersed. In Fizeau imaging, the effect could be corrected by a large low-angle prism located at the entrance aperture, covering all of it, or equivalently by a small prism in the exit pupil, having a larger angle. Whether in the entrance aperture or the exit pupil, the prism could then be segmented into small pieces covering each subpupil. These pieces would have different thicknesses but the same angle.

In practice, the thickness and angle, which change with time, have to be corrected separately. Since the required thickness of glass might become negative as the declination changes, one can insert a fixed piece of glass into one beam and a variable thickness, created by two opposed glass wedges which can slide against one another into the others, as in figure 8.5(a). If the angle is small, the chromatic dependence of fringe positions is then corrected, so that good contrast fringes can be obtained with broad-band light.

But there remains the chromatic dispersion of the image envelope. This is the result of the angular dispersion, which causes the wavefronts at different wavelengths to be mutually inclined. This has to be corrected in Fizeau (image plane) combination, since it causes images at different wavelengths to be mutually displaced. A common way to do this is by the use of a pair of Risley prisms, each of which itself consists of a pair of opposed thin prisms made of glasses with different dispersive powers, their angles being calculated such that in the center of the spectrum there is no deviation (figure 8.5b). The two pairs can then be counter-rotated so that together they give the required dispersion correction.

As an example of the calculation of a dispersion corrector, let us suppose that two types of glass 1 and 2 are used, having refractive indices n_{D1} and n_{D2} for the center wavelength D and Abbe dispersivities ω_1 and ω_2 respectively[†]. When two thin prisms made from these glasses, having angles α_1 and $\alpha_2 = \alpha_1(n_{D1} - 1)/(n_{D2} - 1)$ are cemented in opposition, it is easy to see that the total deviation at the D wavelength is zero. The dispersion angle between λ_C and λ_F is

$$\delta\theta = \alpha_1(n_{D1} - 1)(\omega_1^{-1} - \omega_2^{-1}). \tag{8.1}$$

Now this dispersivity has to annul the effect of the atmospheric prism. The dispersion angle of air for a zenith angle γ is $\gamma(n_{air} - 1)/\omega_{air}$. By using two such pairs of prisms in opposition, the dispersion of the air can be compensated for any value of γ up to a maximum $\gamma_0 = 90°$ by two such prism pairs if

$$\gamma_0(n_{air} - 1)/\omega_{air} < 2\alpha_1(n_{D1} - 1)(\omega_1^{-1} - \omega_2^{-1}). \tag{8.2}$$

The dispersion index of air is about 100, and that for two types of crown glass (BK7 and BaK1) are 64 and 57 respectively, giving a typical prism angle $\alpha_1 = 0.11$ rad $= 6°$.

8.3.4 Path-length equalizers or delay lines

The path equalizers (delay lines) used in long-baseline amplitude interferometers are indeed massive engineering projects, often called "optical trombones". In order that the interferometer be usable at all angles of declination, path differences up to the baseline length have to be compensated to an accuracy better than the coherence length in order to allow interference.

[†] The Abbe dispersivity of a glass is defined in terms of its refractive index n_λ at three wavelengths, $\lambda_C = 656.3\,\text{nm}$, $\lambda_D = 589.3\,\text{nm}$, $\lambda_F = 486.1\,\text{nm}$, as $\omega = (n_D - 1)/(n_C - n_F)$.

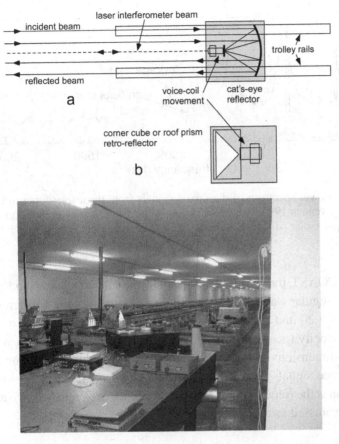

Fig. 8.6. (a) Typical design of a path equalizer, using a cat's-eye reflector. (b) shows the alternative corner-cube reflector. (c) Delay lines at CHARA.

Except by multiple folding, which results in light loss at every fold, there is no way of making a compact equalizer. As a result, there is a long straight track, along which a carriage is moved smoothly by computer-controlled motors, the distance of movement being monitored by a laser interferometer. On the carriage is a retro-reflector; some systems (SUSI, I2T) with only two subapertures use the same track for correcting both optical paths by mounting two retro-reflectors on the carriage, facing in opposite directions, one beam entering from each end; this allows the total length of the track needed to be halved. A typical machine is shown in figure 8.6. The use of retro-reflectors, either cat's eye or roof mirrors, ensures that despite any irregularities in the track, the beam returns in the same direction as it entered. Since the motor-driven movement cannot be accurate to the required fraction of a wavelength, a further movement stage is mounted on the carriage, operated by voice coils or piezoelectric transducers. These elements can also be programmed to provide a path-length sweep for fringe recording. For

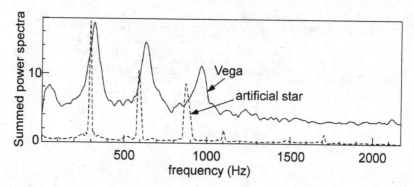

Fig. 8.7. Power spectrum of the mixed signals from three telescopes at COAST observing Vega in 1993. Each peak occurs at the difference frequency corresponding to a particular pair of telescopes. After Baldwin et al. (1994).

example, at COAST, the roof prism position is modulated using a loudspeaker coil fed with a triangular wave at 5 Hz providing a sweep amplitude, different for each beam, between 20 and 60 μm. This allows each path length to be modulated at a constant velocity (positive or negative), conveniently expressed in wavelengths per second (dimensions of frequency). The purpose of this modulation is to allow the interference signals between a given pair of subapertures to be identified by demodulation at the frequency equal to the difference between their two modulation velocities, expressed as above (figure 8.7).

8.3.5 Beam-reducing optics

At some stage, the diameter of the beam from a subaperture must be reduced to about 30–50mm in order to use conveniently sized optics for beam-combination. This may be done in the telescope itself (afocal or high *f*-number Cassegrain) or at a later stage. The optics used is a Cassegrain or Gregorian telescope (SUSI); both require a large paraboloidal mirror with a central hole and a second smaller one with a shorter focal length. A minor advantage of the Gregorian system is that a field stop can be inserted in the real star image plane, in order to cut out light from the surrounding sky (figure 8.8). If the beam-reducer is also used to create a Fizeau image, one mirror might be ellipsoidal.

8.3.6 Beam combiners

The type of beam combiner depends first on the detection system. If image-plane interference is used, as in GI2T and LBT, then the beams from the different

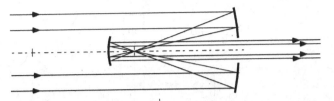

Fig. 8.8. A Gregorian beam reducer for two parallel beams, with a common field stop in the real image plane (SUSI).

telescopes must be focused onto the imaging camera with exact registration[†], where they interfere. The fringe density depends on the angle between the beams. In this case no beam-combining system is necessary. This might seem an enormous advantage to the Fizeau system, but one should take into account the fact that it does need a high-resolution imaging camera to record the fringes faithfully, one which is also fast enough to avoid blurring by atmospheric turbulence unless the telescopes have adaptive optics and the path difference is stabilized. There is, in principle, no reason not to combine many beams this way; if the vector angles between all pairs of beams are nonredundant, the various fringes between pairs of subapertures can then be separated by Fourier analysis of the image as in nonredundant aperture masking (section 6.2). Then the amplitude and phase of the Fourier coefficients indicate the amplitude and phase of γ with, of course, atmospheric errors in the phase part, and the phase of γ itself can then be found by phase closure techniques (section 4.2).

For pupil-plane interferometry a beam combining system can be constructed using beam-splitters. This was originally suggested by Michelson, who called it the "refractometer concept," although he did not in fact use it in his stellar interferometer. Then, if the optics are adjusted correctly, the beams interfere with uniform phase across their diameters, and a single pixel detector is sufficient to record the interference. Since the aperture is imaged onto the detector plane, this uniform phase can be obtained only if the subaperture was less than r_0 in diameter, unless adaptive correction is made to the wavefronts. Otherwise, different regions of the image will have different phases and a pinhole then has to be inserted in the image plane before the detector in order to select a single atmospheric mode or speckle of the beam. This pinhole can be replaced by a single-mode optical fiber which automatically selects one atmospheric mode (COAST, PTI, VLTI, Keck).

Often, a particular beam-combiner is optimized for a certain spectral region only. Light in other spectral bands may be sent to another beam combiner (CHARA, SUSI, NPOI) although there are difficulties in actually observing simultaneously in well-separated spectral regions since the requirements of optimum path-length

[†] "Exact" in this case means to the accuracy of one diffraction-limited image diameter.

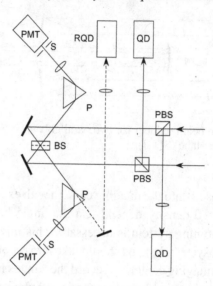

Fig. 8.9. Two-beam combiner at SUSI for shorter visible wavelengths. Polarizing beam-splitters (PBS) are first used to extract one polarization for tip–tilt guidance by the quadrant detectors (QD) and the slits (S) are used for spectral selection. RQD is a reference quadrant detector.

equalization may be different in the different bands. In several interferometers, a "rejected" wavelength band is used for star tracking, by imaging with it and using data from the image as input to a tip–tilt corrector (section 8.3.9) at an earlier stage (COAST, SUSI).

8.3.7 Semireflective beam-combiners

In general, two different types of beam-splitter are used in interferometry: plates, with a semireflecting, semitransmitting coating on one side, and beam-splitter cubes made from two 45° prisms with the coating on the hypotenuse of one, which is cemented optically to the other. All the other faces of the plate or cube are antireflection coated. It seems that plates are more popular for astronomical interferometry. The dielectric multilayer coatings used are designed to have the required reflection coefficient in the spectral region to be used (McLeod 2001). The capabilities of the designer are of course limited; for this reason an interferometer to be used in several wavelength bands will have separate beam-combining systems, each optimized for one band (CHARA, SUSI). When only two beams have to be combined, this can be done with a single beam-splitter. An additional problem with dielectric beam-splitter coatings is to make them polarization independent when used at non-zero angles of incidence. Sometimes, using this to advantage, light with polarization for which the beam-combiner is inefficient can be employed for guiding and tip–tilt correction (figure 8.9). Then there are two outputs; if the input beams

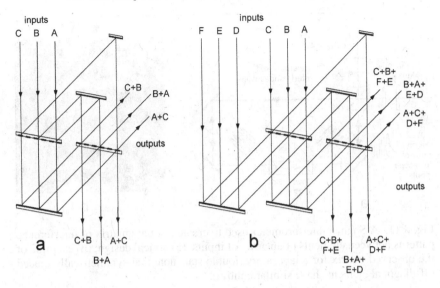

Fig. 8.10. Beam-combining optics designs for NPOI: (a) three inputs and three pairwise outputs; (b) six inputs and three outputs, each combining four of the inputs.

have amplitudes E_1 and E_2, the outputs ideally have amplitudes $(E_1 + iE_2)/2$ and $(iE_1 + E_2)/2$ when the beam-splitters are ideal. Their fringes oscillate in antiphase, and either output can be used to collect data; the difference between signals detected in the two exits has better signal-to-noise.

When there are more than two subapertures, various possibilities arise. Fizeau and densified-pupil imaging are among them, the latter being discussed in section 9.2. Generally, the coherence function $\gamma(\mathbf{r})$ is measured from the visibility and phase of the interference fringes between the pair of beams coming from subapertures separated by $\mathbf{r}$. So the interference scheme must allow the pairwise interference patterns to be separated. In general, the various detectors will receive linear combinations of the amplitudes from some or all of the subapertures. Maybe a basic scheme would have outputs corresponding to all possible pairs, such as that shown in figure 8.10(a) (NPOI) for three inputs with three outputs (each, of course, doubled, with complementary phases). In that case all the data required are read directly. When the number of inputs is larger, it becomes very complicated to design a beam-splitter array which provides outputs from all possible pairs, and it is simpler to combine several inputs (figure 8.10b). If the signal from each telescope is frequency-modulated at a characteristic frequency (section 8.3.4), the pairwise interference fringes can then be separated by demodulation at the specific frequency difference for that pair, as shown in figure 8.7. A simple concept using a Sagnac interferometer which enables pairwise interference of any number of beams (figure 8.11) has been described by Ribak et al. (2005), but has not yet

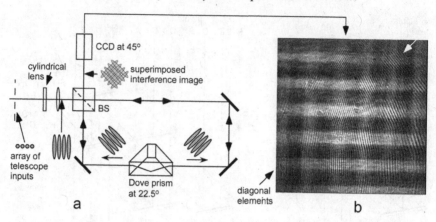

Fig. 8.11. A Sagnac interferometer used to create a square matrix of interference patterns between elements of an array of inputs: (a) optical design; (b) example of the observed matrix for a laboratory double star; note that symmetrically placed off-diagonal elements have similar contrasts.

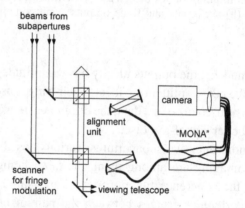

Fig. 8.12. Optical layout of the fiber-linked beam-combiner for the near infrared (FLUOR).

been applied practically. In this system, the telescope inputs are arranged as a linear array which is converted to an array of stripes by means of cylindrical optics. In the interferometer, two images of this array interfere, after they have been rotated respectively by $\pm 45°$ by a Dove prism at $22.5°$. The interference pattern consists of a square matrix of overlapping areas in which region i, j has fringes with contrast $\gamma(\mathbf{r}_{ij})$, the diagonal elements being used for system calibration.

8.3.8 *Optical fiber and integrated optical beam-combiners*

An alternative approach to beam combination uses optical fiber links and integrated optics. A system called "Fiber-Linked Unit for Recombination" (FLUOR) has been

Fig. 8.13. Integrated optic infrared beam-combiner for three inputs (IONIC). Photograph courtesy of Alain Delboulbe, LAOG.

developed for this purpose[†] (figure 8.12) and is in use in several interferometers (CHARA, NPOI, IOTA). Another interferometric beam-combiner using integrated optics technology (IONIC)[‡] is shown in figure 8.13. figure 8.14 is an example of fringes obtained at $\lambda = 1.65\,\mu$m using this combiner at IOTA.

It would be nice to find also a way of constructing a variable delay line using optical fibers, but this does not seem to be possible, so even if fibers instead of evacuated beam pipes were used to communicate signals from the subapertures to the beam combiner, the path-length correction would still have to be made. This is planned for MIRA-I.2. Project OHANA at Mauna Kea is intended to assess the feasibility of using fibers to combine light from the several large telescopes on the Mauna Kea summit in Hawaii and thus to create one big interferometer from them (section 12.1.6).

8.3.9 Star tracking and tip–tilt correction

In small telescopes, with subaperture size comparable to r_0, the most prominent effect of atmospheric turbulence is to create a time-dependent fluctuation in the image position. We see this visually on a hot day as a "heat wave" which causes

[†] Developed by the Laboratoire d'Etudes Spatiales et d'Instrumentation en Astrophysique (LESIA) of the Observatoire de Paris.
[‡] Developed by the LAOG (Laboratoire d'Astrophysique de l'Observatoire de Grenoble), the IMEP (Institut de Microélectronique, électromagnétisme et Photonique) and the LETI (Laboratoire d'électronique et de Technologie de l'Information).

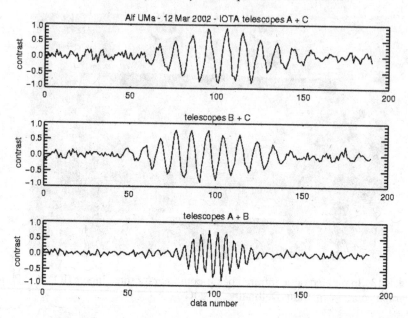

Fig. 8.14. Fringes at $\lambda = 1.65\,\mu$m between the pairs of three telescopes at IOTA obtained using the integrated-optics combiner shown in figure 8.13. Figure courtesy of P. Schuller, IOTA.

distant objects close to the horizon on dry land to appear to be jittering[†]. However, since any two entrance subapertures of an interferometer are separated by a distance much greater than r_0, the motions of both Airy patterns are completely uncorrelated; then, Young's fringes appear only intermittently in the occasionally superposed Airy peaks (see figure 9.12). The other effect of atmospheric fluctuations is a "piston" effect, which describes fluctuations in the mean phase difference between the two waves, and thus causes the fringe position to fluctuate randomly. The center fringe would retain a fixed position if the wavefront disturbance were pure local tilt, with no piston, and conversely the effect of tilt vanishes if the subapertures are shrunk well below r_0, in which case the fringe position responds only to piston fluctuations. The fringe motion observed by Stéphan, Michelson and their followers was caused by piston fluctuations rather than tilt effects.

When the subaperture size is comparable to r_0, the intermittent overlap of the Airy peaks, occasionally split into a few speckles, affects the observing efficiency and the accuracy of fringe visibility measurements, especially if there is no photometric monitoring of both contributing amplitudes in the interference. It is

[†] The astronomical "seeing" effect observed on stars with the naked eye is different since the eye's angular resolution, about an arcminute, is not good enough to detect the angular motion of the image. What is seen visually is twinkling or scintillation, generated by the high layers of the atmosphere.

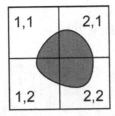

Fig. 8.15. Star image slightly off-center on a quad cell.

then advisable, in order to make full use of the larger subapertures, to stabilize the angular fluctuations. Angular correction is indeed the first order of adaptive optical correction, and is usually corrected independently, since it is considerably larger in amplitude than higher orders. It corresponds to adding linear corrections to the wavefront in the x and y directions, and is accomplished at each subaperture by means of a "tip–tilt mirror," which is usually one of the existing mirrors in the system. The mirror has to provide angular corrections of a few arcseconds[†], ideally at a rate corresponding to τ_0 of the atmosphere, i.e. at up to about 100 Hz. However, because of the paucity of photons available, a bandwidth as low as 10 Hz is often accepted. The error signal controlling the tip–tilt mirror comes from actively monitoring the image of the star being investigated, or maybe one very close to it (within the same isoplanatic patch), in a negative feedback loop. In order to avoid loss of photons, this is sometimes done using light of a wavelength outside the band being investigated, although there is a risk in this because the fluctuations are somewhat wavelength dependent. At long wavelengths, this dependence becomes weaker. In SUSI, light of one polarization is used for tip–tilt correction and the other for measurement.

Two methods are generally used for creating the error signal. One is to image the star on a fast camera and to locate the center of gravity of the image digitally. This has to be completed within less than about 5 ms for the correction to be satisfactory. A second method, requiring simpler engineering, projects the image onto a "quad cell" which is a 2×2 square array of photodiodes. The image is centered in x when the sum of the signals from diodes (1,1) and (1,2) is equal to that from (2,1) and (2,2), and the error signal for x is the difference between these two sums. Likewise for y, evaluating (1,1)+(2,1)–(1,2)–(2,2) (figure 8.15). This system is much simpler than the imaging corrector, but is noisier for weak intensities. We can compare the two quantitatively. Suppose, for example, that the point spread function of the optics is given by $Nf(x, y)$, where N is the number of photons and

[†] If, as is usually the case, the tip–tilt mirror is positioned after beam-size reduction optics, the angular amplitude of the correction needed is increased in proportion to the reduction in beam diameter, the phase difference between points at the edges of the beam being invariant.

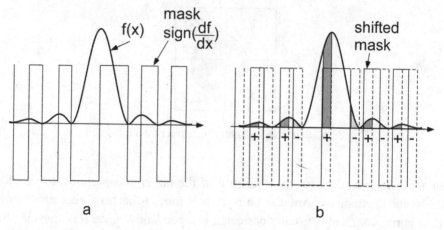

Fig. 8.16. One-dimensional point spread function (sinc$^2 x$) with the masking function sign(df/dx). (a) shows the PSF centered with respect to the mask, and (b) shows the situation after a small movement; the shaded regions indicate signals which contribute to the detected output, with their signs indicated. All the positive signals are greater than the adjacent negative ones.

f is normalized to unity. Then the algorithm representing the relationship between the outputs (s_x, s_y) and the displacement of the origin of the PSF (δ_x, δ_y) is

$$[s_x(\delta_x), s_y(\delta_y)] = \iint_{-\infty}^{\infty} [g(x), h(y)] N f(x - \delta_x, y - \delta_y) \, dx \, dy, \qquad (8.3)$$

where $g(x)$ and $h(y)$ are two masking functions relating the output signal to the light received at the point (x, y). The simplest method to find the center of gravity corresponds to a linear moment calculation, for which $[g, h] = [x, y]$. The quad cell corresponds to $[g, h] = [\text{sign}(x), \text{sign}(y)]$. Both functions must be modified by cut-offs at the edge of the field of view. For a Gaussian point spread function, for example, it is easy to evaluate the integrals and to show that the sensitivity for the quad cell is twice that of the linear moment. The signal-to-noise ratio can then be estimated using photon noise $\sqrt{N}$. However, the function $[g, h]$ can in fact be optimized for any particular point spread function, and clearly a function which reverses sign wherever f is maximum and minimum will take advantage of any linear translations in f. By expanding (8.3) as a Taylor series in (δ_x, δ_y) one can show that the function $f^{-1} \nabla f$ is optimum for small movements, but rather impractical, since it diverges and therefore amplifies noise when $f \to 0$! However, a realistic interpretation of this optimum (e.g. sign(∇f)) can be rather better than the quad cell for experimentally achievable point spread functions (figure 8.16) (Shao et al. 1988), although it is rarely used.

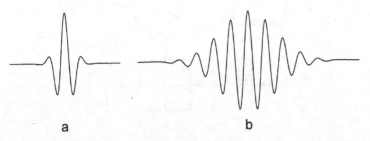

Fig. 8.17. Polychromatic fringe groups with (a) $\lambda/\delta\lambda = 3$ and (b) $\lambda/\delta\lambda = 10$.

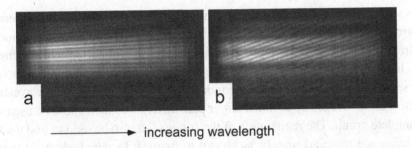

increasing wavelength

Fig. 8.18. Two spectrally dispersed interferograms (wavelength range 2.0–2.4 μm): (a) path-length compensated; (b) with an error in path-length compensation (GI2T: Weigelt et al. 2000).

8.3.10 Fringe dispersion and tracking

In order to equalize the optical paths one has to measure fringes. When using single-pixel detectors (pupil plane interference), this is done by sweeping the path length as described in section 8.3.4. The number of fringes observable depends on the degree of temporal coherence and thus on the bandwidth, and is roughly $\lambda/\delta\lambda$ (figure 8.17). It would therefore appear easiest to use a filter in order to create a very small $\delta\lambda$ and therefore a large number of fringes. But of course this wastes all the photons at other wavelengths. Several systems using a pupil-plane combination (PTI, Keck, SUSI, COAST, IOTA) adopt a compromise with $\lambda/\delta\lambda$ of the order of 10, and sweep the path length by distances of the order of tens of microns. This is not practical for very weak stars, and a close but brighter reference star may be used for this purpose. The path difference to the star of interest, and thus the phase of its fringes, can then be calculated.

Using an imaging detector (Fizeau interference), dispersed fringes can be used to avoid wasting photons (GI2T, CHARA, NPOI). Dispersed fringes were well-known in spectroscopy (Wood 1934) before 1902, and were later used by Michelson and in I2T. Let us suppose that the fringe pattern as observed with broad-band light

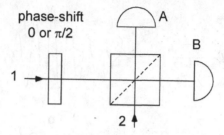

Fig. 8.19. Light from two inputs 1 and 2 interferes at an ideal beam-splitter with an optional additional phase shift of $\pi/2$ and goes to two detectors A and B.

consists of a fringe-group containing a small number of fringes. This is because only around zero path difference do the fringes from all the wavelengths coincide, and already at small deviations from equality the fringes with different wavelengths, which have different spacings, cancel one another. If, however, the fringe group is dispersed so that the different wavelength fringes are separated, each wavelength will show all its fringes up to much larger path differences than could be used with the complete group. The registration of the fringes can also be used to find the zero path difference more accurately, as shown in figure 8.18 which shows dispersed fringes at GI2T. With on-line data processing, the fringe registration information can be used to adjust the path lengths automatically (GI2T and CHARA). An added benefit of dispersed fringes is that if there is a particular wavelength in the band at which the fringe visibility or phase is anomalous, for example because of emission from an extraneous monochromatic source, it may be seen from the display.

8.3.11 Estimating the fringe parameters

The function of a multiple subaperture interferometer is to measure fringe visibilities and phases. Very often these have to be done with very few photons and when designing an interferometer it is important to be aware of the signal-to-noise limitations imposed by the light, the atmosphere and the devices used.

In order to find estimators for visibility and phase, we consider the following experiment. We assume that atmospheric fluctuations define a maximum period of τ during which the measurement must be completed, and that only a single atmospheric cell is involved (subaperture size $< r_0$ if there is no adaptive or tip–tilt correction, and larger if there is), so that the visibility is not corrupted by phase uncertainty. Then, in a typical experiment, N_0 photons are detected during this period. The photons are measured in four bins ($j = 0, 1, 2, 3$), in each of which the phase of one of the interfering waves is changed by added $j\pi/2$. In a simple case this could be done as in figure 8.19, in which there are two input beams 1 and 2. They interfere at an ideal symmetrical lossless beam-splitter (Michelson configuration) and there are two outputs in one of which, A, 1 and 2 interfere with phase difference $-\pi/2$, and at the other, B, 1 and 2 interfere with the phase difference $+\pi/2$ (section 3.1). During the second half of the period of measurement, beam 2 has an additional phase shift $\pi/2$ introduced by a phase-shifter. Thus, between the two detectors and the two halves of the measuring period, there are four output measurements, characterized by phase differences $(-1, 0, 1$ and $2)$ times $\pi/2$. If all the elements (detectors, phase shifter and beam-splitter) are ideal, the average number of photons reaching each detector during $\tau/2$ is $N_0/4$. The individual measured signals of course include noise. To make this account simple, we ignore all noise except photon-counting noise, which is inevitable. Now if the visibility of the fringes is $V = |\gamma|$, the intensities of the fringes in the four bins can be written in terms of numbers of photons detected:

$$n_1 = \tfrac{1}{4}N_0(1 + |\gamma|\sin\phi) + n_{n1}$$
$$n_2 = \tfrac{1}{4}N_0(1 + |\gamma|\cos\phi) + n_{n2}$$
$$n_3 = \tfrac{1}{4}N_0(1 - |\gamma|\sin\phi) + n_{n3}$$
$$n_4 = \tfrac{1}{4}N_0(1 - |\gamma|\cos\phi) + n_{n4} \tag{8.4}$$

where the n_{nj} represents the noise in the jth signal. For Poisson statistics the variance $\sigma_j^2 = n_{nj}^2$ is equal to n_j. Note that ϕ, the fringe phase, is constant during the period of measurement, but is random during a long sequence of M such measurements. Also the sum of the four signals (8.4) is equal to the number of photons N_0 together with a noise term. Consider now the quantity which we might expect to give a good estimate of $\tfrac{1}{4}N_0^2\gamma^2$, namely:

$$q = \left\langle (n_1 - n_3)^2 + (n_2 - n_4)^2 \right\rangle . \tag{8.5}$$

But clearly, even if the values of n_j were quite random, the squaring process would bias this quantity to be positive definite, so it cannot be quite right. However, this bias can be corrected by a standard method in statistics. Given a function $f(x)$ of a variable x whose mean value is $\bar{x}$, $f(x)$ can be expanded around $\bar{x}$ by a Taylor series as

$$f(x) = f(\bar{x}) + \frac{df}{dx}(x - \bar{x}) + \frac{1}{2}\frac{d^2 f}{dx^2}(x - \bar{x})^2 + \cdots \tag{8.6}$$

The expected value of $f(x)$ is thus

$$E(f) = f(\bar{x}) + N^{-1}\frac{df}{dx}\sum_1^N (x - \bar{x}) + \frac{1}{2}N^{-1}\frac{d^2 f}{dx^2}\sum_1^N (x - \bar{x})^2 + \cdots \tag{8.7}$$

$$= f(\bar{x}) + \frac{1}{2}\frac{d^2 f}{dx^2}\sigma^2 + \cdots \tag{8.8}$$

since the second term in (8.7) is zero by definition of $\bar{x}$. In the present case, derivatives of $q = f(x) = x^2$ higher than the second are zero, so (8.8) is exact. From this,

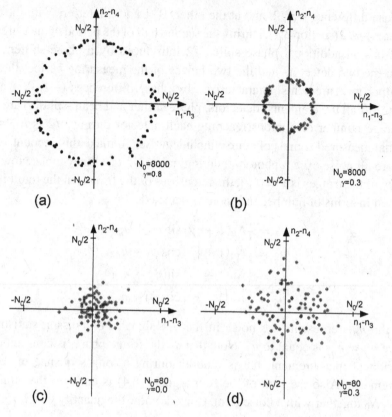

Fig. 8.20. Plots of series of $M = 100$ observations as points in the $((n_1 - n_3), (n_2 - n_4))$ plane. (a) $N_0 = 8000$, $\gamma = 0.8$; (b) $N_0 = 8000$, $\gamma = 0.3$; (c) $N_0 = 80$, $\gamma = 0$; (d) $N_0 = 80$, $\gamma = 0.3$.

we find that the expected value of q is

$$E(q) = \tfrac{1}{4}N_0^2\gamma^2 + n_1 + n_2 + n_3 + n_4 = \tfrac{1}{4}N_0^2\gamma^2 + N_0 , \qquad (8.9)$$

since $d^2q/dn_j^2 = 2$ for each n_j. Thus an estimator for γ^2 is

$$\gamma^2 = \frac{4}{N_0^2}E(q - N_0) = \frac{4}{N_0^2}M^{-1}\sum_{k=1}^{M}(q_k - N_{0k}) \qquad (8.10)$$

for the series of M measurements.

To appreciate the accuracy of this measurement of γ^2 it is instructive to plot a sample of measured values in the $((n_1 - n_3), (n_2 - n_4))$ plane (figure 8.20). The points are concentrated in a ring of radius $\tfrac{1}{2}N_0\gamma$ around the origin. The ring has thickness 2σ, i.e. $\sqrt{2N_0}$. When the ring is well-defined as in figure 8.20(a, b), called the "photon-rich" situation, the standard deviation σ of $N_0\gamma$ is about $N_0^{-\frac{1}{2}}$ and so the standard error on combining M measurements is $(N_0/M)^{\frac{1}{2}}$. However, if γN_0, which is the mean radius of the ring, is of the same order or smaller than its thickness $\sqrt{N_0}$, things get more complicated (figure 8.20 c, d) since the points on diametrically opposed sides of the ring get mixed up. This is the "photon-starved" region, corresponding to $N_0 < \gamma^{-2}$. The results for the photon-rich and photon-starved regions can be combined by adding the variances, giving $\sigma^2 = N_0 + \gamma^{-2}$, from which it follows that the signal-to-noise ratio (SNR) for a measurement of γ is approximately:

$$\mathrm{SNR}(\gamma) = \frac{\sqrt{M}N_0\gamma}{\sqrt{N_0 + \gamma^{-2}}}. \tag{8.11}$$

This relationship (written slightly differently) was proved analytically by Tango and Twiss (1980). One can see that in the photon-rich region, the SNR increases as $\sqrt{N_0}$, whereas in the photon-poor region as $N_0\gamma^2$. But one should be aware of the boundary between the two; if $\gamma = 0.1$, the boundary is at $N_0 \approx 100$ which corresponds to quite a large number of photons per second.

Estimation of the phase of the interference is a more difficult matter. It can be done only in the photon-rich region since the phase changes between measurement periods, and the only value of M is 1. The phase ϕ can be estimated using

$$E(\phi) = \tan^{-1}\left(\frac{n_1 - n_3}{n_2 - n_4}\right). \tag{8.12}$$

8.3.12 *Techniques for measuring in the photon-starved region*

Many of the most important features of a stellar image come from data at long baselines, where the visibility of the fringes is likely to be very small. For example, the diameter of a star is determined from estimating the baseline at which γ first becomes zero, and features of the intensity distribution such as limb darkening can only be elucidated from data at baselines longer than this, where the largest value of $|\gamma|$ possible is 0.13[†]. When γ is small it is necessary to have large N_0 to avoid photon starvation. In order to do this, it is most important to stabilize the fringes against atmospheric piston fluctuations, so that τ and therefore N_0 are as large as possible.

Several methods have been developed to attain this. In general they are called "phase referencing." For example, one stabilizes the interferometer by using high-visibility fringes created with broad-band light to feed a fast path-length corrector so

[†] This is the peak value of $|2J_1(x)/x|$ after the first zero.

that the fringes are stationary, and then measures the visibility with a narrower band in order to get useful-quality information. This has the advantage that even if the system is stabilized on a broad-band fringe which is not the zero order, the error in the measured visibility of narrow-band light, for which the coherence length is much longer, is negligible. Variants of this idea include stabilizing at a wavelength where γ is large for that particular baseline, and measuring at the required wavelength where γ may be much smaller. Usually, this condition is obtained when the stabilizing wavelength is considerably longer than the measurement wavelength, but since the atmospheric fluctuations are a function of wavelength, particularly in the visible, the fringes at the measurement wavelength may not be stabilized as well as one would like. This is not usually a problem when both wavelengths are in the infrared, but an example with the Mark-III interferometer (Quirrenbach et al. 1996) using a broad-band reference in the 600–900 nm for narrow-band measurements at 550 nm shows that it also works well in the visible.

When there are more than two subapertures interfering simultaneously a variant of the technique, called "baseline bootstrapping," is possible. The longest baselines can always be expressed as the vector sum of shorter ones. The fringes on the short baselines, where γ may be quite large at the wavelength of interest, can be simultaneously stabilized. As a result, fringes on the longer baselines must be stable too; this allows integration over long enough periods for the small values of γ for these baselines to be measured accurately.

8.4 Modern interferometers with two subapertures

In this section and the following one, we shall give a brief outline of the multiple aperture amplitude interferometers which have been developed since Michelson and Pease's pioneering work, which finished in about 1930.

The use of two subapertures on a ground-based baseline, together with the diurnal rotation of the Earth, allows a modest sampling of the (u, v) plane. If there were no daylight, nor clouds or other obscuration, the sample would be an ellipse, but these limitations restrict it considerably. Moreover, since phase closure cannot be used with only two subapertures, no real imaging is possible. However, two-aperture interferometers can be optimized to make accurate measurements of particular features, such as stellar diameters and profiles, multiplet separations and distances between stars or proper motions. We shall describe several systems which currently operate by the combination of signals from only two subapertures[†].

[†] Some of these observatories are intended to be upgraded to larger numbers of subapertures in the near future.

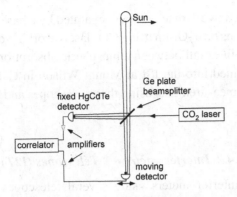

Fig. 8.21. Measurement of spatial correlation of sunlight at 10 μm using heterodyne detection with a CO_2 laser local oscillator (Gay and Journet 1973).

8.4.1 Heterodyne interferometers

Because of the technical difficulty of accurate path compensation, the first amplitude
interferometry systems used heterodyne detection (section 4.4.1). The accuracy
to which the path lengths from the star to the detector have to be equalized is
determined by the length of the coherent wavegroup, $\lambda^2/\delta\lambda = c/\delta\nu$; in heterodyne systems the bandwidth $\delta\nu$ is extremely small, being that of the intermediate
frequency amplifier. Even if this is as large as 100 MHz, the wavegroup has a
length of $c/10^8$ Hz which is about 3 m, and the path lengths only have to be equalized to this accuracy for interference to be observed. The first demonstration of an
optical heterodyne signal, using a laser source as a local oscillator, was made by
Nieuwenhuijzen (1970) who observed interference between the optical radiation
from several stars and a He–Ne laser at 6328 Å. Subsequently, the same idea was
extended to a spatially separated pair of detectors by Gay and Journet (1973) in
the infrared region using a CO_2 laser at 10.6 μm (P20 line) as the local oscillator,
in an experiment in which they measured the Sun's diameter. At this wavelength,
the coherence radius r_c of sunlight is $1.22\lambda/\alpha \approx 1.5$ mm, and their experimental
apparatus is shown in figure 8.21; it looks very similar to that used by Hanbury
Brown and Twiss to measure intensity correlations at spatially-separated points
(figure 4.2b) except for the addition of the laser source and the different detector types. The fact that this experiment showed a clear dependence of correlation
on detector separation, including the expected negative correlation region when
$r_c < r < 2r_c$ encouraged astronomical development of the method. Two groups set
up heterodyne stellar interferometers in this spectral region. Johnson et al. (1974)
at Kitt Peak used a baseline of 5.5 m and measured the diameter of the planet
Mercury (6 as) at 10.6 μm. Gay, Journet and colleagues (Assus et al. 1979.) at

Plateau de Calern used two 1-m telescopes separated by a baseline of 15 m to measure spatial correlations on α-Orion using CO_2 lasers working on the 11 μm region, where the CO_2 laser lines fall between atmospheric absorption peaks. The former group's work blossomed into the ISI at Mount Wilson in California, which now makes simultaneous measurements using three telescopes and will be described in section 8.5.5.

8.4.2 *Interféromètre à 2 Télescopes (I2T)*

This forerunner of interferometers using several telescopes was built at Nice by Labeyrie (1975) and consists of two coudé reflecting telescopes with mirror diameter 25 cm separated by 13.8 m. The telescopes were supported on alt–alt mounts, resulting in two antiparallel $f/3000$ output beams which were fed into a beam-combiner on an optical table which could be translated linearly by a mechanical motor and cam system in order to equalize the path lengths (figure 8.22). Each of the telescopes was individually trained onto the star so that the two Fizeau images overlapped in the focal plane as completely as possible. Since the subaperture diameter was not much greater than r_0, the image was essentially diffraction-limited. On the optical table, the beams impinged on two mirrors in a roof-like configuration which reflected them with a small angular separation onto an imaging camera in the focal plane, resulting in parallel interference fringes modulating the combined image (figure 8.23). Using a 5-Å bandwidth filter, giving a coherence length of 0.3 mm, fringes could occasionally be observed despite inevitable mechanical vibrations. An alternative slit and direct-view prism allowed dispersed fringes (section 8.3.10) to be observed by eye; as Michelson found, this considerably facilitates fringe acquisition. Fringes were seen on 13 August 1974 and demonstrated for the first time the practical possibility of observing interference between light waves from spatially separated telescopes. The interferometer was subsequently moved to the CERGA observatory on the Plateau de Calern, where the baseline was extended to 35 m and it was used for more than a decade for measuring angular diameters and effective stellar temperatures in the visible and the 2.2 μm region (for example Di Benedetto and Rabbia 1987).

8.4.3 *Grand interféromètre à deux télescopes (GI2T)*

Following the success of the I2T system, a larger version, GI2T, has been developed on the Plateau de Calern (Mourard et al. 1994) (figure 8.24). It has two telescopes in spherical mounts, with subapertures 1.52 m and secondaries 8 cm. Since the subaperture diameter is larger than the atmospheric r_0, the combined speckle image obtained when the two images overlap will have each speckle

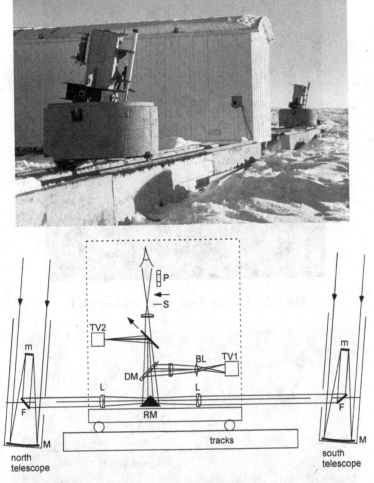

Fig. 8.22. I2T. In the drawing of the optical layout, M is a 250-mm primary mirror, m is a Cassegrain secondary, F a coudé flat, L a field lens, RM a roof mirror in the pupil plane, D a dichroic mirror, TV1 a guiding camera, BL a bilens to separate the two guiding images; S and P are slit and prism which can be inserted to observe dispersed fringes and TV2 a photon-counting camera with 500–700 nm filter.

individually modulated by parallel fringes with random phase (see section 5.9.1). For maximal illumination of the camera pixels, the number of fringes per speckle, which depends on the angle between the interfering beams, was arranged to be about 3. The maximum baseline is 65 m in the N-S direction. Like in the I2T, path-length equalization was made by moving the beam-combiner on a rolling table, but using, instead of a mechanical programming device, a computer-driven screw with optical rotary encoder. A trombone delay line was later added to increase the tracking time

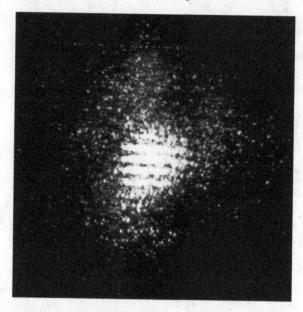

Fig. 8.23. Fringes observed on Vega with I2T.

Fig. 8.24. GI2T.

beyond 2 hours before and after transit. The interferometer is designed to work in the V band, 400–850 nm, with two array detectors optimized at 570 nm and 700 nm, and in the K band around 2μm.

In the dispersed fringe mode, a slit across the speckle image allows a one-dimensional cut to be spectrally analyzed (figure 8.18). Thus the instrument has the capability of high spatial and spectral resolution at the same time: e.g. the diameter of a star as a function of wavelength. Alternatively the entire combined speckled image can be recorded simultaneously in four different narrow wavelength bands, providing clear fringes within each speckle. The limiting magnitude has been limited to 6 or 9 in practice, but can in principle be much higher, especially if adaptive optics are installed to concentrate the energy of about 100 speckles into a single Airy peak.

The GI2T was intended as a precursor for an "Optical Very Large Array" or OVLA (section 12.1.2), involving tens of telescopes arrayed across one or several kilometers. A prototype 1.5-m telescope, built and tested by Dejonghe et al. (1998), was constructed sufficiently compactly and light of weight to be movable during the observation, thereby making path equalizers unnecessary. Variants of the OVLA have been proposed for the Magdalena Ridge interferometer (section 12.1), and for the Dome C in Antarctica, where excellent seeing conditions prevail (section 5.1).

8.4.4 The Mark III Interferometer

The Mark III interferometer was built at the Mount Wilson Observatory in order to test basic ideas involved in long-baseline optical interferometry without attempting to reach outstanding performance. It operated between the years 1986–1990, with siderostats feeding two 5-cm subapertures and a baseline which could be varied from 3.0 to 31.5 m. The star images were stabilized (tip–tilt) using light between 450 and 600 nm wavelength, and the light between 600 and 900 nm was used for fringe tracking. The fringe-tracker modulated the vacuum path-length equalizer at 500 Hz with 800 nm amplitude and the output signal was used to track and compensate movement of the fringes due to atmospheric effects at 20 Hz. This allowed fringes to be stabilized for periods between several seconds up to a few minutes, depending on the seeing. Then narrow-band quantitative interferometry was carried out in three separate channels at 500, 550 and 800 nm (figure 8.25). The narrow bands ensured that several clear fringes would be present in each band, so that identification of the zero order would not be critical to getting accurate visibility magnitudes; this meant that a single fringe-tracker would be adequate for all the bands.

8.4.5 Sydney University stellar interferometer (SUSI)

In 1986, Davis and Tango constructed a basic amplitude interferometer to make a quantitative comparison between the intensity interferometer and the newly developing amplitude interferometry technique. They used it to remeasure the angular diameter of Sirius. While the quantitative results from the two interferometers were

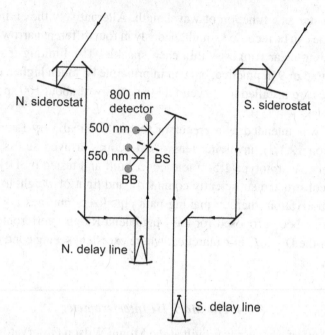

Fig. 8.25. Schematic optics of the Mark III interferometer. BB indicates the broadband detector used for fringe tracking.

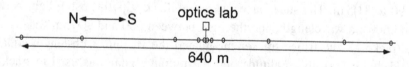

Fig. 8.26. Schematic linear layout of SUSI.

in excellent agreement, they showed that the same degree of accuracy could be obtained by amplitude interferometry in 1/40 of the time required by intensity interferometry, and that with a subaperture 10-cm diameter compared with 6.5 m in the intensity interferometer. This was essentially the justification for discontinuing intensity interferometry (section 7.8). Following this, SUSI was built to use pairs of 14-cm subapertures, whose diameter approximately equals r_0 in visible light. Although it has in fact 11 subapertures, they are used in pairs (figure 8.26). The maximum baseline planned is 640 m. Each subaperture is preceded by a plane 20-cm diameter siderostat and the reflected parallel beams travel along evacuated pipes to the beam-combining table. Path-length equalization is provided by a single carriage moving along a guide track with cat's-eye retro-reflecting systems on both sides and a laser metrology system to measure the total path difference and monitor the deviations from smooth tracking of the carriage. Since the path equalization is in air and not vacuum, correction for the refractive index dispersion is made by additional glass plates. Path-length equalization is followed by beam-reducing optics which reduce the diameter of the beams from 14 cm to 5 cm by the use of two opposed paraboloidal mirrors, between which a field stop can be placed (figure 8.8), and atmospheric dispersion correction is applied by the use of Risley prisms (figure 8.5b). Pupil-plane interference is achieved using a single dielectric plate beam-splitter, and both mixed beams reach photomultipliers via prism dispersers, which allow any wavelength within the full visible wavelength region to be investigated (figure 8.9). Because beam-splitters which are achromatic over the whole spectrum are too difficult to design, there are actually two independent beam-combining systems, "red" and "blue," covering the spectral ranges of 400–550 nm and 550–900 nm. Before combining, the beams are polarized parallel to one another by the use of polarizing beam-splitters; the rejected light is used for tip–tilt guiding.

8.4.6 The large binocular telescope (LBT)

The LBT illustrates a different approach to the use of interferometry, closer to the original Michelson stellar interferometer concept. This instrument is in the process of construction at Mount Graham in SE Arizona. It has two large mirrors, 8.4-m diameter, whose centers are separated by 14.4 m, and are mounted together so that the whole structure containing both mirrors can be rotated. Each mirror will have adaptive optical correction for the atmosphere in the H band. The telescope optics and beam combination are shown in figure 8.27; since the baseline is constant and the whole structure is rotated about an axis pointing toward the star, there are no gross changes in the path length and it only has to be actively corrected to compensate mechanical flexure and atmospheric fluctuations. Using the data from a sequence of orientations of the baseline, each of which gives a high-resolution image crossed by interference fringes, a single image with resolution equivalent to a 22.8-m aperture can be obtained. For a 180° rotation of the structure, the (u, v) diagram is a ring (figure 8.28), whose transform gives a point spread function little different from that of a filled aperture and, according to simulations, deconvolution can be applied quite successfully to this type of problem. In addition, the two mirror images can be combined in antiphase so as to carry out Bracewell nulling for planet detection. If the adaptive correction is done well, a good intensity zero can be obtained (see section 10.6). The baselines covered with big-aperture telescopes by the LBT and the VLTI are designed to be complementary (0–23 m and 30–100 m, respectively), so for brighter, complex objects which are visible to both, better images will come by combining data from both telescopes.

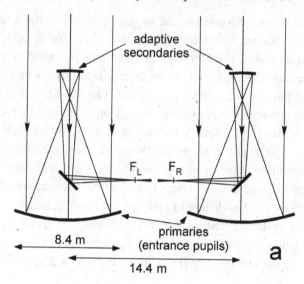

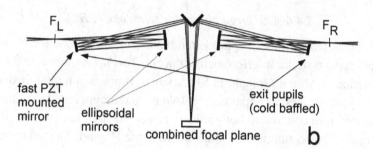

Fig. 8.27. LBT optics: (a) the binocular telescope; (b) detail of the beam-combining region.

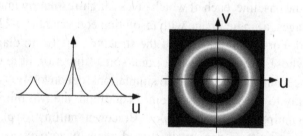

Fig. 8.28. The (u, v) plane coverage of LBT for one complete rotation: (a) u-section of the autocorrelation function; (b) grayscale representation.

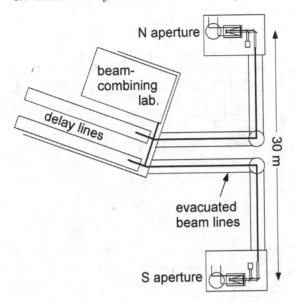

Fig. 8.29. Optical layout and beam-combination at MIRA-I.2.

8.4.7 The Mikata optical and infrared array (MIRA-I.2)

MIRA-I.2, in Japan, is a two-beam interferometer with fixed telescopes of 30-cm diameter observing through siderostats. The baseline is 30 m, without laser metrology, so that measurements of significance which can be made include diameters of large stars and binary separations. The limiting magnitude is estimated as 4.5. Observations of fringes in the pupil plane are made at 800 nm using avalanche photodiode (APD) detectors. With subapertures 30-cm diameter, the only atmospheric correction necessary at that wavelength is tip–tilt, which is applied to the collimated beams before they enter the vacuum beam lines. The two delay lines are also operated *in vacuo*. Fiber links are being considered to replace the vacuum lines. In the beam-combining system, wavelengths outside the range of interest are focused onto quadrant detectors to provide the input to the tip–tilt correctors and to monitor the beam intensities. Figure 8.29 shows the optical layout. Second and third stages with longer baselines are planned.

8.4.8 Palomar testbed interferometer (PTI)

The PTI is designed both for interferometric measurements and for differential astrometry by observing two stars simultaneously. If the two stars are within the same isoplanatic patch (about 4 arcsec in the visible) the atmospheric fluctuations cancel one another in the two measurements, and very accurate values for their

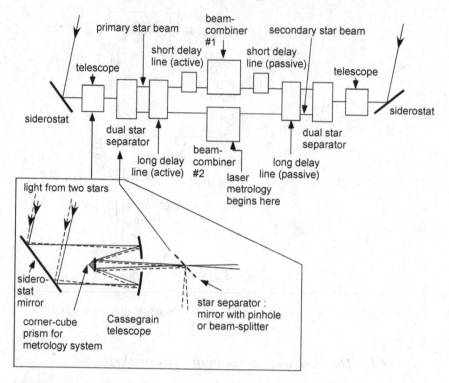

Fig. 8.30. Optical layout of PTI. The metrology system uses laser interferometry traversing the same optics as the star beams, returning from the corner-cube reflectors in the shadow of the Cassegrain secondaries (lower drawing).

separation can be obtained. Even if they are not in the same patch, improved accuracy is obtained by simultaneous measurement. The interferometer has three 40-cm collecting subapertures which can be used two at a time, with a maximum baseline of 110 m and active fringe tracking working in the near infrared (H and K bands). It is possible to switch pairs of telescopes fairly quickly in order to improve (u, v) coverage for imaging. The optics are shown schematically in figure 8.30.

In the "larger separation" mode, the image plane of each subaperture can be split by a beam-splitter or mirror with a pinhole so that the light can be sent down two separate beam paths to two beam-combiners. One star (primary) is observed on-axis, while a second one can lie at distance 8–60 arcsec from it. Both beams have one common optical delay line, while the light from the primary star is further corrected by an additional short delay line whose length then measures the angular separation between the two. Tip–tilt correction for the atmosphere is made by stabilizing the image of the primary star, which is why there is a minimum separation of 8 arcsec required between them, because otherwise the system cannot separate

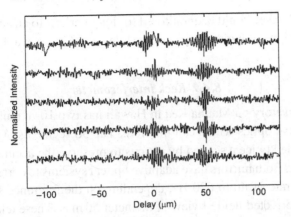

Fig. 8.31. Examples from PTI of five consecutive fringe trains containing groups from two stars (Lane and Muterspaugh 2004).

the two. After path-length correction, the signals from the two subapertures are combined at beam-splitters and fed to photodectors for fringe-tracking and spectrometry. PTI can operate in three ways: in one, fringes are tracked and measured on a single star, as in other interferometers; in the second, the fringes are tracked simultaneously on the two stars and the delay lines act independently; in the third, fringe tracking on the brighter star is used to correct the interference phases of both the primary and secondary star, a mode which is called "phase-referencing" and allows fringe-tracking for short periods (up to 0.1 s) on secondary stars as dim as sixth magnitude.

In the "small separation" mode, one observes for example the two components of a binary star, which are simultaneously in the field of view. If the separation of the binary components is small, the fringe packets may overlap, and the separation can then be deduced from the visibility variations caused by their interference. A single beam combiner is used and the fringe positions are measured by modulating the instrumental delay with an amplitude large enough to record both fringe packets (figure 8.31). This would seem to cancel atmospheric fluctuations, but since the fringe position measurements of the two stars are no longer truly simultaneous, it is possible for the temporal variation of the atmosphere nevertheless to introduce path-length changes (and hence positional error) in the time it takes for the scan to go from one fringe group to the other. To reduce this effect, a fraction of the incoming starlight is split off and directed to the second beam-combiner which is used for continuous phase referencing on one of the stars. This has the effect of stabilizing the fringes measured by the astrometric beam-combiner. As a result, a direct and very accurate measurement of the separation between the two stars is obtained. For example, with the binary star HD 1779 identified by speckle interferometry, the separation 0.25 arcsec could be monitored for long periods to an accuracy of 9 μas in a search for signs of a planetary system (Lane and Muterspaugh 2004).

8.4.9 Keck interferometer

The Keck observatory on Mauna Kea in Hawaii has two 10-m multi-element mirrors with adaptive optics which can interfere. The 36 segments of the mirrors are actively controlled twice a second by pistons to preserve the accurate figure of the complete mirror. Both mirrors have adaptive optical systems to correct atmospheric turbulence at a rate of up to 670 Hz, depending on the reference star magnitude, and provide a corrected field of view of diameter 50 mas. These telescopes are not dedicated to interferometry, but the interferometric mode is one option that is available. As discussed in chapter 6, Fizeau interferometry has also been carried out on the Keck-I telescope alone. Four-to-six additional "outrigger" telescopes with apertures 1.8 m are planned specifically for interferometry, which will be included in the interferometric array and will then improve the (u, v) coverage for imaging. The Keck-I and Keck-II telescopes provide an 85-m baseline oriented at 38° E of N (see figure 12.3). A single beam from each telescope is routed through the telescope coudé train to the beam-combining laboratory. Optical path delay is implemented in two stages, with long delay lines which remain stationary during an observation, and fast delay lines up to 15 m to track sidereal motion.

The images are stabilized by tip–tilt correction in the beam-combining laboratory provided by data from a near-infrared array camera (1.2 μm) operated at 100 Hz frame rate. A second similar camera is used at 500 Hz for dispersed fringe tracking and science measurements, being coupled to the beam-combiner by means of a single-mode fiber. The adaptive optics are critical here, because otherwise the image would be a speckle pattern and a single-mode fiber could only sample the light statistically in one diffraction-limited region, the rest of the photons collected being wasted (see section 8.5.6).

One of the purposes of the interferometer is to search for extrasolar planets in relatively close stars by detecting the very small periodic movement of the star about the common center of gravity (see section 8.2). To do this interferometrically, the principle is to measure the position of the target star with respect to one or two reference stars. While the target star will generally be nearby and bright, the reference stars will be distant and relatively faint. Phase referencing is used, which tracks the bright target star to stabilize the interferometer in order to obtain sufficient sensitivity to detect the fainter reference stars. Two reference stars are needed in principle to resolve ambiguities in the astrometric signature; with two reference stars, two measurements are available, and positional variations are attributed to the target star only if they are seen simultaneously in both astrometric measurements.

Another mode of operation with the two Keck telescopes is nulling interferometry, which will be discussed in more detail in chapter 10. In this mode of operation, the beams from the two telescopes are combined in antiphase in the Fizeau mode, with an angle of intersection between the two beams calculated to produce destructive interference at the center of the star, and constructive interference (first fringe) at the radius of a suspected planet. Note that complete destructive interference requires an *achromatic* phase difference of π between the beams, which cannot be produced simply by an additional path difference, although within a restricted bandwidth quite good destructive interference (of order 1:2000) can be obtained using glass plate phase retarders (section 10.4.3).

8.5 Interferometers with more than two subapertures

When an interferometer is built with more than two subapertures, various combinations between them can be employed to allow the (u, v) plane to be filled more completely. Moreover, when light from three subapertures interferes simultaneously, providing a hexagonal or "honeycomb" pattern of fringes, phase closure can be used in order to find the phase of $\gamma(u, v)$. The result is that more complete images can be synthesized. Optimization of the layout of the telescopes becomes important, either to obtain a nonredundant array (for the best point spread function) or to get a partially redundant array sufficient to determine the phases of γ unambiguously. In this section we describe briefly several interferometers in this class.

8.5.1 The Cambridge optical aperture synthesis telescope (COAST)

COAST is a five-telescope interferometric array working in the visible and near infrared, where the telescope positions sample a set of fixed positions lying on a "Y" structure. The telescope array is located on a level site close to sea-level a few km from the city of Cambridge, UK. Although it has rather poor seeing, it has nevertheless produced several new astronomical results (Young et al. 2002) since its first operation in 1995. The longest baseline achievable is 67 m. The beam-mixing optics allow the signals from any four of the five telescopes to interfere simultaneously and, since the interchange between two telescopes can be made relatively easily, two different groups of four telescopes can be used on one particular night. Visibilities and closure phases are measured by temporal fringe scanning (section 8.3.10) using single-element detectors and pupil-plane inerference. One group of four telescopes gives six nonredundant baselines and therefore four closure phases can be measured; after exchanging one telescope with a fifth one, the number of baselines is increased to nine and the number of closure phases to seven (since one group of three is common to both setups) (Haniff et al. 2002).

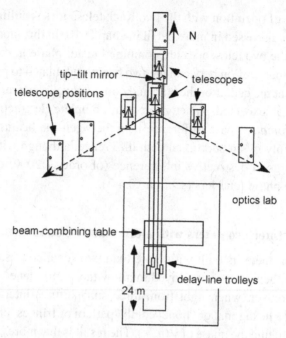

Fig. 8.32. Layout of the telescope stations and optics laboratory of COAST.

In figure 8.32 one sees the layout of the telescope stations and in figure 8.33 the (u, v) coverage for a typical star is shown. On the ground, each telescope position is defined by a three-leg kinematic mount[†]. Each telescope comprises a 50-cm diameter siderostat feeding a 40-cm $f/5.5$ Cassegrain telescope with a magnification 16. The 40-cm aperture was chosen to be about $3r_0$ at $\lambda = 1\,\mu m$, since $2.75r_0$ has been shown (see section 5.8.3) to give optimum signal-to-noise for detecting fringes when tip and tilt are corrected actively (section 8.3.9). The magnification 16 leads to a beam diameter of 25 mm with negligible diffraction effects, and an angular field of view of 10–15 arcsec.

The path equalizing system used at COAST has already been described as an example in section 8.3.4. After path equalization, the beams from the four inputs are combined by plate beam-splitters, and there are four output beams each of which contains one quarter of the input from each telescope (figure 8.34). The relative phases of the elements in each combination are of course different (see section 3.1.1). The result is that fringes between any pair of beams within the total detector

[†] A kinematic mount provides a well-defined mechanical constraint using the minimum number of contact points. Suppose one wishes to have n degrees of freedom. Since a rigid body has six degrees of freedom (three translations, x, y and z, and three rotations about these axes), the number of points of contact between the body and the mount must be $6 - n$. In the present case, absolute reproducibility is required, so that $n = 0$ and the mount must therefore provide six points of contact; commonly used designs provide two on each leg, or three, two and one on the three legs, respectively.

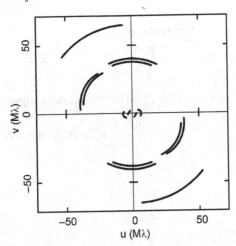

Fig. 8.33. The u, v coverage diagram at $\lambda = 1\,\mu$m for one configuration of COAST observing a source at declination $45°$ (Haniff et al. 2002).

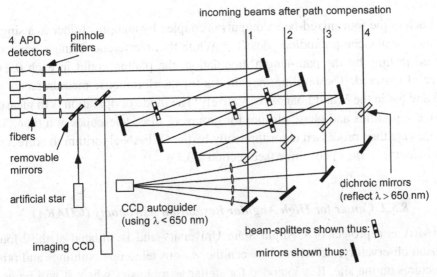

Fig. 8.34. The beam-combining optics of COAST. The four detectors each receive one-quarter of the light from each telescope.

signal are swept at a rate depending on the difference between the modulation velocities of those two beams; since each telescope beam has its own characteristic modulation velocity (amplitude/frequency), the fringes between the different pairs of telescopes can be distinguished by their frequencies. Light at wavelengths not used for interferometry ($< 650\,$nm) is used for guiding the telescopes and as input to the adaptive tip–tilt correction apparatus.

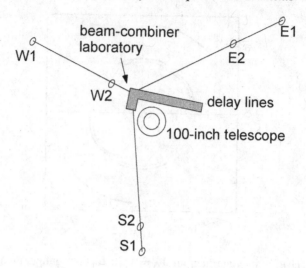

Fig. 8.35. Schematic layout of CHARA at the Mount Wilson Observatory. The longest baseline is S1-E1 = 331 m.

Each of the four mixed-beam outputs is coupled by an optical fiber to a single-element avalanche photodiode detector. While the interference fringes are modulated in time by the path-length modulation, the photon count in each 0.2 ms interval is stored. The limiting magnitudes for interferometric measurements are 8.5 and 6.4 in the V and I bands respectively. The mode of operation is to measure fringe visibilities and phase closure for groups of three telescopes at a time. The images are then processed algorithmically by the CLEAN algorithm to correct for the incomplete (u, v) coverage (see section 4.3.1).

8.5.2 Center for High Angular Resolution Astronomy (CHARA)

CHARA is a project of Georgia State University and is situated at the Mount Wilson observatory, stretching between the various telescope buildings and other structures on the site. It is intended for stellar astrophysics where it will be used to measure the diameters, distances, masses and luminosities of stars as well as to image features such as spots and flares on their surfaces. There are six alt–az mounted afocal telescopes with 1-m apertures, arranged on a Y (figure 8.35) for which the longest baseline is 350 m. The telescopes all have active secondaries for tip–tilt control and focus. Each light beam exiting a telescope provides a 125-mm diameter collimated beam which passes folding mirrors designed to preserve polarization as far as possible. This is followed by an afocal beam compressor to form a 25-mm diameter beam which is transported through evacuated beam-line tubes

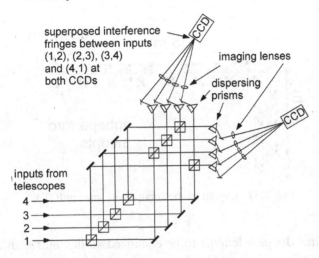

Fig. 8.36. Simplified schematic optical layout for the fringe-tracking subsystem at CHARA, as if there were just four telescopes (in fact there are six). The CCDs record four (six) superimposed fringe patterns, each with its own period. The reflections are shown to be at 90°; in the real system these angles of reflection are much less, in order to minimize polarization problems.

(about 1 torr pressure) to a central laboratory. Each beam is reflected 12 times en route to the laboratory; the minimum number necessary is seven, but additional mirrors have to be included so as to ensure the same sequence of reflections in each beam and thus to match the polarizations of all the beams on their arrival (figure 8.4). The laboratory is within a building with accurate climatic stability and it houses path-length equalizers up to 345 m long (figure 8.6c), beam combination optics, and detection systems. Of the 15 possible baselines from six subaperture positions, a subset of seven baselines is sampled by choosing the appropriate pairs within the fringe-tracking optics (figure 8.36). The pairs are combined by a series of plate beam-splitters, a separate set being used for each of two spectral operating regimes: one regime contains visible and near infrared up to 1.1 μm, and the other from 1.1 to 2.2 μm. The former band undergoes a second dichroic split, wavelengths shorter than 0.6 μm being reflected to the tip–tilt sensors. The rest of the spectrum is transmitted to the visible beam-combiner and the combined beams are divided equally between a boresight camera and a spectrograph for fringe tracking. Since polarization inequalities result in apparent loss of correlation, polarizing beam splitters are used for this division.

In the fringe-tracking system, the fringes from each pair of combined beams are spectrally dispersed by a prism and imaged onto a bare CCD. In general, although fringes are observed independently at each wavelength, group fringe

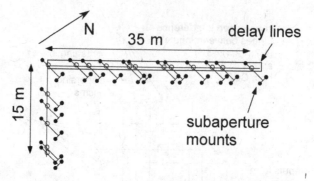

Fig. 8.37. Layout of the subaperture sites at IOTA.

tracking requires the path lengths to be equalized so that all the different wave-lengths give fringes with the same phase (see section 8.3.10). The fringe-tracking optics are based on a modified Mach–Zehnder interferometer, where two initially parallel beams are combined. The seven sets of spectra are imaged onto a single detector. Different wavelength bands can be selected for tracking by adjusting the position of the prisms. Since there is less dispersion in the longer wave infrared band, fringes in this region are observed by scanning the complete fringe packet. A fiber-optic based beam-combiner, FLUOR, has been designed for this purpose (section 8.3.7).

The limiting resolutions, accuracies and magnitudes estimated for CHARA are 200 μas, $\pm 20\,\mu$as and +11 mag in the V band and 1 mas, $\pm 35\,\mu$as and +15 in the K band.

8.5.3 *Infrared optical telescope array (IOTA)*

IOTA, situated on Mount Hopkins in southern Arizona, began with an agree-ment in 1988 among five institutions, the Smithsonian Astrophysical Observatory, Harvard University, the University of Massachusetts, the University of Wyoming, and MIT Lincoln Laboratory, to build a two-telescope stellar interferometer for the purpose of making fundamental astrophysical observations, and also as a prototype instrument with which to perfect techniques which could later lead to the devel-opment of a larger, more powerful array. The first fringes were seen in Decem-ber 1993. Since 2002 it has three 45-cm diameter collectors that can be located at different stations at multiples of 5 m and 7 m respectively on the arms of an L-shaped array 15 × 35 m, thus reaching a maximum baseline of 38 m (figure 8.37). Each light collector consists of a siderostat feeding a stationary afocal Cassegrain telescope which produces a 45-mm diameter parallel beam. Each beam is next reflected from a piezo-driven active tip–tilt mirror that makes a first-order correction for atmospheric turbulence. The beams then enter evacuated tubes and proceed to the corner of the "L" array, where path compensation is implemented *in vacuo*. There are three beam-combination tables at the IOTA, and all of them implement pupil-plane (Michelson) beam combination. In two cases, at visible and near-IR wavelengths, the combination occurs at a beam-splitter. The third table houses the FLUOR and IONIC experiments (section 8.3.7), in which beam combination occurs in single-mode fibers. For near-IR operation, a pair of dichroic mirrors at 45° transmit wavelengths less than 1 μm toward the CCD based tip–tilt servo system, and reflect the near infrared light toward the beam-combining optics and science detector. For visible operation only part of the visible light is directed toward the star tracker CCD, and most of the light is directed toward the visible table for combination. The near infrared detector is based on a 256×256 HgCdTe array and the visible light detector is a thinned back-side-illuminated 512×512 CCD.

8.5.4 Navy prototype optical interferometer (NPOI)

NPOI is a joint project between the Naval Research Laboratory and the US Naval Observatory. It is situated at the Lowell Observatory, near Flagstaff, Arizona, and received its first images using phase closure in 1996. A good description of the basic design and considerations leading to it is given by Armstrong et al. (1998). It has six movable subapertures for imaging, which can be placed at any of 10 defined positions on each arm of a Y-formation, forming baselines between 2 m and about 60 m, to be enlarged eventually to 437 m. There are another four subapertures at fixed positions for astrometry, with baselines between 19 and 38 m, monitored by a laser metrology system. All the subapertures have 50 cm siderostats to track the object, the imaging set being followed by 12-cm apertures, and the astrometry set by 35 cm apertures. A schematic drawing of the layout is shown in figure 8.38. Many of the features evolved directly from the Mark III interferometer (section 8.4.4).

The imaging array is intended for stellar surface imaging of stars up to 7th or 8th magnitude. The baselines available are partially redundant, which allows baseline bootstrapping to be used to stabilize low-visibility fringes on long baselines when they can be observed simultaneously at higher visibility on the shorter component lines; the resultant image quality may not be theoretically as good as that obtainable with a non-redundant array, but this is a trade-off against the improved accuracy of low visibility data. On the other hand, if the source contains a bright and small component such as a flare, which gives rise to good visibility at all baselines so that fringe-locking is not a problem, a non-redundant array may be chosen for observing the circumstellar surroundings with better resolution. The detection is in the visible region between 570 and 850 nm, and images with resolution of 0.2 mas can be

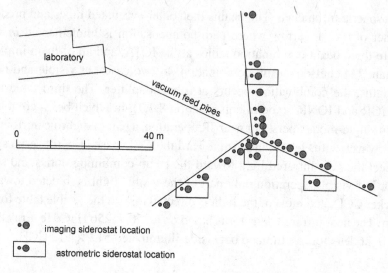

laboratory

vacuum feed pipes

0   40 m

•● imaging siderostat location

|•●| astrometric siderostat location

Fig. 8.38. Layout of the NPOI subaperture stations. The relative positions of the astrometric substations are measured by an independent laser metrology system which is not shown.

achieved, or stellar diameters of this order can be measured with accuracy about 1%. Figure 8.39 shows images of the triple star η-Virginis made with six telescopes using phase closure (Hummel et al. 2003). The close pair was previously known as a spectroscopic binary with a period of about 72 days, whereas the wide pair has been extensively observed by speckle interferometry and the orbital period determined to be about 13 years.

The astrometry array makes accurate measurements of stellar positions on both large and small field scales. This involves using the fringe data together with very precise knowledge of the geometry of the array and the lengths of the path compensators. For this reason the interferometer incorporates very accurate laser interferometric metrology for determining the geometrical parameters on a real-time basis; one reason for using plane mirror siderostats is that their points of rotation can be determined more accurately than those of telescope systems. The goal of accuracy for wide-field astrometry is 2 mas. In the case of stars which are close enough to be measured simultaneously, an astrometric accuracy of 0.01 mas can be achieved; this should be compared to the reflex motion of the Sun due to Jupiter, which would appear as 1 mas from a distance of 10 parsec. The metrology system provides relative but not absolute subaperture coordinates, but it is possible to obtain these from the wide-field astrometric data itself, when the positions of several stars are measured with the same baseline array. Suppose we have n subapertures at $\mathbf{r}_0, \ldots, \mathbf{r}_{n-1}$, giving $n-1$ three-vectors $\mathbf{r}_{i0}$ and $n-1$ path-length corrections: a

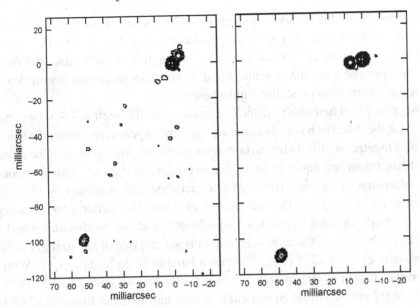

Fig. 8.39. Synthesized images from NPOI of the triple star η-Virginis on February 15 and May 19, 2002, after processing with CLEAN (section 4.3.1) (Hummel et al. 2003).

total of $4(n-1)$ parameters. N stars have $2N$ angular coordinates, and so the total number of degrees of freedom is $4(n-1)+2N-1$. The number of measured data for the N stars is $N(n-1)$ path lengths. Now the total number of measurements must be at least as great as the number of degrees of freedom. For four subapertures, this gives 11 stars as the minimum number that must be measured together.

To reduce atmospheric fluctuations and dispersion to a minimum, the beam lines and the coarse and fine path-length equalizers are all in vacuum. The beams are combined using a combiner shown earlier in figure 8.10. This creates three pairs of outputs from six inputs, and each output contains mixtures of two independent pairs of signals, which can be separated by their different time modulations. Compared with systems where all the inputs are mixed in all the outputs (e.g. COAST), smaller modulation amplitudes are needed and there is less cross-talk between the interference patterns. The output beams are separated into 16 wavelength channels and the dispersed fringe technique (section 8.3.10) is used for fringe tracking.

8.5.5 The Berkeley infrared spatial interferometer (ISI)

ISI is a heterodyne interferometer working at a wavelength around 10 μm, where the Earth's atmosphere is highly transparent, using heterodyne detection (Johnson et al. 1974; Hale 2000, 2003). The development of ISI was based on earlier experiments on CO_2-laser-based heterodyne systems described briefly in section 8.4.1. Results obtained by Weiner et al. (2003) emphasize the fact that stellar diameters and structure are very much functions of the wavelength of observation, and the fact that mid-infrared emission is a molecular rather than an atomic feature leads to important information on stellar atmospheres.

The principle of heterodyne detection was described in section 4.4.1. One should recall that the detector has a nonlinear square-law response, and when the starlight falls on it together with light from a laser local oscillator, a component of the received signal has frequency equal to the difference between the two. This component, the intermediate frequency (IF) signal, is detected and amplified by electronics having a cut-off at f_{max}. The wavelength at which the stellar interferometry is done is clearly defined by the laser wavelength, and the bandwidth around this wavelength by f_{max}. If the laser wavelength is λ_0, and that of the starlight is λ, the requirement $c|\lambda^{-1} - \lambda_0^{-1}| < f_{max}$ defines a bandwidth $\delta\lambda = 2f_{max}\lambda_0^2/c$. With the center wavelength at $10\,\mu$m and $f_{max} = 2.6 \cdot 10^9$ Hz, we have $\delta\lambda \sim 2 \cdot 10^{-3}\,\mu$m (2 nm) or 0.2 cm^{-1}. This is equivalent to a very narrow-band filter; its width is in fact determined by the response time of the cooled HgCdTe detectors used. A larger bandwidth would, of course, allow more energy to be collected, and weaker sources investigated. On the other hand, such a narrow bandwidth means that the temporal coherence time τ_c is very long, and so there is no call for very accurate compensation of time delays between the interfering signals. The coherence length $c\tau_c = \lambda^2/\delta\lambda$ is about 0.5 m, which means that if the optical paths from the two telescopes to their detectors are equalized or compensated to about this accuracy, there will be little loss of fringe contrast. One should compare this to similar considerations in intensity interferometry, and contrast it with the very stringent requirements for path-length compensation in direct interferometry. Although any of the laser lines at $10\,\mu$m can be used as the local oscillator frequency, that at $11.15\,\mu$m is the most suitable because the atmospheric transparency is greatest at that wavelength.

The interferometer uses three mobile telescope stations which can be situated on any of an array of seven predetermined positions forming a nonredundant array (figure 8.40). It should be appreciated that at $10\,\mu$m the atmospheric turbulence problems are considerably less severe than at visible wavelengths; for example the Fried parameter r_0 depends on $\lambda^{\frac{6}{5}}$ which means that, using tip–tilt correction, a mirror diameter as large as 5 m could be profitably employed, although such a construction would be a major undertaking, and the mirrors used in practice are 1.65 m diameter. The size is of course crucial in determining the weakest sources measurable, in particular since emission at $10\,\mu$m is considerably weaker than that in the visible. Each telescope consists of a siderostat, which follows the stellar position, feeding a 1.65-m diameter $f/3.1$ parabolic mirror. This is followed by a second smaller paraboloid and hyperboloidal mirror (Schwartzschild combination) providing an output $f/89$ (almost parallel) beam. All four mirrors have aligned axial holes,

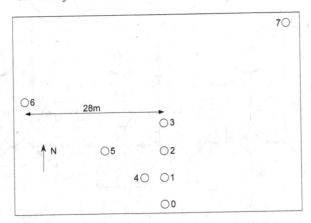

Fig. 8.40. Layout of the eight ISI telescope sites at Mount Wilson.

amongst other things to allow laser alignment. The smallest mirror provides tip–tilt correction provided by feedback from an InSb imaging camera using near infrared (2.0–2.4 μm) light, separated off from the telescope beam by a dichroic mirror. The atmospheric fluctuations at this wavelength and 10 μm are almost identical, although the star intensities may be very different. Each telescope station contains its own CO_2 laser, whose wavelength can be finely tuned by piezoelectric control of one resonator mirror. Part of each laser beam (a reference beam) is separated off and transmitted through an air path to a master laser, and a mixing signal between the two lasers (the same idea as heterodyne) is used to actively control the telescope laser to a constant phase and frequency relationship to the master; this signal is modulated at 20 Hz by a dither mirror in order to facilitate the electronic detection (figure 8.41). The control includes correction for variations in the air path between the lasers, obtained by receiving back part of the master laser signal, after it has traveled the path twice, and interfering it with itself to monitor the path length. Each laser is tuned to a different offset from the master; one is at 1.000 000, one at 1.000 193 and one at 1.000 107 MHz. As a result, when the IF signals interfere, the fringes between the pairs have temporal frequencies 193, 107 and 86 Hz. The use of three telescopes simultaneously allows phase closure (section 4.2.1) to be used to obtain real two-dimensional images. Path-length compensation is done by changing the communication cable lengths by steps. Provided that the compensation remains within the 0.5 m mentioned earlier, the modulated fringes are observed continuously; there is no fringe scanning involved and therefore no "dead time" during which the fringes are out of range. This all contributes to detection efficiency.

The detectors used are liquid-nitrogen cooled single-element photovoltaic Hg_x-$Cd_{(1-x)}$-Te detectors with the value of x optimized for maximum signal-to-noise in the 10 μm band. The signal is measured relative to that from a background

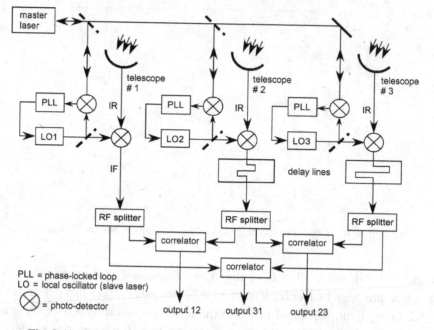

Fig. 8.41. Schematic flow diagram of the optical and RF signals in ISI.

at 77 K by means of a mechanical chopper at 140 Hz which alternately allows the detector to see the star and a reflection of the cold source as seen in its reflective vanes. The output from the detectors is amplified first by a cold FET amplifier and then by a further room-temperature preamplifier. Further background correction is obtained by switching the observation from star to sky on a 15 s cycle (compare a similar technique used with intensity interferometry, chapter 7).

As we pointed out, the heterodyne technique in the mid-infrared allows spectral images to be obtained in a spectral region where molecular spectroscopy is important. First, the use of different isotopes of the C and O in the lasers and rotation of the grating in the laser resonator allows different spectral regions to be selected; then, within each spectral line, further high-resolution spectroscopy can be done by the use of RF filters selecting regions within the 2.8 GHz IF band. The resulting spectral resolution of 10^5 corresponds to a Doppler shift of 0.7 km s^{-1} (Monnier et al. 2000).

8.5.6 Very large telescope interferometer (VLTI)

The European Space Organization's Very Large Telescope observatory is situated on the Paranal mountain platform in Chile. It consists of an array of four 8.2-m telescopes, each one having active optics to reduce aberrations by controlled

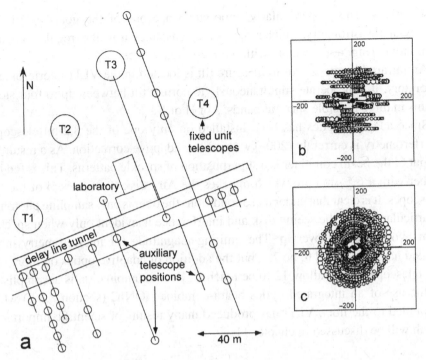

Fig. 8.42. (a) Layout of the VLT observatory, showing the four 8.2-m telescopes T1–T4 and 30 positions for 2-m auxiliary telescopes, joined by rail tracks. (b) and (c) show (u, v)-plane coverage for T1–T4 and three optimally chosen auxiliary telescopes, for source declinations of $0°$ and $-35°$, respectively. The u and v are in units of $10^6\lambda$. After von der Lühe et al. (1994).

bending of the flexible primary mirror and movement of the other mirrors. The time-scale is 30 s for major corrections and 1 s for minor ones. There is an adaptive optical system working at 2.2 μm on one telescope (2005). The four telescopes work independently but can also be used in an interferometric mode to provide high-resolution imaging. The useful wavelength range extends from the near ultra-violet up to 25 μm in the infrared. In the interferometric mode, light from the coudé focus is sent to the interferometric focus. The layout of the observatory is shown in figure 8.42. The figure also shows examples of (u, v) diagrams calculated for the four large telescopes and three auxiliary telescopes at positions where they optimize the areal coverage of the plane.

Although the ultimate sensitivity of the VLTI will be obtained when combining the four VLT 8.2-m telescopes, there are in addition four auxiliary telescopes, 1.8-m diameter, which can be placed on any of 30 possible stations and provide many interferometric baselines up to 202 m. The auxiliary telescopes are dedicated to VLTI, while the 8.2-m telescopes will be only intermittently available for interferometric observations. This is particularly important since one of the goals of VLTI is narrow-angle astrometry, which requires long baselines as well as regular and long-term monitoring, not available with the 8.2-m telescopes.

An infrared image sensor to measure tilt is located in the VLTI interferometric laboratory. It will operate simultaneously to correct tilt between up to four stellar beams in three possible spectral bands, J or H or K.

Since adaptive optics has been installed on only one of the 8.2-m telescopes, interferometry is currently (2005) without any adaptive correction. As a result, the output of the beam combiner is a superposition of speckle patterns. This is fed into a fiber with a 6.5 μm core which matches the Airy disk (0.06 arcsec) of the 8-m telescopes. It is clear that under these conditions the fiber is only sampling photons in a particular part of the seeing disk and interference is evident only when speckles from two telescopes overlap. The limiting magnitude for interferometry in the K band has been found to be 7.7, but the addition of adaptive optics on more than one telescope should allow 12 to be reached. Beam combination is accomplished by the use of an integrated optics beam-combiner IONIC (section 8.3.8) which is coupled to the fiber. VLTI has produced many results of scientific importance, which will be discussed in chapter 11.

References

Armstrong, J. T. *et al.* (1998). *Astrophys. J.* 496, 550–71.

Assus, P. H. Choplin, J. P. Corteggiani *et al.* (1979). *J. Opt.* (Paris), **10**, 345.

Baldwin, J. E. *et al.* (1994). *Amplitude and Intensity Spatial Interferometry,* II, ed. J. B. Breckenridge, Proc. SPIE, **2200**, 627.

Dejonghe, J., L. Arnold, O. Lardiére *et al.* (1998). *Advanced Technology Optical and IR Telescopes*, Proc. SPIE, **3352**, 603.

Di Benedetto, G. P. and Y. Rabbia (1987). *Astron. Astrophy.*, **188**, 114.

Gay, J. and A. Journet, (1973). *Nature Physical Science*, **241**, 32.

Hale, D. D. S. *et al.* (2000). *Astrophys. J.*, **537**, 998.

Hale, D. D. S. (Nov 2003). *Sky and Telescope.*

Haniff, C. A. *et al.* (2002). *Interferometry for Optical Astronomy, II*, ed. W. A. Traub, Proc. SPIE, **4838**, 19.

Hummel, C. A. J. A. Benson, D. J. Hutter *et al.* (2003). *Astron. J.*, **125**, 2630.

Johnson, M. A., A. L. Betz and C. H. Townes, (1974). *Phys. Rev Lett.*, **33**, 1617.

Labeyrie, A. (1975). *Astrophys. J.*, **196**, L71.

Lane, B. F. and M. W. Muterspaugh (2004). *The Astrophysical Journal*, **601**, 1129.

Lawson, P. R. ed. (2000). *Principles of Long Baseline Stellar Interferometry*, Course Notes from the 1999 Michelson Summer School, JPL, NASA.

von der Lühe, O. *et al.* (1994). *Amplitude and Spatial Interferometry*, II, ed. J. B. Breckenridge, Proc. SPIE, **2200**, 168.

McLeod, H. A. (2001). *Thin Film Optical Filters*, Bristol: Institute of Physics Publishing.

Monnier, J. D., W. C. Danchi, D. S. Hale *et al.* (2000). *Astrophys. J.*, **543**, 868.

Mourard, D., I. Tallon-Bosc, A. Blazit *et al.* (1994). *Astron. Astrophys.*, **283**, 705.

Nieuwenhuijzen, H. (1970). *Mon. Not. R. Astron. Soc.*, **150**, 325.

Noll, R. J. (1976). *J. Opt. Soc. Am.*, **66**, 207.

Proceedings of the SPIE meetings on Stellar Interferometry, **1237** (1990), **2200** (1994), **4006** (2000), **4838-40** (2003), **5491** (2004).

Quirrenbach, A. *et al.* (1996). *Astron. Astrophys.*, **312**, 160.

Ribak, E. N., M. Gai, D Gardiol *et al.* (2005), *Multiple anamorphic beam combination*, Proc. ESO-EII Workshop on "The Power of Optical/IR Interferometry: Recent Scientific Results and 2nd Generation VLTI Instrumentation", Garching, Germany, Ed. A. Richichi, in press.

Shao, M. *et al.* (1988). *Astron. Astrophy*, **193**, 357.

Tango, W. J. and R. Q. Twiss, (1980). *Progress in Optics*, ed. E. Wolf, Amsterdam: North Holland, **17**, 241.

Weigelt, G. *et al.* (2000). *Interferometry in Optical Astronomy*, ed. P. J. Léna and A. Quirrenbach, Proc. SPIE, **4006**, 617.

Weiner, J., D. D. S. Hale and C. H. Townes, (2003). *Astrophys. J.*, **589**, 976.

Wood, R. W. (1934). *Physical Optics*, New York: Macmillan.

Young, J. S. *et al.* (2002). *Interferometry for Optical Astronomy*, II, ed. W. A. Traub, Proc. SPIE, **4838**, 369.

9

The hypertelescope

9.1 Imaging with very high resolution using multimirror telescopes

Recent years have seen very large aperture telescopes constructed by piecing together smaller mirrors and carefully mounting them on a frame so that the individual images interfere constructively at the focus. This way the two multimirror Keck telescopes on Mauna Kea are constructed, each from 36 hexagonally shaped mirrors which together form a paraboloidal mirror about 10 m in diameter. The frame is constructed very rigidly, but since each segment weighs half a ton, it still distorts significantly when the telescope is pointed, so that the mirror positions have to be actively corrected to compensate for the small movements. Then, together with adaptive optics correction for atmospheric turbulence, diffraction-limited images are obtained since, for a small enough field of view, the off-axis aberrations of the paraboloidal mirror are insignificant. However, the maximum aperture which can be operated this way is expected to be of the order of 100 m, the size of the "Overwhelmingly Large Array" (OWL) being studied by the European Southern Observatories. The resolution achievable for a pointable direct imaging telescope is therefore limited to a few milli-arcseconds at optical wavelengths.

By using synthetic imaging, interferometry represents one way around this problem, when angular resolution rather than light-gathering power is the dominant aim. Effective apertures of hundreds of meters have been achieved, but the number of subapertures is still quite modest. As we have seen in the preceding chapters, interferometers with even tens of apertures and optical delay lines become extremely expensive and complicated to operate. By extrapolation from operating interferometers, the problems of continuously phasing a substantially larger number of individual apertures individually would be prohibitive.

When the feasibility of larger interferometric space arrays with formation-flying spaceships began to be explored in 1983, the potential application to imaging

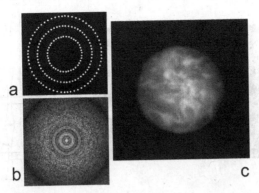

Fig. 9.1. A simulated raw image of an exo-Earth as would be recorded using a hypertelescope, with contrast enhancement. The aperture (a) has 150 subapertures equally spaced around three rings, the outermost one having diameter 150 km. The central peak and rings of the interference function (b) resemble the Airy pattern from a filled disk of identical outer size, but the outer rings are broken into speckles. (c) The simulated image of the Earth as seen from 10 light-years distance, using this hypertelescope. The central peak of (b) has been weakened by a factor of 4 in order to bring out the surroundings.

planets around nearby stars (exoplanets) was considered as a major objective. Resolving an "exo-Earth" in sufficient detail to detect vegetated areas, for example, and therefore to deduce the presence of life, requires an array size larger than 100 km. More than 100 mirrors would be needed to obtain snapshot Fizeau images providing usable detail of the planet's morphology, and their diameter must be 3 m at least to obtain usable exposures in 30 minutes, a period short enough to freeze the diurnal rotation of an exo-Earth. A sizable problem, however, with such Fizeau imaging with a sparsely filled aperture is that its point spread function contains a broad halo of diffracted light surrounding the central interference peak (figure 9.1 a, b). It leaves only a very small part of the collected energy in this peak, about 10^{-7} in the above case. Conceivably, the full halo could be recorded through a narrow-band filter, in order to allow all of its complicated speckle structure to be deconvolved and thereby obtain a cleaner image. But this would require as many as 10^{10} detector pixels and for full sensitivity, their performance would need to be photon-limited. To avoid wasting the unused part of the spectrum, a multichannel spectro-imager arrangement could be needed, but its 10^5 channels would then require 10^{15} pixels.

A solution appeared in the form of "densified pupil imaging," also called "hyper-telescope imaging," a principle which can provide direct images of exo-Earths such as the simulated raw image shown in figure 9.1(c), and will be described in this chapter.

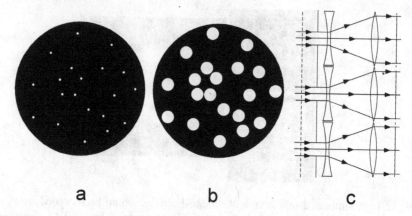

a   **b**   **c**

Fig. 9.2. (a) A sparse array aperture. (b) A densified copy of (a) in which the pattern of subpupil centers is conserved with respect to the entrance pattern, while the size of the subpupils relative to their spacing is increased. (c) Densification achieved by the use of inverted Galilean telescopes.

9.2 The physical optics of pupil densification

A hypertelescope can be defined as a multi-aperture Fizeau interferometer which is modified by attaching a pupil densifier. More generally, it may be defined as a multi-aperture interferometer where the detecting camera is illuminated through an exit pupil which is a densified copy of the entrance aperture. "Densified copy" implies that the pattern of exit pupil centers is conserved with respect to the entrance pattern, while the size of the exit pupils relative to their spacing is increased by an array of linearly magnifying optical devices such as inverted Galilean telescopes aligned to each aperture individually (figure 9.2). One assumes that the phasing, which is done to an accuracy better than a quarter of a wavelength, is ensured passively, actively or adaptively, and will remain stable over long periods.[†] The array of mirrors may be either random (nonredundant) or periodic. Both possibilities will be considered, since they lead to rather different characteristics. For simplicity, this discussion will be limited to the paraxial approximation.

9.2.1 A random array of apertures

For the random array, exploited in the Fizeau imaging mode, the point spread function is the diffraction pattern of N randomly situated circular apertures of diameter d within the bounding locus, ideally a circle of diameter D. We discuss this problem in Appendix A. The diffraction pattern consists of a central peak,

---

[†] Nonphased hypertelescopes can be operated in a speckle interferometry mode, of interest in the absence of adaptive optics.

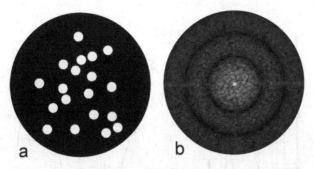

Fig. 9.3. Point spread function for 20 randomly spaced circular apertures of diameter D within a circle of radius $20D$. Notice the *interference function*, consisting of a sharp central point on a weaker speckle background, multiplied by the *diffraction function*, the coarser ring pattern which is the diffraction pattern of the individual apertures.

with intensity proportional to N^2, and profile $[4J_1(uD/2)/uD]^2$, at the center of a noisy speckle halo with mean intensity proportional to N. The peak and speckle pattern together are called the "interference function," using terminology borrowed from crystallography. The mean envelope profile of the speckle halo depends on the diameter of the individual apertures as $[4J_1(ud)/ud]^2$, and is called the "diffraction function" (figure 9.3). If N is large enough, the speckle background might be negligible, and the point spread function is essentially the same as that of the bounding locus, which provides angular resolution $1.22\lambda/D$. As we saw in section 5.9.1 and Appendix A, the ratio $N^2/N = N$ between the intensity of the central peak to that of the background is only correct if the individual apertures are correctly phased, and of course approaches unity as the phase errors approach 2π. However, the speckle background is an embarrassment, and takes energy away from the central peak.

Labeyrie (1996) proposed to shrink the halo by the use of "densification," in which the small wavefronts from the mirrors are individually magnified by inverted Galilean telescopes, so that they fill the spherical front converging on the focus as completely as possible. In this system, each of the beams propagating from an individual mirror is magnified laterally by a factor g by means of an inverted Galilean telescope aligned with its axis (figure 9.4a). The wave which converges on the focal point now has an interesting structure. Consider the wavefront incident on the Galilean from a star at angle α from the axis. Compared to the wavefront from a star on the axis, it has an angle α at each point. After the Galilean, which demagnifies the angle by the same factor as it increases the extent of the wavefront, it is at angle α/g. On the other hand, if all the Galileans have identical optical thicknesses, the average wavefront from all of them is unchanged. So the output

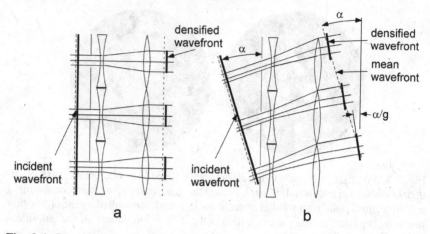

Fig. 9.4. Densified pupil configuration using inverted (demagnifying) Galilean telescopes, and the composite wavefront formed: (a) normal incidence; (b) incidence at angle α. $g = 1.7$ in this figure.

wavefront is piecewise continuous through each exit aperture, and has a stair-case shape as shown in figure 9.4(b).

We now ask what is the point spread function resulting from such a wavefront. Following the method described in Appendix A and shown in detail below, we find this to have the same form of interference function as before, still centered on angular position α, consisting of an interference peak and speckle halo. Both are modulated by the diffraction function, but this function is now centered on a different position α/g and has the form $[4J_1(udg/2)/udg]^2$, which is laterally shrunken by a factor g. As α changes, the sharp interference peak follows the star, but the diffraction function lags behind. When the peak reaches the edge of the diffraction function, it essentially disappears. Its side-lobes located within the halo remain visible, and become intensified to conserve the energy.

The hypertelescope concept has been tested with up to 78 apertures in a miniature-sized global aperture of 10 cm, needed to preserve phasing through the atmosphere in the absence of adaptive optics (Pedretti et al. 2000; Gillet et al. 2003). Large versions, under construction, use ground-supported mirrors and an imaging camera supported by a helium balloon with computer-controlled tethering (section 9.4.3).

Analysis of an undensified array

We shall now consider more formally the theory of the point spread function of a densified wavefront in one dimension. When necessary, there will be no difficulty in extending this to two dimensions. First, for an axial point star, an array of N undensified apertures, each one being represented by the individual aperture function $f(x)$ centered on position x_n may be described as a convolution of the aperture function with an array of δ-functions. The image, its Fourier transform, is therefore a product of their Fourier transforms, which are respectively the halo envelope or diffraction function, and the interference function. The combined wavefront is $f_U(x) = \sum_1^N f(x - x_n)$ which gives an image:

$$F_U(u) = F(u) \sum_1^N \exp(-iux_n). \tag{9.1}$$

For an off-axis point at α the wavefront becomes $\sum_1^N f(x - x_n)\exp(-ik_0\alpha x_n)$ which has transform

$$F_U(u) = F(u - k_0\alpha) \sum_1^N \exp[-i(u - k_0\alpha)x_n]. \tag{9.2}$$

The intensity $I_U(u) = |F_U(u)|^2$ is

$$I_U(u) = |F(u - k_0\alpha)|^2 \sum_{m=1}^N \sum_{n=1}^N \exp[-i(u - k_0\alpha)(x_n - x_m)]. \tag{9.3}$$

The double sum can be divided for convenience into groups:

(i) for $n = m$, N unit terms, summing to N;
(ii) for $n \neq m$, $N(N - 1)/2$ random cosine terms; this is noisy but averages to zero except around $u = k_0\alpha$ and has value between $\pm N$;
(iii) when $|(u - k_0\alpha)| \times \max(x_n - x_m) < \pi$ the exponential term is of order unity for all terms in the sum, and so the sum has value N^2.

Together (i) and (ii) give a positive definite speckle pattern with mean intensity N, to which is added the coherent image (iii) of height N^2 at $u = k_0\alpha$ with width about $\pi/\max(x_n - x_m)$, i.e. π/D. Notice the diffraction function envelope $|F(u - k_0\alpha)|^2$, which multiplies everything.

An estimate of the fraction of the total energy which is in the central peak for a two-dimensional array can now be made. The central peak has height N^2 and radius π/D in u space. The speckle pattern, which contains the rest of the energy, has height N and radius π/d. The fraction in the former is therefore $N^2/D^2 \cdot (N^2/D^2 + N/d^2)^{-1} \approx Nd^2/D^2$ which is about 10^{-7} for the parameters in section 9.1.

Analysis of a densified array

Now for the case of a densified pupil, the linear scale of $f(x)$ is increased by the factor g and its amplitude is reduced by g, making it $g^{-1}f(x/g)$, and the local wavefront slope is reduced from α to α/g so that the total wavefront excursion is unchanged (the Galilean is assumed to be optically perfect). The center-point of the wavefront (i.e. $x = x_n$) lies on the wavefront at angle α. The amplitude of the wave therefore becomes

$$f_D(x) = \sum_1^N \frac{1}{g} f\left(\frac{x - x_n}{g}\right) \exp\left[\frac{ik_0\alpha(x - x_n)}{g}\right] \exp(ik_0\alpha x_n)$$

$$= \exp\left[\frac{i(k_0\alpha x)}{g}\right] \cdot \sum_1^N \frac{1}{g} f\left(\frac{x - x_n}{g}\right) \exp[ik_0\alpha x_n(1 - 1/g)]. \quad (9.4)$$

which is identical with $f_U(x)$ when $g = 1$. The transform of this is evaluated term by term as

$$F_D(u) = \delta\left(u - \frac{k_0\alpha}{g}\right) \star \left[F(gu) \sum_1^N \exp(iux_n) \exp[-ik_0\alpha x_n(1 - 1/g)]\right]. \quad (9.5)$$

Since convolution with $\delta(u - k_0\alpha/g)$ creates a linear shift, implemented by replacing u by $u - k_0\alpha/g$ in the function following the $\star$, we have:

$$F_D(u) = F\left[g\left(u - \frac{k_0\alpha}{g}\right)\right] \sum_1^N \exp\left[i\left(u - \frac{k_0\alpha}{g}\right)x_n\right] \exp[-ik_0\alpha x_n(1 - 1/g)]$$

$$= F\left[g\left(u - \frac{k_0\alpha}{g}\right)\right] \sum_1^N \exp[i(u - k_0\alpha)x_n]. \quad (9.6)$$

The intensity is then

$$I_D(u) = |F[g(u - k_0\alpha/g)]|^2 \sum_{n=1}^N \sum_{m=1}^N \exp[i(u - k_0\alpha)(x_n - x_m)]. \quad (9.7)$$

The double sum can be visualized as before and the result is shown in figure 9.5. It follows that the diffraction function, the envelope, has shrunk by the factor g and its center-point is at angular position α/g instead of α. On the other hand, the interference function, the coherent point image and the associated speckle background, remain at angular position α and have unchanged width. But the interference function is still modulated by the diffraction function, and in particular the interference peak now has intensity

$$I_P(\alpha) = N^2 |F[gk_0\alpha(1 - 1/g)]|^2. \quad (9.8)$$

This means that images outside the region of the diffraction function have essentially zero intensity.

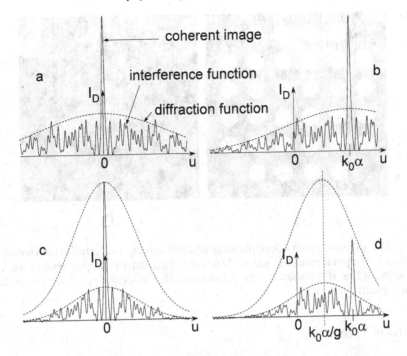

Fig. 9.5. Schematic profiles of undensified and densified images of a point source for a random array of apertures: (a) and (b): undensified, with object at angles 0 and α; (c) and (d): densified, $g = 2$, with object at angles 0 and α.

We can now repeat the estimate of the fraction of the energy in the central peak of a two-dimensional array which we made in the previous section. For a source on-axis, the central peak has height N^2 and radius π/D in u space, but the speckle pattern, which contains the rest of the energy, now has height N and radius π/gd. The fraction in the former is therefore $N^2/D^2 \cdot [N^2/D^2 + N/(gd)^2]^{-1} \approx Ng^2d^2/D^2$. We shall see in section 9.3 that for a random array of apertures, $g < D/Nd$, which means that for large N the fraction can be about $1/N$. This is an improvement of 10^5 over the undensified array for the parameters of section 9.1.

9.2.2 A periodic array of apertures

For a densified periodic array, the speckle halo in the interference function (9.6) is replaced by the periodic function which is its Fourier transform (section 3.1.3). We can call it the "reciprocal array" using the crystallographic term. This consists of a periodic array of peaks, of which the interference image is the zero order, with regions between them which are completely dark in the limit of an infinite number of apertures. For a finite number N spaced by a in a one-dimensional array,

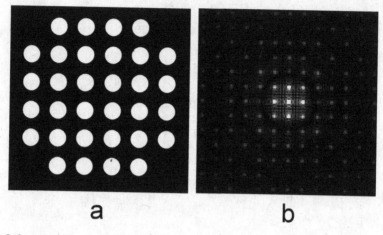

Fig. 9.6. (a) Aperture of a periodic array and (b) the reciprocal array of interference peaks in the point spread function. The scale of the latter is proportional to wavelength, so that if the source is polychromatic, the off-center peaks are dispersed into spectra.

(9.6) and (9.7) become

$$F_D(u) = F\left[g\left(u - \frac{k_0\alpha}{g}\right)\right] \sum_{n=1}^{N} \exp[i(u - k_0\alpha)na]$$

$$= F\left[g\left(u - \frac{k_0\alpha}{g}\right)\right] \frac{1 - \exp[i(u - k_0\alpha)Na]}{1 - \exp[i(u - k_0\alpha)a]} \tag{9.9}$$

$$I_D(u) = F\left[g\left(u - \frac{k_0\alpha}{g}\right)\right]^2 \frac{\sin^2[(u - k_0\alpha)Na]}{\sin^2[(u - k_0\alpha)a]} \tag{9.10}$$

as shown in figure 9.6. For polychromatic light, the zero order is white, and the other peaks are spectrally dispersed, the blue ends of the dispersion streaks pointing toward the center. Again, this array of orders is multiplied by the diffraction function corresponding to the individual apertures.

In the two-dimensional case, there are N apertures on a square lattice with spacing a, so $\pi D^2/4 = Na^2$ and the higher orders of the interference function for an axial source are at angles which are multiples of λ/a along both axes of the array. On densification, the extent of the diffraction function is compressed by the factor g, thus having its first zeros at angles $1.22\lambda/gd$ (figure 9.6). In the limit of complete densification, the apertures of diameter d are each expanded till they touch the sides of the square of side a, i.e. $gd = a$, from which it follows that $g^2 = \pi D^2/4d^2N$, and the first zeros of the diffraction function approximately coincide with the first interference order[†]. Then, for an axial source there is just the zero-order peak in the field of view, the next orders being on its periphery and having close to zero intensity. As the source moves away from the axis by angle α, as before the array of interference orders moves by α and the diffraction function by α/g. This means that the higher orders on the side opposed to α start to move into the field of view, and the first order will reach the center just as the zero-order peak disappears at the edge of the peak of the diffraction function. This is similar to the "umklapp" process in phonon theory, and also to the blazed diffraction grating. As a result, no information is actually lost, and in principle the image can be found algorithmically. Moreover, for simple objects, the fact that the regions around the peaks are dark (as opposed to speckled) the contrast ratio might be better. However, one should take into account the fact that the dark regions between the orders of a *finite* periodic array contain some well-defined peaks, most prominently the first minor peak flanking each interference peak, which has relative intensity $(2/3\pi)^2$, and might be mistaken for a planet image. This value is stronger than the speckle peaks, whose relative intensity can reach $2/N$, if $N > 45$.

Once again, we can estimate the improvement in efficiency of collection of light in the central peak of the interference function, when the source is on-axis. In this case, we saw that the maximum $g^2 = \pi D^2/4d^2 N$. Then the fraction of light in the central peak is $N^2/(N^2 + 4/\pi^2)$ which is unity for large N. This is rather obvious since the shrunken diffraction function has completely obliterated the higher order peaks for an axial source. Some laboratory experiments to illustrate these ideas using a periodic array are discussed in section 9.5.

9.3 The field of view of a hypertelescope and the crowding limitation

In section 4.3 we discussed the question of how many individual sources can be recognized by an aperture synthesis system, in which the (u, v) plane is partially sampled. It is interesting to reconsider this question for the case of the hypertelescope, which is not an aperture synthesis system, but where the interference between waves from a finite number of subapertures results in similar limitations. The densification also adds a new feature. We shall show that complete densification does not affect the crowding limits already determined. The direct imaging field of view is now defined as the angular region of the sky within which the interference peak has a non-zero value. As we have seen, when a source moves in the sky, the interference peak moves faster than the envelope diffraction function, and thus crosses it. As it approaches the edge, it is attenuated by the envelope's multiplicative effect;

[†] If the apertures were square, the approximation would be exact.

since energy is conserved, the background speckle pattern becomes intensified, mostly in the central part of the envelope. As the peak moves beyond the edge of the envelope it vanishes, and only the speckles remain. We have seen in (9.8) that the direct imaging field has angular radius $\lambda/d(g-1)$, which becomes very small when g is large. The resolution limit is given by the angular radius of the interference peak, which remains as λ/D, and so the number of resolvable sources is then

$$S = \left[\frac{\lambda}{(g-1)d} \frac{D}{\lambda} \right]^2 = \frac{D^2}{d^2(g-1)^2}. \qquad (9.11)$$

But the largest possible value of g comes when the densification is complete and the magnified beams from the individual mirrors touch. For the periodic array, this is occurs when $g = D/\sqrt{N}d$. For large g, it then follows that $S = N$, which is the same result as we obtained for aperture synthesis. Likewise, for a random array which is optimally non-redundant, as in section 4.3.2, we find that the minimum distance between mirrors (there the minimum baseline) is D/N from which it follows that g has a maximum value D/Nd and consequently $S \approx N^2$ as before.

Another way of looking at the limitations for a densified system is by requiring the interference peak signals to dominate the speckle noise. Consider the case of a nonredundant undensified array with N very small apertures of diameter d. It can be visualized as a Fizeau interferometer, having a point spread function with an envelope of angular size λ/d crossed by many fringes created by the array (this is the speckle). If phasing is perfect, all the fringes are in phase at the center, causing the envelope to have a central peak N times more intense than the average local background B of speckles. Now suppose this array is used to image several point stars simultaneously. The crowding limit is the number of such stars that can be within one such envelope, without degrading catastrophically the contrast of their direct image. Suppose again that there are S stars within the domain. Their envelopes are wide, and add without appreciable shift, but the speckles and peaks are shifted. The average speckled background intensity level thus becomes BS, while the speckle noise level becomes $B\sqrt{S}$. The stellar peaks remain detectable if their level NB is higher than the speckle noise, a condition which may be expressed as $NB > B\sqrt{S}$, or $S < N^2$. The crowding limit in this case is therefore N^2, as shown above.

The N^2 limit can be improved by repeating the exposure with a different non-redundant array, which can be achieved either by moving the subaperture mirrors or, more simply, by letting the Earth's diurnal rotation change the sky projection on the aperture. Since each exposure then creates a different speckle pattern, but the interference peaks are invariant, the signal-to-noise ratio improves. If E exposures are taken with significantly different configurations, the background speckle noise increases by factor $\sqrt{E}$ while the signal as E, resulting in a crowding limit $S < \sqrt{E}N^2$. This improvement would also apply to aperture synthesis systems, since it is essentially allowing a larger field of view to be obtained by the exploitation of the larger number of baselines sampled.

In the case of the periodic densified array, the interference peak is replaced by the reciprocal array of peaks. Spectral information can then be used to improve the crowding limit. For a white source, the zero order is white and the other orders are spectrally dispersed, with the blue end pointing toward the zero order. On densification, the whole pattern is multiplied by the shrunken diffraction function. As densification is maximized, the higher orders disappear, leaving the interference peak alone. But, just as a single source moves out of the direct imaging field as the source moves off-axis, higher orders move in, and the position of the single source can be inferred from the spectral dispersion (direction and magnitude) of these orders. As a result, the periodic array has potential for algorithmically imaging a wide field of view. With a random array, similar algorithmic reconstruction might also be possible by exploiting the speckle pattern recorded from a source located beyond the direct imaging field. The source's offset can also be calculated since the detailed speckle map is an offset part of the interference function, which is known *a priori*.

Related to the crowding phenomenon, although somewhat different, is the problem of the spurious background illumination which may be caused by zodiacal light (the solar light diffused by dust in the solar system), by the thermal emission of infrared radiation from mirrors when working at longer wavelengths, or the equivalent noisy background introduced by a camera which is not photon-limited. Such contaminations in the image cannot be completely uniform, if only due to photon noise, and thus degrade the detectability of faint sources. Densifying the pupil similarly intensifies both the low-resolution image of the sky background and the point source image. But this intensification improves the ratio of the peak to the background's photon noise: since the former grows as g^2 while the latter grows as $\sqrt{g^2} = g$; so the ratio grows as g. If, however, the numerous side-lobes in a broad Fizeau image, in monochromatic light, can also be exploited by a photon-limited camera having enough pixels, then the signal-to-noise ratio is improved by these side-lobes and reaches the same level achieved with the densified pupil. In practice this would require about 10^{15} pixels for a highly diluted imaging array such as the neutron star imager (section 12.4), and in fact many more to accommodate multiple monochromatic images. Unless so many pixels become available, the densified-pupil array can be expected to reach higher limiting magnitudes in practice. A similar advantage arises for densified-pupil imagers if the camera is not photon-limited; imperfections such as a response threshold or a background "read-out noise" generate degradations comparable to zodiacal or thermal contamination.

9.4 Hypertelescope architectures

9.4.1 Michelson's stellar interferometer as a hypertelescope, and multi-aperture extensions

The 20 and 50 foot interferometers built by Michelson and Pease at Mount Wilson (section 8.1.1) and the Large Binocular Telescope (section 8.4.6) can be considered as two-element hypertelescopes since the periscopic arrangement of mirrors densifies the pupil. The scheme can in principle be extended to include more apertures, instead of just two, but this would be very expensive.

9.4.2 Hypertelescope versions of multitelescope interferometers

Multitelescope interferometers such as the OVLA and the VLTI can be equipped with a pupil densifier within their beam combiner. At the VLTI, the entrance aperture is already rather dense, and the intensification gain achievable by fully densifying the exit pupil is moderate, of the order of 30. This can, however, appreciably improve the limiting magnitude if the detector has a sensitivity threshold.

9.4.3 Carlina hypertelescopes

The Carlina architecture for hypertelescopes represents a new approach to the problem of achieving a very large aperture by using many small mirrors in a fixed sparse array. Lack of steering capability requires a solution that is equally good for any angle of incidence; this indicates that the individual mirrors must be parts of a large spherical surface which, however, they fill only sparsely. A spherical mirror has the important property that its axis is not unique; any source lies on an axis of the mirror, and forms an image on the same axis at the paraxial focal distance (half the radius). As the night proceeds, the image of a given source will therefore move on a spherical surface with radius half that of the mirror sphere (figure 9.7) and a moving camera can capture this image continuously.

On the other hand, the fact that the surface is not paraboloidal means that the image suffers from spherical aberration, as shown enlarged in figure 9.7. In small telescopes, this same problem is classically solved by the use of an aspheric secondary mirror or, in the incoming beam, a Schmidt plate at the curvature center. These classical types of correcting elements compensate the difference between a spherical mirror and a paraboloid. The Schmidt plate does it equivalently for any axis, but has diameter nearly equal to its aperture, which is obviously impractical for large telescopes. Instead, an aspheric secondary mirror, or a pair of them arranged in "clam-shell" fashion near the focus (figure 9.8), can have relatively

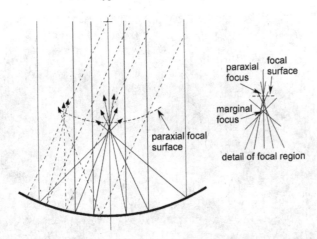

Fig. 9.7. The focal surface of a spherical mirror, with rays incident from two directions. The expanded view of the focal region indicates the geometrical origin of spherical aberration.

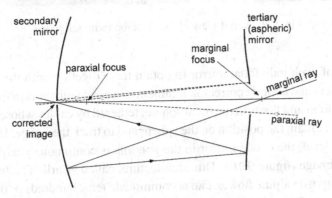

Fig. 9.8. The principle of a Mertz ("clam-shell") corrector, which compensates the difference between the sphere and paraboloid at a position close to the focus. Only one marginal and one paraxial ray are shown, but all intermediate rays focus to the same stigmatic image point.

small dimensions and it can be moved as part of the imaging camera, thus correcting the problem of spherical aberration equivalently at any image point. As discussed by Mertz (1996), aspherical clam shell correctors can also compensate the lower-order coma aberration arising when the star moves off-axis.

Using this solution, analogous to the Arecibo radio telescope (figure 9.9), and also similarly applicable to filled apertures, we can envisage a large fixed spherical basin (the Caldera de Taburiente crater on the Canary Islands has been suggested for this purpose) in which a sparse array of small spherical mirrors is mounted so that their centers of curvature coincide accurately. In principle, it requires a one-time

Fig. 9.9. Aerial view of the Arecibo radio telescope.

adjustment of each individual mirror to obtain this. Together with the appropriate pupil densifiers and a Mertz corrector, a camera or a coudé mirror is suspended from a helium balloon and floats in the focal sphere, tethered by cables whose lengths are adjusted to maintain the position on the sphere and to track the image. If the camera is at ground level, the coudé mirror in the gondola is continuously adjusted[†] so as to relay the image (figure 9.10). This architecture, called "Carlina" after a ground-hugging composite alpine flower, can accommodate tens, hundreds or thousands of primary mirrors. Several such balloon-suspended gondolas can make simultaneous observations in different parts of the sky. As described below, a prototype system with a 17-m effective aperture is under construction and has provided fringes with the first pair of mirrors installed (Le Coroller et al. 2004) (figure 9.11). Proposals for space-based hypertelescopes of much larger size will be discussed briefly in chapter 12.

9.4.4 A fiber-optical version of the hypertelescope

The principle of the hypertelescope remains applicable if single-mode optical fibers are used as densifiers. In this case each fiber samples the incident wavefront at a

---

[†] If the axis joining the coudé mirror to the camera is parallel to the Earth's rotation axis, the mirror orientation is invariant.

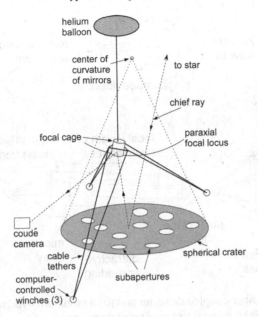

Fig. 9.10. Hypertelescope concept using a balloon-supported coudé mirror and Mertz corrector in a focal cage, and computer-controlled tethering.

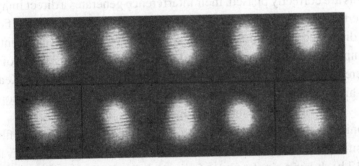

Fig. 9.11. Sequence of fringes observed on Vega during a 200 ms period with a two-subaperture hypertelescope.

point, so that $f(x)$ would correspond to the core of the fiber, although concentrating telescopes are needed in practice. Here, the fibers are inserted between the subapertures and the beam-combiner. The bundle of emerging fibers can be packed tightly together, while respecting the pattern of the entrance apertures, but with the cores now filling a larger fraction of the wavefront than at their entrance end (figure 9.12a). Their exit openings behave like subpupils, although the notions of image and pupil planes vanish in fiber optics. A camera, located in the far field of the

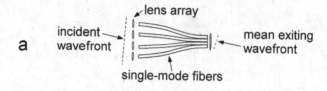

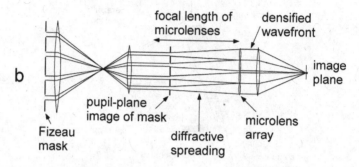

Fig. 9.12. (a) A fiber-coupled densifier and (b) a miniature hypertelescope due to Pedretti et al. (2000) using diffractive pupil densification.

densified pseudo-pupil thus obtained, receives superposed lobes from all fiber exits. If the beams are correctly phased, their interference generates a direct image, like in other hypertelescopes. The limitations to the size of the direct imaging field and to crowding discussed in section 9.3 also apply. As of 2005, this has apparently not yet been attempted, even in laboratory simulations, but an imaging mode of this kind is considered for the OHANA fiber-linked interferometer at Mauna Kea (section 12.1.6), which is intended to link interferometrically all the seven telescopes on the summit having apertures greater than 3-m (Perrin et al. 2000).

The miniature hypertelescope built by Pedretti et al. (2000) has no fibers but a pupil densifier which exploits the diffractive divergence of light to densify the pupils (figure 9.12 b). It has a single array of microlenses and the densified pseudo-pupil obtained in this case makes it somewhat analogous to the fiber version.

9.5 Experiments on a hypertelescope system

Pending the completion of the first hypertelescope having dimensions useful to astronomers, some small-scale experiments in the laboratory and on the sky have confirmed the approach. Gillet et al. (2003) verified the imaging performance on the sky, using a miniature version of a hypertelescope with a 10-cm aperture containing 78 subapertures of 1-mm size, arrayed periodically on a square grid. This aperture is small enough to avoid seeing limitations, so that no adaptive optics were needed to

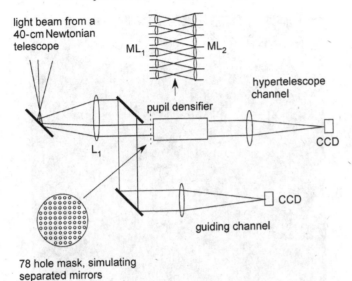

Fig. 9.13. Hypertelescope experimental set-up used in miniature form for preliminary testing. The incoming light beam from a Newtonian telescope is collimated by lens L_1. A Fizeau mask installed for convenience in the pupil plane following L_1, rather than at the primary mirror, has $N = 78$ holes of $100\,\mu$m size each. It defines in the entrance aperture a virtual "diluted giant mirror" of 10 cm size with 1 mm subapertures. The densification is achieved with two microlens arrays (ML_1 and ML_2). (Gillet et al. 2003).

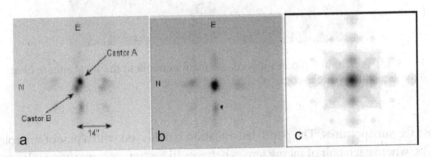

Fig. 9.14. (a) Image of Castor made using the miniature hypertelescope, showing the resolved binary A-B, spaced 3.8 arcsec. The half direct imaging field is about 14 ± 0.6 arcsec wide. (b) Image of Pollux, obtained with a 10-min exposure. It matches the theoretical pattern, with residual first orders due to incomplete pupil densification. With respect to the laboratory images and the numerical simulation, the peaks are however somewhat widened by seeing and exceed the theoretical arcsecond resolution limit of the 10-cm array. (c) Numerical simulation of a monochromatic point source image with the 78-aperture hypertelescope.

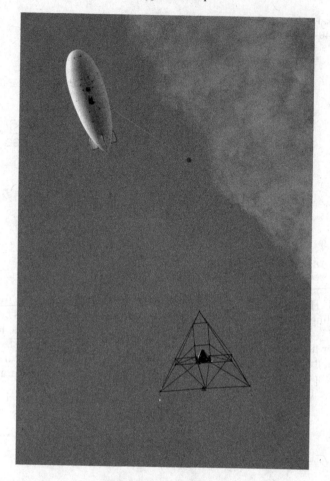

Fig. 9.15. A helium balloon supports the focal gondola in the focal sphere of an experimental hypertelescope (see figure 9.10).

phase the subapertures. The pupil densification is achieved with a pair of microlens arrays, where each pair of facing lenses behaves like a tiny demagnifying telescope. The experiment was used to observe laboratory-simulated multiple stars and also the stars Castor (binary α-Gem) and Pollux (β-Gem). The experimental set-up and results are shown in figures 9.13 and 9.14. The densification factor is $g = 6$ in this experiment; since complete densification would imply $g = 11$, the first diffraction orders are not eliminated by the diffraction function and are visible in figure 9.14.

A Carlina-type hypertelescope with 17-m effective aperture and 35-m primary focal length is under construction. A tethered balloon at 140-m altitude carries the focal optics and camera (figure 9.15). The cable suspension is arranged to allow equatorial tracking of the star's motion with a single computer-controlled winch. The initial testing has shown that the primary mirrors at ground level can be co-spherized within one micron before observing, so that fringes are immediately obtained in a star's image. Under conditions of low wind, the focal optics have enough stability with passive tracking, and this can be further improved with servo tracking corrections. A larger version called Carlina II is to be built at the Plateau de Calern observatory, in a sink hole. The effective aperture diameter which can be reached is 40–50 m, enough for many astrophysical programs requiring rich snapshot images. The mode of operation will initially use speckle interferometry; adaptive optics will be added later.

References

Le Coroller, H., J. Dejonghe, C. Arpesella, *et al.* (2004). *Astron. Astrophys.*, **426**, 721.
Gillet, S., P. Riaud, O. Lardière, *et al.* (2003). *Astron. Astrophys.*, **400**, 393.
Labeyrie, A. (1996). *Astron. Astrophys. Suppl. Ser.*, **118**, 517.
Mertz, L. (1996). *Excursions in Astronomical Optics*, New York: Springer.
Pedretti, E., A. Labeyrie, L. Arnold *et al.* (2000). *Astron. Astrophys. Suppl. Ser.*, **147**, 285.
Perrin, G., O. Lai, P. Léna (2000). *Interferometry in Optical Astronomy*, ed. P. J. Léna and A. Quirrenbach, Proc. SPIE, **4006**, 708.

10

Nulling and coronagraphy

10.1 Searching for extrasolar planets and life

The emerging possibility of searching for life outside the solar system creates a challenging situation for today's civilization. Maybe it is somewhat analogous to that of our Paleolithic ancestors when, roaming along a shoreline, they saw inaccessible off-shore islands and wondered what life forms they could carry. Today's inaccessible islands, which may also carry life forms, are the extrasolar planets, dubbed as "exoplanets," and man's curiosity about them has pushed NASA and ESA to encourage efforts toward observing exo-Earths. When images of exoplanets are obtained with large telescopes in space, a subsequent step will be the construction of even larger versions designed to obtain images with enough resolved detail to search for life signatures. There are so many possible answers to the question of where life might be sustainable, and therefore where to look for such signs, that it is necessary to define what is called a "habitable zone": for the purposes of present-day efforts this is considered to be the region where liquid water can exist for at least some part of the time. This defines a rather small shell around any star, which in the case of our solar system includes the planets Venus, Earth and Mars. Of course this definition excludes the possibility of life which has evolved independently of water, but somewhere a line has to be drawn in order to contain the investigation.

The following sections discuss the observational problems and potential solutions. Essentially, the biggest problem is the extreme contrast ratio between the star and its planet, which has been likened to observing a firefly next to a searchlight. The solutions go in two directions, both directed at attenuating the residual star light appearing at the field position where planets may exist. One is to weaken the image of the star selectively, together with its diffracted halo (nulling and coronagraphy), and the other is to ensure that, bright as the star's image may be, the halo region surrounding it is dark enough for the planet to be observable (apodization). The two approaches can then be combined.

Optical systems are conveniently modeled by calculating Fourier transforms of complex vibration amplitudes, and much insight is thus gained into their performance and possible improvements. We should remark, however, that the Fourier transform model of diffraction phenomena is only an approximation, itself based on the scalar wave approximation for electro magnetic propagation, where the vectorial structure of the electromagnetic field is ignored (section 3.1.2). The edges of masks, mirrors and refracting materials used in actual coronagraphic telescopes or interferometers do introduce weak nonscalar effects which vary with polarization. These are usually negligible in ordinary telescope images, but the extreme nulling levels needed for exoplanet detection require a more rigorous analysis of coronagraphic performance, using Maxwell's equations and accurate data on the polarizing properties of aluminum coatings, the edges of masks, etc. Fortunately, careful analysis has found no significant degradation due to polarization even in extreme coronagraphy experiments, so the scalar diffraction model will be used exclusively in the following discussion.

10.2 Planet detection methods

Among the celestial objects, those located near much brighter sources can be difficult to see amidst the light scattered from the bright source. Such is the case for the Sun's corona, for the planets of other stars, and for more remote objects such as distant galaxies containing a very bright nucleus, or emitting localized light flashes during gamma ray bursts. More than 160 extrasolar planets have been discovered by means of indirect methods since Mayor and Queloz (1995) first detected periodically shifting lines in the spectra of the star 51 Peg. Interpreted as radial Doppler shifts, these indicated a motion of the star toward and away from the observer, with period four days in that case, presumably resulting from the reflex orbital motion induced by a planet which could not be seen. Another indirect method by which planets have been inferred is eclipsing, in which a periodic dip in the intensity of a star results from the passage of a planet in front of it (Pont et al. 2005); quantitative measurements allow the areal ratio to be calculated, but of course the need for the observer to be quite accurately in the plane of the orbiting planet makes such detection somewhat rare. But the first direct observations of large companions to nearby stars have recently been made at the VLT observatory using adaptive optics in the infrared (Chauvin et al. 2004; Neuhäuser et al. 2005), in both cases the companions having temperature of 1000–2000 K, masses of 1–10 Jupiters and orbital radii 55 and 200 times that of the Earth. Although these are far from being exo-Earths, their discovery is very encouraging in the search for closer and more elusive planets.

10.2.1 The relative luminosities of a star and planet

Observing directly such exoplanets requires detecting the weak light which they receive from their parent star and partially reflect toward the observer. Their thermal emission is also potentially detectable in the infrared. Other possible forms of light emission, caused by fluorescence, auroral or thunderstorm activity and bioluminescence are expected to be much fainter if present at all. Somewhat indirect biological emissions, such as laser beams emitted toward the Earth, light pollution, or the brief flashes from an exoplanetary nuclear war, could be much brighter, but of shorter duration.

An estimate of the ratio between the light fluxes received from a star and its planet depends only on the planet's radius r, its orbital radius a, and its albedo[†] x, which can be assumed to be of order 0.5 from our planetary system. For a star with radius R and irradiance I (in units of watt m^{-2} str^{-1}), an observer on Earth at distance D receives luminous flux $J(D) = \pi R^2 I / D^2$ and the planet receives $J(a) = \pi R^2 I / a^2$. The irradiance of the planet is thus $J(a)x/2\pi$, assuming it scatters uniformly into the solid angle 2π, from which the luminous flux at Earth due to the planet is $J_\text{p}(D) = \pi r^2 J(a)x/2\pi D^2$. The ratio between the fluxes due to the planet and the star is therefore

$$\frac{J_\text{p}(D)}{J(D)} = \frac{xr^2}{2a^2}.$$

(10.1)

For Jupiter's parameters, this ratio has the value $2.2 \cdot 10^{-9}$ and for the Earth's, $3.2 \cdot 10^{-10}$, indicating 22 and 24 magnitudes relative attenuation respectively. If we add self-emission by the planet, which peaks around $10\,\mu$m for the habitable zone, the ratios become smaller in the infrared (figure 10.1).

Another problem facing the direct imaging of extrasolar planets is zodiacal light, caused by dust and larger particles which accumulate in the plane of a planetary system, and also scatter light from the star. The magnitude of the problem can be estimated by looking at the solar system, where the total flux from the zodiacal disk at $10\,\mu$m is two orders of magnitude greater than that from Jupiter. There is little data about the magnitude of this problem around other stars, although the images showing planet-like companions around two stars obtained using adaptive

---

[†] Albedo is the fraction of received light which is scattered diffusively by a rough surface.

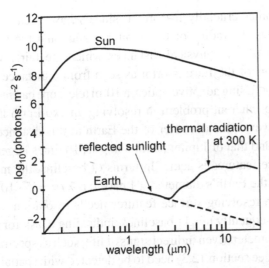

Fig. 10.1. Light flux spectra received from the Earth and Sun at a distance of 10 parsec. The ratio between the two graphs is independent of the distance.

optics could also give some information about exo-zodiacal luminosities closer to the stars.

Detecting the weak planet light amidst the parent star's stellar glare requires cleaning techniques known as *coronagraphy*, *apodization* and *star nulling*. The first of these closely related techniques dates back to the 1940s when Bernard Lyot invented his solar coronagraph and succeeded in observing the faint solar corona with it, for the first time without a total solar eclipse. Lyot's coronagraphic telescopes were small refractors of 10 or 20 cm diameter, larger apertures being catastrophically affected by the seeing. A stellar version of the Lyot coronagraph (Bonneau et al. 1978) was built by ESA for the Hubble Space Telescope, but was rendered inoperable by the fabrication defect of the primary mirror. NASA is now planning TPF-C, a coronagraphic telescope modernized and dedicated to exoplanet coronagraphy (section 12.2.3).

Realizing that the planet/star contrast can be 1000 times higher in the thermal infrared than at visible wavelengths, Bracewell (1978) proposed a space version of Michelson's beam interferometer for resolving planets from their star at 10 μm, with a "nulling" device which behaves analogously to a coronagraph by strongly attenuating the star's light selectively.

10.2.2 *Requirements for imaging planet surface features*

The ratio between the luminosity of a star and its planet is independent of its distance from us. However, when it comes to separating the images of the two, the distance becomes crucially important, since what we can control is angular resolution, whereas the radius of the habitable zone and the expected size of a life-supporting planet are physical distances which are fairly well defined. For example, the radius of the Earth's orbit as seen from a distance of 1 parsec is by definition[†] 1 arcsec. Using adaptive optics, a 10-m telescope can resolve 0.01 arcsec, so that there is no inherent problem in resolving an exo-Earth from its star from that distance. However, the diameter of the Earth at this distance is about 83 μas $(4.1 \cdot 10^{-10}$ rad), and that of Jupiter about 830 μas, 10 times larger. So the problem of imaging surface features is acute. In terms of baseline, the minimum baseline needed to resolve the Earth's diameter at 1 parsec is $\lambda/\alpha = 5 \cdot 10^{-7}/4.1 \cdot 10^{-10} = 1.2$ km. Moreover, resolving surface features needs resolution many times better than this, and consequently longer baselines, indicating a task for a hypertelescope or a space interferometer. Even indirect signs of life, such as spectral signatures with yearly variations (see section 12.3) need to be detected with spatial resolution, since the variations in opposite hemispheres would be expected to be in antiphase. The sobering conclusion is that we should be searching for habitable planets by imaging only within our immediate surroundings (less than 10 parsec from us) within which there are a few hundred candidate stars.

10.3 Apodization

If an image, focused by a lens or a telescope, contains light from a bright source concentrated in a small region, the bright source always contaminates the surrounding pixels, beyond its geometric image, with diffracted light, thus affecting the detection of images from adjacent fainter sources. Part of this light is diffracted by the abrupt edges of the imaging lens or mirror and appears in the form of the concentric rings in the classical Airy pattern. Obscurations such as a spider structure supporting a secondary mirror, etc. disturb the ring structure. Periodic secondary peaks also appear if the primary mirror is a mosaic of segments. An addition to the contaminating diffracted light, appearing in the form of random speckles, is generated by phase disturbances such as the residual bumpiness of the main mirror or lens.

Apodization, literally "removal of the feet" of the point spread function, typically uses a continuous attenuation pattern with graded edges across the entrance aperture or a conjugate pupil, in order to reduce the contaminating light in the pixels surrounding the central peak of the diffraction-limited image. Such soft-edged apertures can be fabricated by degrading the reflectivity of the aluminum,

---

[†] The parsec is defined as the distance at which the parallax of a star, with reference to a star at infinity, is one arcsec between points separated by the Earth's orbital radius. 1 parsec = 3.3 light-years.

or coating it with partially absorbing material. Various such patterns of radially graded intensity can be applied to the wavefront. A general effect is to attenuate the Airy diffraction rings, at the expense of somewhat enlarging the central peak. Apodization degrades somewhat the angular resolution and the light throughput, but improves the detectability of faint features near a star. For some applications, the resulting removal of the Airy rings in the spread function is so efficient that no additional coronagraphic mask might be needed, although an absorbing mask in the bright central peak is often useful to protect the camera from saturation or destruction. An important advantage of apodization, compared to the techniques of nulling and coronagraphy, is that it applies equally well to point sources anywhere in the field of view, and therefore very accurate pointing of the telescope toward the star is not a prime requirement.

In order to attenuate the "feet" of the point spread function isotropically, an apodization mask rotationally symmetric around the optic axis is required. But the use of an aperture mask which is not rotationally symmetric, and apodizes excellently along certain directions at the expense of others, is also possible and . often easier in practice. This proved its worth in the discovery of the star Sirius B, and remains a good illustration of the sensitivity gain achievable with apodization. In the 1840s, the German astronomer F. W. Bessel observed a periodic motion of the bright star Sirius, and deduced from it the existence of a stellar companion Sirius B. This turned out to be a white dwarf companion about 10,000 (10 magnitudes) times fainter and was observed in 1972 by using a hexagonal mask on a telescope in order to attenuate the Airy rings around the image of Sirius A. The Fraunhofer diffraction pattern formed in the focal plane from a hexagonal aperture has six radial spikes on a background which is darker than the Airy pattern of rings from a circular aperture. When the companion star was in one of the dark spaces, it was bright enough to be visible. However, this example represents a weak form of apodization. Better results are obtainable with graded attenuation at the edge of the aperture. A solution to the general problem of finding the optimum linear and radial attenuation functions for a finite aperture, to give an arbitrarily dark surrounding field around a point image, was found by Slepian (1965) using prolate functions, and an example is shown in figure 10.2. This solution can in principle achieve the required background of 10^{-10} of the peak intensity, with an acceptable reduction of the resolution limit by a factor of about 4, but the manufacture of an accurately controlled gradation which is also resistant to degradation with time remains a problem. A combination of a square aperture and graded absorption was suggested by Nisenson and Papaliolios (2001), who also checked the sensitivity of their solution to inevitable wavefront errors (figure 10.3)

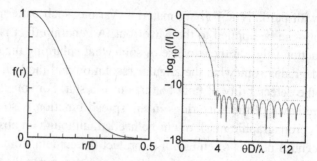

Fig. 10.2. An example of Slepian's prolate function apodization mask (intensity attenuation factor as function of radius) and the cross-section of the point spread function, shown on a logarithmic scale. The abscissa angle θ is in units of λ/D, so that the first zero of the Airy function for the full aperture would be at 1.22 (Kasdin et al. 2003).

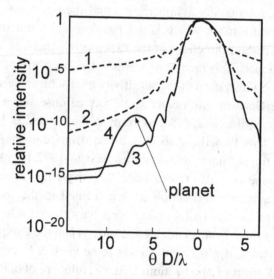

Fig. 10.3. Nisenson and Papaliolios (2001) considered apodization of a square aperture with the sonine function $[(1-x^2)(1-y^2)]^3$. The figure shows diagonal cuts through the PSF in polychromatic light for a circular aperture, without apodization (1) and with sonine apodization (2), and a square aperture with sonine apodization (3) and with the addition of a planet of relative intensity 10^{-9} of the star (4). Absicissa angle θ as in figure 10.2.

10.3.1 Apodization using binary masks

In order to solve the apodization problem without using graded attenuation, it is necessary to limit the search for suitable masks to functions which have either zero or unit value, and therefore can be manufactured by cutting holes in an opaque sheet. Ideally, the opaque part of the mask should be contiguous so as to avoid the need for a supporting structure, although a mask could be printed onto the primary mirror to solve the problem of mechanical support. For example, if a slit is cut out of such a sheet with variable width $y(x)$ proportional to the value of Slepian's function for the graded attenuation mask in one linear dimension, the point spread function intensity has along its u axis the value of $|\int y(x) \exp[-iux] \, dx|^2$ which is similarly dark outside the central peak. This mask would have to be rotated in its plane to allow scanning of the complete surroundings of a star. The scanning problem can be considerably alleviated by using an array of narrow slits of this type of construction which broadens the angular region within which the point spread function is darker than the required value. This type of mask has been suggested by Kasdin et al. (2003) and is illustrated in figure 12.6, where the overall shape of the mask is restricted by the demands of NASA's TPF-C project (section 12.2.3).

Some other possible solutions have been found for this problem with rotational and quasirotational symmetry (Vanderbei et al. 2003). For example, Slepian's solution can be "digitized" as a set of annular apertures. In this case, the fact that the set of rings is finite in number results in a finite outer limit to the dark part of the field of view, but this is hardly a problem if the limit is far enough out (figure 10.4). In another solution, it is digitized as a set of radial spoke-like slits with angular width suitably varying as a function of radius, which creates a point image surrounded by a completely dark field in the limit of an infinite number of spokes. Compromises result from using a finite number of spokes; apparently 120 are necessary to achieve the required background darkness.

10.3.2 Apodization using phase masks

A problem with aperture masks is that a significant amount of light is absorbed by the opaque part, more than 50% in the examples in section 10.3.1. An alternative is to make a rotationally symmetric phase distortion of the wavefront. Zernike's phase contrast, the classical technique used with microscopes to see transparent biological cells, transforms phase patterns into intensity patterns by means of a $\pi/2$ phase change at the zero order. This means that a phase mask corresponding to one of the designs in section 10.3.1 can be used together with a $\pi/2$ stop in the image plane. This type of approach has been developed by Martinache et al. (2004). A major problem is to carry out such an operation achromatically, taking into account the fact that every wavelength must be apodized differently. An alternative method of phase apodization, called "phase-induced amplitude apodization," was proposed by Guyon (2003) who calculated the figuring distortions of the primary and secondary mirrors of a Cassegrain telescope needed to create an intensity modulation in the pupil plane with properties similar to those of Slepian's function.

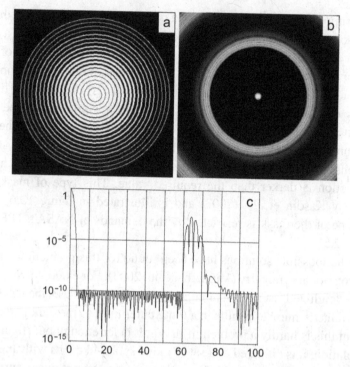

Fig. 10.4. Rotationally symmetric apodization mask providing an extended region of intensity below 10^{-10}: (a) the mask, (b) and (c) calculated PSF. Courtesy of R. J. Vanderbei.

An advantage of this system is that the intensity variation is created by geometrical optics and is therefore achromatic. However, the system introduces considerable off-axis aberrations and the contrast of 10^{-10} is only achieved in a very small field around the axis. The principle has been tested in the laboratory by Galicher et al. (2005).

10.4 Nulling methods in interferometers

The idea of nulling, originally suggested by Bracewell (1978) is to use destructive interference in a stellar interferometer to selectively delete the starlight, without affecting that of the planet. It can be carried out in different ways, depending on whether the planet separation is resolved only as a result of the long baseline between the subapertures of the interferometer, or whether the subapertures are individually capable of resolving the planet from its star. For example the former might be the situation when using the two Keck telescopes jointly as an interferometer, since the baseline is much longer than the telescope aperture diameters, but the latter

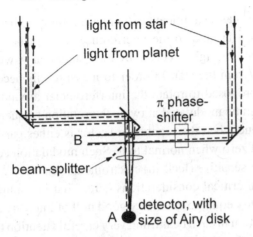

Fig. 10.5. Bracewell's concept of a Michelson interferometer with small subapertures used as a nulling interferometer. As a result of the phase shift, the waves
from the two subapertures interfere destructively when the source is on the axis
of the interferometer, but when the source is at a non-zero angle to the axis, constructive interference may be obtained. Because the requirements for nulling are
less stringent in the infrared, this is practical in the mid-infrared region.

case would be obtained with the Large Binocular Telescope, where the separation
between the individual apertures is not much greater than their diameters.

10.4.1 Bracewell's single-pixel nulling in nonimaging interferometers

Bracewell's concept of a single-pixel nulling interferometer is based on a variant
of the Michelson stellar interferometer (section 8.1.1), which has two subapertures
separated by a long baseline. The sub-apertures are small enough that the star is
not resolved, so that it behaves as a point source. The plane waves transmitted from
the two apertures are combined using a beam-splitter (pupil-plane configuration[†]),
so that the interfering wavefronts are parallel and can be detected by single-pixel
detectors in the two outputs A and B. The geometry is such (figure 10.5) that in
one output of the interferometer (A) there is a phase difference π between the two
waves from the star. Moreover, the two waves are arranged to have equal amplitudes
so that there is complete destructive interference at A, and no light is received from
the star.

On the other hand, the baseline is assumed long enough to resolve a suspected
planet when the vector separating it from the star is parallel to the baseline. At
this length of baseline, the light from the planet reaching the beam-splitter from

---

[†] Michelson suggested this form in his 1890 paper, where he called it the "refractometer" configuration, but he
did not build it.

the two subapertures has a significant extra phase difference because of its angular separation from the star, and so the interference will not be quite destructive at output A. As a result, all light received by the detector at A will have come from the planet, and not from the star. In order to measure this necessarily very weak signal, Bracewell proposed to rotate the interferometer around the star axis. The planet light will then be modulated at twice the rotation frequency, because it will have maximum value when the separation vector is either parallel or antiparallel to the baseline, and zero when normal to it. Such modulation could be sensitively detected by a phase-sensitive (lock-in) electronic amplifier.

There are several critical considerations here. First, the amplitudes of the two waves must be exactly equal to get a very good null of intensity at A. This requires some active attenuation in each beam, and very careful attention to polarization (see section 8.3.7, where there are similar but less stringent requirements). Second, the phase shift must be π at all wavelengths of observation, so it must be achieved by geometrical means (section 10.4.3). Unless the waveband is very narrow, it is not sufficient to use a glass phase-shifting plate or a controlled path-length correction, which give the required phase shift at discrete wavelengths only. Third, pointing the rotation axis exactly at the star is critically important, since any slight offset would result in the star itself giving a spurious signal which is also periodic at twice the rotation frequency. Ground-based precursor versions of Bracewell's interferometer have been tested at the Multimirror Telescope in Arizona (Hinz et al. 2001) and at the Keck telescope. In these cases, each telescope aperture is obscured by a mask having two large holes, and the beams from these two subapertures are combined in antiphase, resulting in a system which is similar to the LBT (section 8.4.6). In addition, the outputs from the two Keck telescopes can be combined in antiphase. Variants on Bracewell's principle have also been proposed for space experiments and will be discussed in chapter 12.

10.4.2 Bracewell nulling in imaging interferometers

If the subapertures are large enough, we have two superimposed Fizeau images of the star–planet system, each of which individually resolves them. Suppose that by using a Michelson stellar interferometer a small angle is introduced between the two wavefronts in the image plane so that the image is crossed by interference fringes. Introducing a π phase shift between them at the center of the field of view creates a deep black fringe across the image, and the system is pointed at the star so that its image lies exactly under this fringe. On the other hand, the angle is such that a planet in the habitable region will lie on the bright fringe; since its location is not known, it is necessary to rotate the whole system about the star image so as to cover all possibilities.

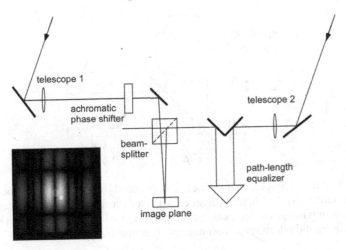

fringes multiplying the image

Fig. 10.6. Nulling in an imaging interferometer. The picture sketches the sort of image expected, and the origin of starlight leakage.

This sounds good, but where has the starlight gone? Destructive interference can't destroy its energy. In fact, as described above, the system cannot achieve its purpose. Suppose that each image is formed by a cone of light with semi-angle β. Then the resolution limit in the image plane is λ/β, and this will be the diameter of the star image (assuming it is itself unresolved). But if the two star images are superimposed without the use of a beamsplitter, as in Michelson's instrument, the angle between their wavefronts must be at least 2β, otherwise their cones will overlap. The fringe spacing is therefore less than $\lambda/2\beta$, and so there are at least one bright and one dark fringe crossing the unresolved star image, so that the star is not nulled; its energy is just redistributed.

As a result, the two Fizeau images have to be combined at an angle considerably smaller than β, and such a combination once again requires a beam-splitter, as in section 10.4.1. There are then the two outputs and one of them can have destructive interference whereas the other will have constructive interference. In other words, the starlight goes to one exit and the planet light to the other, and all is well with conservation of energy. An advantage of this arrangement is that the starlight alone can be used for tracking the star and maintaining good alignment on it (figure 10.6).

10.4.3 Achromatic nulling in Bracewell interferometers

An aspect of nulling interferometry which has inspired a great amount of ingenuity is the achievement of an achromatic π phase shift. There are several approaches which can be employed. The first is to use the fact that the complex amplitude

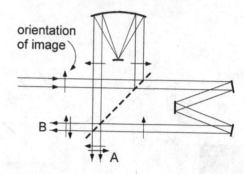

orientation
of image

B

A

Fig. 10.7. An interferometer in which a π phase shift at the A exit is achieved using the Gouy effect. When an image is projected through this interferometer, the two interfering images at the exits are mutually rotated by 180°; this effect is used in the achromatic interference coronagraph (section 10.5.4).

reflection coefficients r and $\bar{r}$ from opposite sides of an ideal nonabsorbing beam-splitter are related by $\bar{r} = -r^*$ (see, for example, Lipson et al. 1995). If the beam-splitter is symmetrical, so that in addition $\bar{r} = r$, then r is pure imaginary, i.e. the beam-splitter introduces $\pi/2$ phase shift into each reflection. Incorporated into a Michelson interferometer (section 3.1) the two exiting waves at the usual A exit, having amplitudes rt and $\bar{r}t$, have zero phase difference, and this is not useful for achromatic nulling. But if the beam-splitter is asymmetric, and r is real, the required condition $\bar{r} = -r$ is achieved. For example, using reflection from one face of a transparent plate with refractive index such that $r = 1/\sqrt(2)$ at 45° incidence for one polarization would give the required effect (but only for that polarization). Actually, certain multilayer nonpolarizing achromatic beam-splitter cubes have been found to have the right property, and a Michelson interferometer built with this component shows in white light a good black zero-order fringe at the A output.

A second approach uses the π phase shift arising when a wave goes through its focus, called the Gouy (1890) effect. In this case a symmetrical Michelson type of interferometer is built with a cat's-eye retro-reflector in one arm, in which the reflected wave goes through the focus once on the surface of the secondary mirror, and an ordinary mirror combination in the other arm. The phase difference achieved at zero path difference is therefore π; this is achromatic and polarization independent. This is the basis of the achromatic interference coronagraph (section 10.5.4) (figure 10.7).

A third type of approach uses out-of-plane interferometers. The description usually given is in terms of following the direction of the wave-field vectors as they are successively reflected from the various mirrors. This is most easily done by assuming boundary conditions $\mathbf{E} = 0$ on all metal reflecting surfaces (figure 10.8). However, it is in fact most instructive to consider the phase shifts in these interferometers as arising from the geometrical phase (Berry 1984) which is found by

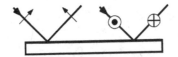

Fig. 10.8. Electric field vectors before and after reflection at a perfectly conducting mirror. Note that there is a change in sense of rotation if the incident wave is circularly or elliptically polarized.

following the routes of the waves on the sphere of wave-vector orientations, or Poincaré sphere (Chaio and Wu 1986; Lipson 1990), as shown in Appendix A. The geometric phase introduced into each wave is given by the solid angle subtended by its route at the center of the sphere. A simple symmetrical case is shown in figure 10.9. One then requires π solid angle difference between the two beams. In addition, each reflection at a mirror introduces a phase shift of approximately π[†]; but the difference from π is not achromatic, although usually small, being zero for a perfect conductor. It is therefore necessary to have an even number of reflections, arranged symmetrically in pairs so that the errors cancel, to which is added the geometrical phase, for a nonplanar interferometer to give an achromatic phase shift. Thus an interferometer of the type shown in figure 10.10, which also has the advantage of the Sagnac common-path interferometer's exceptional mechanical stability, provides the necessary phase difference (Tavrov et al. 2002). Essentially, this arises because one of the interfering beams traces a right-hand helical path, while the other, its mirror reflection in the beam-splitter, traces a left-handed helix. Each one provides $\pi/2$ phase difference, but in opposite senses. This type of interferometer can be used as part of an imaging nuller when an additional phase plate is included in order to add a phase shift which is zero at normal incidence, but has dependence on the angle of incidence such that there is constructive interference (which does not need to be strictly achromatic) at the planet radius.

Two further methods of achieving achromatic π phase shifts have been considered, but have rather low throughput efficiencies. One is based on using half-wave plates with orthogonal axes sandwiched between two quarter-wave plates (Pancharatnam phase). A second uses a sequence of diffraction gratings such that the total diffraction angle is zero, but a phase change can be introduced by mutual linear translations.

10.4.4 Starlight leakage in nulling interferometers

Another problem compromising the completeness of nulling is the effect of the finite diameter of the star, which means that the whole of its disk cannot be obscured

---

[†] It should be noted that this approach can easily be seen to result in the reversal of the sense of rotation at each reflection of a circularly or elliptically polarized wave.

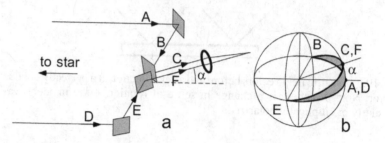

Fig. 10.9. (a) An out-of-plane Michelson stellar interferometer in which an arbitrary phase shift 2α is achieved using the geometrical phase shift. (b) The route traced on the sphere of propagation vectors for the two waves in (a).

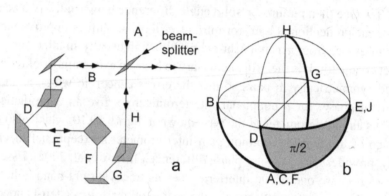

Fig. 10.10. Sagnac-type interferometer creating π phase difference at the output (Tavrov et al. 2002). The two routes through the interferometer introduce geometric phases $\pm\pi/2$, respectively.

completely by the null interference fringe. The starlight which is not nulled completely then competes with the planet image. We consider a one-dimensional model of this, remembering that limb darkening and the fact that the star is circular will improve the situation somewhat, but only by about 10%. The basic idea of nonimaging nulling interferometer is that a source at angle θ to the axis of the interferometer has its intensity modulated by $\sin^2(k_0\theta B/2)$, where B is the baseline. Then B is chosen so that, while the star of flux J_0 at $\theta = 0$ is nulled, the planet with flux J_p at angle β is at the first fringe maximum, i.e. $k_0\beta B/2 = \pi/2$. If the star has finite angular diameter α, we represent it by the top-hat function $I(\theta) = J_0/\alpha \, \text{rect}(\theta/2\alpha)$. The flux from the star after nulling by the sinusoidal fringe is thus

$$J_s = \frac{J_0}{\alpha} \int_{-\alpha/2}^{\alpha/2} \sin^2(k_0\theta B/2) \, d\theta \approx \frac{J_0(k_0\alpha B)^2}{48} \tag{10.2}$$

since $k_0 \alpha B \ll 1$. This is small, but not zero. The ratio between the planet flux and that of the star has thus been increased, under optimal conditions when the planet is exactly at $\beta = \pi/k_0 B$, by a factor $J_0/J_s = 48/(k_0 \alpha B)^2 = 48\beta^2/(\pi\alpha)^2$. Note that this result is independent of wavelength, but of course the optimum baseline B does depend on wavelength. We also learn from this that the baseline should be as short as possible[†], since it appears squared in the numerator of (10.2); this is one reason for the rather short baselines proposed for Darwin and TPF-I (section 12.2.2–3).

We can examine the hypothetical case of a solar system similar to ours, seen from afar (section 10.2.1). For Jupiter and the Sun, $\beta/\alpha = 560$ so that the contrast would be improved by $1.5 \cdot 10^6$. Compare this with the natural contrast of $2.2 \cdot 10^{-9}$ in the visible region, but maybe 10^{-6} in the thermal infrared, to see that only in the latter region would nulling with sinusoidal fringes help to make the planet visible. For an exo-Earth, the situation is even worse; $\beta/\alpha = 108$, and the natural contrast ratio is $3.2 \cdot 10^{-10}$.

Starlight leakage can be reduced by using non-sinusoidal fringes, which are a property of interference between more than two waves (Mieremet and Braat 2003). We shall illustrate this by an example. If two pairs of apertures having areas A_1 and A_2 are used, with baselines B_1 and B_2, the nulled fringes have amplitude proportional to $A_1 \sin(k_0\theta B_1) + A_2 \sin(k_0\theta B_2)$. At small angles, retaining powers of θ up to the third, this is proportional to $A_1[(k_0\theta B_1) - \frac{1}{6}(k_0\theta B_1)^3] + A_2[(k_0\theta B_2) - \frac{1}{6}(k_0\theta B_2)^3]$. The first-order term can be canceled by making $A_1 B_1 + A_2 B_2 = 0$, which implies that the phases of the two fringe sets are opposite. The residual amplitude is proportional to $A_1 B_1(B_1^2 - B_2^2)\theta^3$ and the zero-order fringe has θ^6 variation in intensity, which reduces starlight leakage very considerably. As before, the planet is best sought at the first maximum of the fringe pattern. If, for example, $B_2 = 3B_1$, the first fringe maximum is found at $\theta = \lambda/4B_1$ which is the same angle as would be obtained from the shorter baseline alone, but the addition of the second baseline has reduced the starlight leakage. We leave it as an exercise for the reader to show that this example would suffice to make both Jupiter and the Earth stand out with respect to the Sun at visible wavelengths. The fringe profiles for this example are shown in figure 10.11. Interferometer designs for Darwin, using similar principles, are described in section 12.2.2.

10.5 Imaging coronagraphy

Coronagraphy was invented as a method of observing the solar corona between solar eclipses. In a solar image formed by a perfect telescope, the geometric image of the bright photosphere is surrounded by a halo of diffracted light. This halo is a

† i.e. using the second or higher order fringes is counter-productive.

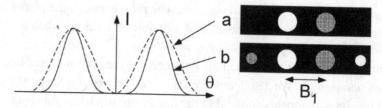

Fig. 10.11. Fringe profiles using (a) two small subapertures with equal areas A_1 and phases 0 and π separated by B_1; (b) four small subapertures at positions $(0, 1, 2, 3)B_1$ with phases respectively $(\pi, 0, \pi, 0)$ and areas $(\frac{1}{3}A_1, A_1, A_1, \frac{1}{3}A_1)$. The maxima have been normalized to unity. In the subapertures, white indicates phase 0 and gray indicates π.

convolution of the Airy rings in the point spread function with the photosphere's apparent disk. Because its intensity exceeds that of the faint corona, it seemed, until the 1930s, that the corona would never be observable except during a total solar eclipse.

10.5.1 The Lyot coronagraph in its original and stellar versions

Bernard Lyot proved otherwise by removing the diffraction rings. Using a rather small refractor with a superpolished lens, he imaged the Sun and masked the solar disk image with a circular field occultor. When looking back at the entrance lens from behind the field occultor, he noticed that the residual sunlight appeared to be mostly concentrated in a bright double ring appearing at the edge of the lens. He concluded that this light concentration produced the residual halo of solar light in the image, and thus arose his idea of removing the double ring by means of an image relay lens with a diaphragm, now called the "Lyot stop," installed in the pupil plane (figure 10.12). This stop removed much of the stray light in the image, making the solar corona visible. He also had to develop superpolishing methods to reduce the microbumpiness and resultant light scattered at the lens surface. Lyot's coronagraph started a revolution in solar physics; one no longer had to wait for rare total eclipses to briefly observe the solar corona. Lyot himself, observing at the Pic du Midi in the Pyrenees, was able to produce films showing the growth and evolution of the spectacular coronal "flames" or proturberances.

The Lyot coronagraph is also applicable to unresolved sources such as stars, in which case the occulting mask must hide the central peak and at least the first few rings of the Airy pattern. The outer Airy rings which are not masked become greatly attenuated in the output image when the double pupil-edge ring is removed by the Lyot stop. A complete removal of these rings is possible by also using apodization, as discussed in section 10.3. This generally leaves a fainter speckle pattern of residual scattered light, which results from disturbances on the wavefront, such as those

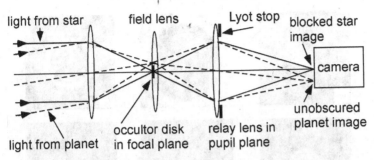

light from star · field lens · Lyot stop · blocked star image

camera

occultor disk in focal plane · relay lens in pupil plane · unobscured planet image

light from planet

Fig. 10.12. The Lyot coronagraph uses an opaque occultor disk in the focal image to mask the central Airy peak and a few rings in the diffraction pattern of the brighter source. A "Lyot stop" located in a pupil relayed by the field lens has an aperture slightly smaller than the geometric pupil. It masks the rings where light from the non-occulted Airy rings is mostly concentrated. In the image then relayed onto the camera C by the relay lens, the star's Airy pattern is strongly attenuated. The image of an off-axis planet is little affected.

produced by atmospheric turbulence or the bumpiness of the telescope's optical surfaces. This is illustrated in section 5.9. If the phase errors ϕ are much smaller than 2π, the complex amplitude distribution $\exp[i\phi(u, v)]$ on the wave approximates as $1 + i\phi(u, v)$ multiplied by the aperture, where the first term accounts for the Airy pattern in the focal plane. The second term on its own behaves like a real distribution, globally phase-shifted by $\pi/2$, thus providing in the image a centrosymmetrical speckle pattern added in quadrature to the Airy pattern (see figure 10.18a). When the Airy pattern is removed by a perfect coronagraph, the speckle pattern remains. A simulation showing the way in which an exoplanet image is enhanced by a Lyot coronagraph is shown in figure 10.13[†].

Only in recent years has Lyot coronagraphy been applied to stars, using large telescopes, since it is adversely affected by the atmospheric turbulence and requires efficient adaptive optics to operate with apertures larger than Fried's parameter r_0. It is also affected by the shadow pattern generated on the aperture by high-altitude turbulence (section 5.5.3). This is not corrected by a standard adaptive optical system, which only senses phase errors (section 5.8.1). For these reasons, space provides a more attractive environment for coronagraphic telescopes. The Hubble Space Telescope's coronagraphic camera built by the European Space Agency was prevented from operating by the well-known fabrication error which caused spherical aberration and required the in-orbit installation of COSTAR, a

[†] It will be noticed that the ring of light in figure 10.13(c), to be obscured by the Lyot stop, appears single, and not double as observed by Lyot. This is an artifact of digitization in the simulation; as will be seen in section 10.5.5, the amplitude switches abruptly from positive through zero to negative at the radius of the pupil aperture, but the simulation shows the same value of the intensity. By convolving the amplitude with a small smearing matrix (slight defocus), the double nature of the ring can be brought out (figure 10.14).

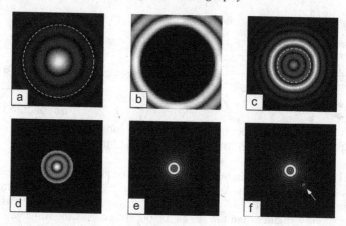

Fig. 10.13. Simulation of imaging a star and planet (intensity ratio 10^{-5}) by a Lyot coronagraph. (a) shows a magnified picture of the central "Airy disk" of the telescope image, and the dotted circle represents the edge of the occultor disk. (b) shows the same image after occulting, with contrast enhancement by 500 with respect to (a), so that the outer diffraction rings now become visible. (c) shows the re-imaged pupil, with the Lyot stop (dotted circle) and (d) the masked aperture pattern, with contrast enhancement 10 with respect to (c). The final coronagraph images (e) and (f), on a scale eight times smaller than (a) and (b), show the star respectively without and with the planet at the position indicated by the arrow. The intensity ratio between the planet and the star image in (f) is now about 0.2, an enhancement of 2×10^4.

Fig. 10.14. Detail of the ring in figure 10.13(c), showing its double structure.

three-mirror corrector located before the focal plane. The COSTAR made the focal ratio somewhat slower, thus reducing the pupil size in the Lyot stop and destroying its cleaning effect. More recently, coronagraphic cameras have been proposed for the HST's successor, the James Webb Space Telescope, in the infrared and for the TPF-C for the visible (section 12.2.3). On Earth, a coronagraphic camera is being considered by the European Southern Observatory for one of its VLT 8-m telescopes which is equipped with high-performance adaptive optics.

The prospect of detecting extrasolar planets has stimulated much effort in recent years toward inventing new variants of the older apodization and coronagraphy schemes. Some of them are discussed below.

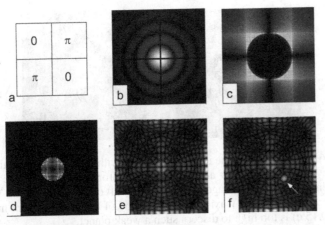

Fig. 10.15. Simulation of the four-quadrant phase mask coronagraph. The star and planet have intensity ratio 10^{-6}. The phase changes due to the mask are shown in (a). (b) shows the telescope image with the mask superimposed (magnified eight times with respect to (e) and (f)). This diffracts most starlight outside the relayed geometric pupil, shown in (c). (d) shows the field transmitted by the Lyot stop. (e) and (f) show, respectively, the final images without and with a planet along the diagonal.

10.5.2 The Roddier–Roddier phase-dot coronagraph

This variant of the Lyot coronagraph has the occulting mask replaced by a transparent phase-shifting "dot." The dot, a thin transparent disk appreciably smaller than the Airy peak, has its thickness and size adjusted to produce a π phase shift over half of the image. Destructive interference then results in zero intensity at the center of the geometric beam, the energy from which is diffracted outside the pupil[†]. A Lyot stop in the pupil therefore removes most of the stellar residual light, without much affecting the light from a planet if its focal image impacts the focal plane outside the phase-shifting dot.

Unlike the basic Lyot coronagraph, this variant is critically affected by the apparent size of the star. It must be very nearly unresolved. Also, achromatization of the dot size and thickness, both of which are proportional to wavelength, is needed to accommodate a broad spectral band.

10.5.3 Four-quadrant phase-mask and phase-spiral coronagraphs

Another variant of the Lyot coronagraph (Abe et al. 2003) uses a π-shifting phase mask shaped as four quadrants (figure 10.15 a). In the relayed pupil plane, most

[†] Phase-shifting dots, of submicron size, are also found in billions on compact discs and DVDs. Inscribed on a disc as submicron dips, their nulling effect serves to modulate the reflected intensity of the laser beam focused onto them.

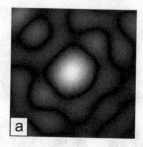

Fig. 10.16. Comparison of the images of planets with intensity 10^{-6} of their stars, as seen by the four-quadrant and phase-spiral coronagraphs when the planet is along a diagonal. The star image is off to the top left of the field of view. When the planet is close to the x or y axis, the background of the four-quadrant mask field (figure 10.15 e) is too high to discern such a weak planet.

residual stellar light is again rejected from the geometric beam toward the four corners of the pupil. Simple analysis shows that the star image, again relayed through a Lyot stop, is fully nulled if the mask and field lens are of infinite size. Like the phase-dot coronagraph, this variant is affected by the star's angular size. Achromatization is also needed for broad-band operation, but unlike the phase dot, only the thickness needs to be achromatized in the case of four quadrants. The stellar light is most effectively rejected along the diagonals of the field, since if the planet is close to an x or y axis, its light is also partially canceled by the phase mask. A variant on this idea which overcomes this disadvantage is to use a mask in the form of the second-order phase spiral, $\exp(2i\theta)$, where θ is the angle in the (x, y) plane (Foo *et al.* 2005). The rejection in this case is isotropic. A comparison of planet images for the quadrant and phase-spiral is shown in figure 10.16.

10.5.4 *The achromatic interference coronagraph*

Proposed by Gay and Rabbia (1996), the achromatic interference coronagraph is really a form of nulling interferometer, since it uses a beam splitter to split the wavefront and then recombine it (figure 10.7), flipped and phase-shifted by π. This nulls much of the star's wavefront, incident on-axis, but splits the planet's wavefront into two oppositely tilted waves. The corresponding planet images thus become spatially separated in the image, appearing symmetrically on each side of the nulled star. Whereas other coronagraphic schemes (except Lyot's), work in a narrow spectral range, the achromatic interference coronagraph uses the achromatic Gouy shift (section 10.4.3).

10.5.5 *Elementary modeling of mask coronagraphs*

The various mask coronagraphs mentioned above can be understood in simple terms if described by using the image occultor to achieve a subtraction of complex

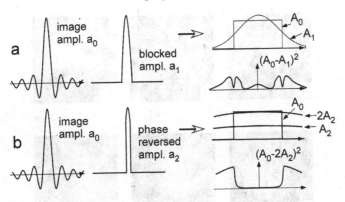

Fig. 10.17. Illustrating schematically the effect of coronagraphic field masks described as a superposition: (a) Lyot mask; (b) phase-dot mask.

amplitudes, rather than a multiplication (Martinache et al. 2004; Bonneau et al. 1978). For instance, the complex amplitude transmitted by a Lyot occultor mask and propagating toward the image plane may be represented as the unmasked distribution minus the distribution occulted by the mask. In the re-imaged pupil plane, the former is just the pupil aperture disk while the latter, being the Fourier transform of the central Airy peak and only a few of the outer rings, is necessarily a disk of the same size but with smeared edges. When the subtraction is performed, most of the light is concentrated around the edge of the disk. Figure 10.17(a) shows the process schematically when the Airy disk alone is obscured, in which case the transform of the latter function approximates a Gaussian of width about equal that of the aperture. In the pupil plane, the difference between the complex amplitudes is therefore restricted to the region around the edge of the disk.

With a phase disk mask, the idea is that the mask is smaller and it transmits integrated amplitude equal to exactly half the total in the image. Since the mask is small, it can be approximated by a δ-function of this strength. In the pupil plane, the transform is a constant with half the amplitude of the incident wave. This is subtracted, and then returned with opposite phase, which subtracts it again (figure 10.17 b). Thus the result is to cancel the wave amplitude over the pupil, but to create a light intensity outside the pupil which can be completely blocked by the Lyot stop. Clearly, the performance is critically dependent on the mask transmitting *exactly* half the integrated amplitude, and the phase shift being *exactly* π, neither of which is easy.

10.5.6 Mirror bumpiness tolerance calculated with Maréchal's equation

Some of the coronagraphic schemes mentioned above can reach theoretical perfection, preserving the planet's image while completely suppressing the light from

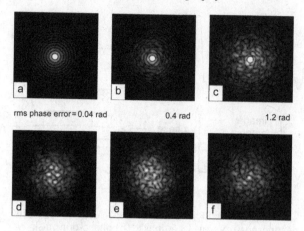

Fig. 10.18. (a–c) Simulated diffraction patterns of a circular aperture with wavefront degraded by different levels of wavefront bumpiness; (d–f) the same at the exit of an ideal coronagraph. The central peaks in (a–c) are overexposed in order to empasize the surrounding speckle patterns. The coronagraph removes the theoretical Airy pattern of the aperture, and retains the contribution from the wave bumpiness. If the bumpiness is weak (d), this contribution is a centrosymmetrical speckle pattern, but not if the bumpiness is strong (f). There is no centro-symmetry in the speckles of (a–c) since the antisymmetric speckle phase interferes with the symmetric ring phase.

the star if it is unresolved and perfectly guided. This unfortunately assumes perfect optical surfaces; it appears that for adequate suppression, the bumpiness tolerances for the main mirror of a coronagraphic telescope are much tighter than those for imaging telescopes. A smooth optical wavefront reflected from a bumpy mirror indeed becomes similarly bumpy, with twice the mirror bump amplitude. If a lens of refractive index n is used, as did Lyot, instead of a mirror, then the wavefront bumpiness per surface is only $(n-1)$ times the surface bumpiness amplitude; with two surfaces, this is still about a two-fold improvement with respect to a mirror. In the focal plane, the resulting effect is a degradation of the diffraction pattern, with the Airy rings of increasing order becoming progressively broken into speckles. The diameter of the annular transition zone between the rings and speckles shrinks for increasing levels of bumpiness (figure 10.18). With bumpiness amplitudes much smaller than a quarter wave, the chromatic dependence of the focal pattern is proportional to wavelength. At higher levels, particularly beyond the Rayleigh tolerance of $\lambda/4$, where the central peak becomes markedly attenuated and eventually vanishes, the speckle pattern becomes spectrally decorrelated; but such large amplitudes are completely unsuited to coronagraphy.

A useful expression for the amount of diffracted light caused by wavefront bumpiness was given by Maréchal. This diffracted light materializes as the speckled background observed in the image plane of an ideal coronagraph, which is a coronagraph where the "organized" diffraction rings are entirely removed together with the star's Airy peak. The peak-to-halo ratio G is analogous to a signal-to-noise ratio and is defined as the ratio of the intensity in the unmasked star peak to the average residue level, and can be related to the peak and halo visible in figure 5.12(a).

We consider the case of light of unit amplitude reflected by a mirror of diameter D from which it is also scattered by N bumps with average diameter d and random height h at random positions within the aperture diameter D, where $d \ll D$ and $h \ll \lambda$. A bump scatters light with phase $\phi = 2\pi h/\lambda$. The reflected wave is described by an amplitude function

$$f(\mathbf{r}) = \exp[i\phi(\mathbf{r})] \approx 1 + i\phi(\mathbf{r}) \tag{10.3}$$

where $\phi = k_0 h$. The spatial frequency spectrum of the scattered light is given by the Fourier transform of (10.3), which is

$$F(\mathbf{u}) = \delta(\mathbf{u}) + i\Phi(\mathbf{u}). \tag{10.4}$$

Parseval's theorem (Appendix A) states that the integrated power of the function and its Fourier transform are equal, i.e. $\int |f(\mathbf{r})|^2 d^2\mathbf{r} = \int |F(\mathbf{u})|^2 d^2\mathbf{u}$, the integrals being over the complete $\mathbf{r}$ and $\mathbf{u}$ planes, respectively. Thus, integrating (10.4) over the whole $\mathbf{u}$ plane,

$$\frac{\pi}{4}D^2 = \frac{\pi}{4}D_0^2 + \int |\Phi(u)|^2 d^2\mathbf{u} \tag{10.5}$$

in which D_0^2 is the zero (unscattered) order, which is the integrated power of the δ-function in (10.4). Since the scattering is weak, $D_0^2 \approx D^2$. The ratio between unscattered and scattered light G is thus given by

$$G = \frac{\pi D^2}{4\int |\Phi(u)|^2 d^2\mathbf{u}} = \frac{D^2}{Nd^2 k_0^2 \overline{h^2}} \tag{10.6}$$

in which $\overline{h^2}$ is the mean square bump height.

Maréchal's equation also relates to the Strehl ratio, which is just $1 - G^{-1}$, since the light energy missing in the Airy peak, at the first focus before the coronagraph, is distributed among the surrounding speckles. The halo region extends to angles of order λ/d, while the resolution angle is about λ/D, so that the number of resels affected by the halo is $(D/d)^2$. For planet detection, the residual speckle per resel after the coronagraph must be of order 10^{-9} of the star intensity, implying $G \approx 10^{-9} D^2/d^2$. If the whole surface is covered by bumps, so that $Nd^2 \approx D^2$, then $k_0^2 \overline{h^2} \approx 10^{-9} N$. The expression can also be applied to an adaptive correction system required to annul the scattered light to the required level. For example, if the surface is modulated by $N = 10^4$ piston actuators, then $k_0^2 \overline{h^2} \approx 10^{-5}$, implying an accuracy of control of the actuators of about $5 \cdot 10^{-4}\lambda$, or about 0.5nm.

10.6 Further cleaning, coherent and incoherent, for high-contrast coronagraphy and apodization

Some of the apodization and coronagraphic schemes described above in section 10.3 and section 10.5.1 can approach perfect performance in terms of removing the star's diffraction rings and other organized forms of diffracted light. But we have seen in section 10.5.6 that the performance of an ideal system at visible wavelengths will be degraded by the bumpiness of the main mirror even if this is of the order of ångstroms. The bumpiness adds residual stellar light in the image, in the form of a speckled halo which disturbs the Airy rings at the coronagraph entrance, and then reappears at the coronagraph exit. Several methods have been proposed to attenuate this halo. These fall into two classes depending whether they operate predetection, using interference nulling, or post-detection by subtracting a reference image. We will first concentrate on pre-detection cleaning methods, also called *coherent cleaning*, which can in principle remove the photon noise if the star is a point source, and then discuss post-detection, or incoherent, cleaning, which is limited by Poisson photon noise.

The basic image cleaning achieved by any type of coronagraph can also be considered as coherent cleaning, when applied to those organized or expected stray-light patterns, such as Airy rings. Here we are more concerned with the additional stray light generated by mirror bumpiness or a shadow pattern which is to be annulled by a second layer of coherent cleaning using adaptive optics.

10.6.1 Adaptive coherent correction of mirror bumpiness

To enhance the planet detection performance of HST's initial coronagraph, KenKnight (1977) considered adaptive optics to further remove the residual stellar speckles from the coronagraphic image. With a perfect coronagraph, which has completely removed the diffraction rings of the Airy pattern, the actuators only have to correct the bumpiness of the incoming wave. But in fact, actuators can serve more generally to clean, at least partially, the residual image speckles, whether they be generated by imperfect coronagraphy or by a shadow pattern combined with bumpiness or dust on the mirror. A "dark hole" in a region of the observed field of a coronagraphic image can be created by adding suitable weak corrections to the actuators located in the aperture which shape the incoming wavefront. This was demonstrated by Green et al. (2003) with a laboratory experiment. A typical algorithm (Malbet et al. 1995; Boccaletti et al. 1998) finds the most intense speckle of residual starlight in the field, then nulls it by adding to the wavefront in the aperture a phase function calculated to create a properly located "antispeckle." Nulling a speckle is a matter of using this added spot destructively. Its phase is adjusted by trial and error for optimal destruction. The algorithm can be iterated to deal with the brightest speckle then remaining at each stage. If the wavefront has a shadow pattern, in addition to bumpiness, optimizing the coronagraphic image within part of the field by creating a "dark hole" may however result in some bumpiness remaining in the corrected wavefront. Using this kind of method, Green et al. (2003) demonstrated in their laboratory experiments a very encouraging attenuation of the residual halo by a factor approaching 1000. They only had to process the image detected by the science camera, and calculate corrections which were then applied to a piezoceramic deformable mirror. With 32×32 actuators, finally adjusted with 20 nm accuracy, they reached the 10^{-9} level of residual starlight needed to detect exo-Jupiters in the visible range. The iterative correction cycles however take a long time to converge, typically 10 hours, which is somewhat demanding for the stability of the instrument. The other methods of coherent cleaning discussed in the following sections require no such iterations and respond much faster.

10.6.2 Adaptive hologram within the coronagraph

Another proposed method of coherent cleaning uses a hologram inside the coronagraph to similarly fabricate a copy of the stellar residue and subtract it coherently. As sketched in figure 10.19, the hologram is the recorded interference pattern of the stellar residual light with a flat reference wave. In a second step, the recorded hologram reconstructs a copy of the recorded stellar residue. According to the classical principle of holographic interferometry, the copy interferes with the true residue, destructively if the holographic material has a negative response. The planet's light does not contribute to forming the hologram since it does not reach the reference beam, and is incoherent with the star's light. The planet image transmitted through the hologram by zero-order diffraction is therefore unaffected by the destructive interference. Its contrast is therefore improved with respect to the stellar residue.

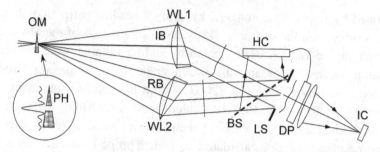

Fig. 10.19. Lyot coronagraph containing hologram-like adaptive optics for nulling the residual star light. The focal occultor mask OM is a small optical wedge (inset) with pinhole PH, which deviates the main stellar light out of the imaging beam IB containing the planet's light, to provide a cleaned reference beam RB. Both beams, collimated and deviated by wedge lenses WL1 and WL2, intersect in the pupil plane, within the aperture of a Lyot stop LS. Their interference produces a hologram, recorded directly on a photosensitive plate or indirectly by a camera HC. The camera is fed by the beam-splitter BS and displays the recorded image as a phase pattern on the deformable plate DP which then behaves as a phase hologram. When it is transmitted through the hologram, the stellar wavefront in beam IB subtracts coherently from the copy of it reconstructed by the hologram as the first-order diffraction of the reference beam RB. The cleaned image of circumstellar features, including planets, is recorded by camera IC.

The holographic recording plate can be replaced by a camera, fed by a beam-splitter as sketched in the figure, which captures the fringed pupil pattern and displays it onto a deformable plate or mirror, also located in the pupil. This dynamic form of the hologram would be adapted continuously to a variable wave-front bumpiness. Also, the diffractive efficiency and fringe phase can be controlled electronically to finely tune the holographic nulling. Modern cameras can be sensitive enough to exploit the fringed residue of starlight with nearly photon-limited performance.

Codona and Angel (2004) have analyzed a similar system to the above, and show simulations for a star–planet system imaged with the Hubble Space Telescope. They suppress the speckled halo (including diffraction rings and streaks from edge scattering) by the use of adaptive phase modulation in a pupil plane. The wavefront modulator feedback loop is controlled by iterative phase retrieval (Fienup 1982) based on the interference pattern or hologram between the halo image and the spatially filtered starlight as reference wave. Since the planet light is incoherent with that from the star, its image would be unaffected by the cleaning. They point out that phase modulation alone can suppress the halo from the star completely only over half the field of view. Amplitude modulation would be necessary as well in order to deal with the whole field, resulting in serious attenuation of the light.

This can be seen from figure 10.18, where for small-amplitude phase modulations the speckle halo is seen to have centrosymmetric intensity, and can therefore match the halo to be suppressed over at most a 180° segment of the field of view.

10.6.3 Incoherent cleaning of recorded images

Post-detection, also called incoherent, image processing is also of interest to further clean the signal. Several methods have been proposed, but only two will be discussed. The first method, which was to be carried out with the coronagraph of the Hubble Space Telescope, requires it to be rotated around the star axis. The speckle pattern, which results from the properties of the instrument itself, rotates together with the telescope and attached coronagraph, but the planet image does not. Thus the difference between two images in the frame of reference of the coronagraph will cancel the speckle pattern, while leaving positive and negative planet images at two points in the field, separated by the rotation angle. The method can equally well be applied to an apodized telescope. However, because image subtraction is not coherent, the result is limited by the Poisson noise (the square root of the photon count recorded in each exposure) before the subtraction. Repeated sequences further improve the sensitivity if the pattern drifts. It is important that the rotation be exactly around the star axis, but the method is tolerant of partial resolution of the star.

"Dark speckle" analysis (Labeyrie 1995) involves a more elaborate statistical treatment of multiple exposures which exploits the drifting or "boiling" of the speckles due to guiding errors or servo noise in the adaptive optics. Many exposures are made, each shorter than the speckle lifetime, milliseconds on Earth and minutes or hours in space. Since the planet peak adds its intensity to the residual stellar speckles, one looks for resels where the speckles never become fully dark.

10.6.4 Comparison of coherent and incoherent cleaning

Coherent and incoherent cleaning methods have complementary qualities. Coherent cleaning has the theoretical advantage of evading the photon statistical limitation of incoherent subtraction, for which the residual photon count per speckle amounts on the average to $\sqrt{N_p}$, where N_p is the number of photons detected in the same speckle during the double exposure. With coherent cleaning, wave amplitudes are directly subtracted by interference, and the photon fluctuations arise only in the subtracted residue, which can be very low if the amplitudes and phases are finely tuned for a highly destructive interference. Bracewell's basic form of nulling (section 10.4) can in principle give a complete null if the tuning is achieved exactly, and the same is true with a hologram inside a coronagraph. In both cases, however, the accurate tuning itself requires many photons to measure the amplitude and phase with a corresponding level of accuracy. With a hologram in the coronagraph, using a 50/50 beam-splitter for exposing a hologram leaves, on the average, one photon per speckle as a final residue. There is little need to further null the residue, since a planet's image must have several photons for a clean detection, but in principle it could be achieved by pointing a brighter reference star or a laser star, the latter being particularly convenient to record a dynamic hologram within the corona-graph. Adaptive cleaning within a coronagraph thus evades the $\sqrt{N_p}$ limitation of incoherent cleaning.

If the star is partially resolved by the aperture, the result is a slight blurring of the stellar speckles at the entrance of the coronagraph, which produces a more complicated distortion of the residue at the coronagraph exit. Guiding jitter during exposures, even amounting to a fraction of the Airy radius, introduces a comparable distortion. Pure apodization is highly tolerant of resolved stars and poor guiding; some coronagraph types are also tolerant, as are some of the coherent cleaning methods. Incoherent cleaning is not affected by the star's partial resolution. This comparison therefore suggests that it is of interest to use coherent cleaning first and then, if needed, to apply incoherent cleaning to the recorded image.

References

Abe, L., A. Domiciano de Souza Jr., F. Vakili and J. Gay, (2003). *Astron. Astrophys.*, **400**, 385.

Berry, M. V. (1984). *Proc. Roy. Soc.* London A, **392**, 45.

Boccaletti, A., C. Moutou, A. Labeyrie *et al.* (1998). *Astron. Astrophys. Suppl. Ser.*, **133**, 395.

Bonneau, D., M. Josse and A. Labeyrie, (1978). *Image Processing Techniques in Astronomy*, Proceedings of a Conference, held in Utrecht, March 25–27, 1975, ed. C. de Jager and H. Nieuwenhuijzen. Astrophysics and Space Science Library, Vol. 54, 403.

Bracewell, P. N. (1978). *Nature*, **139**, 274, 780.

Chaio, R. Y. and Y.-S. Wu, (1986). *Phys. Rev. Lett.*, **57**, 933.

Chauvin, G., A.-M. Lagrange, C. Dumas *et al.* (2004). *Astron. Astrophys*, **425**, L29-L32.

Codona, J. L., and J. R. P. Angel (2004). *Astrophys. J.*, **604**, L117.

Fienup, J. R. (1982). *Appl. Opt.*, 21, 2758.

Foo, G., D. M. Palacious and G. A. Swartzlander, Jr. (2005). *Opt. Lett.* **30**, 3308.

Galicher, R., O. Guyon, M. Otsubo *et al.* (2005). *Proc. Astron. Soc. Pac.*, **117**, 411.

Gay, J. and Y. Rabbia, (1996). *C. R. Acad. Sci.* Paris, **322**, séries IIb, **265**.

Gouy L. G., (1890). *C. R. Acad. Sci.* Paris, **110**, 1215 (1890); see M. Born and E. Wolf, *Principles of Optics*, 6th ed. Pergamon, Oxford (1980) p. 448.

Green, J. J., S. A. Basinger, A. Scott *et al.* (2003). *Techniques and Instrumentation for Detection of Exoplanets* ed. Daniel R. Coulter, Proc. SPIE, **5170**, 38.

Guyon, O. (2003). *Astron. Astrophys*, **404**, 379–387.

Hinz, P. M., W. F. Hoffman and J. L. Hora (2001). *Astrophys. J.*, **561**, L131.

Kasdin, N. J., R. J. Vanderbei, D. M. Spergel (2003). *Astrophys. J.*, **582** 1147.

KenKnight, C. (1977). *Icarus*, **30**, 422.

Labeyrie, A. (1995). *Astron. Astrophys.*, **298**, 544.

Lipson, S. G. (1990). *Opt. Lett.* **15**, 154.

Lipson, S. G., H. Lipson and D. S. Tannhauser, (1995). *Optical Physics* 3rd ed,
Cambridge: Cambridge University Press.

Malbet, F., J. Yu and M. Shao, (1995). *Proc. Astron. Soc. Pac.* **107**, 386.

Martinache, F. (2004). *Astronomy with High Contrast Imaging II*, C. Aime and R.
Soummer (eds) EAS Publications Series, **12**, 311–316.

Mayor, M. and D. Queloz (1995). *Nature*, **378**, 355.

Mieremet, A. L. and J. J. M. Braat (2003). *Appl. Opt.* **42**, 1867.

Neuhäuser, R., E. W. Guenther, G. Wuchterl *et al.* (2005). *Astron. Astrophys*, **435**, L13–16.

Nisenson, P. and C. Papaliolios (2001). *Astrophys. J.*, **548**, L201.

Pont, F., C. H. F. Melo, F. Bouchy, S. Udry, D. Queloz, M. Mayor and N. C. Santos
(2005). *Astron. Astrophys.*, **433**, L21-L24.

Slepian, D. (1965). *J. Opt. Soc. Am.*, **55**, 1110.

Tavrov, A., R. Bohr, M. Totzeck *et al.* (2002). *Opt. Lett.*, **27**, 2070.

Vanderbei, R. J., D. N. Spergel and N. J. Kasdin (2003). *Astrophys. J.*, **590**, 593.

11

A sampling of interferometric science

11.1 Interferometric science

Interferometry has already provided many scientific results that could be obtained in no other way. In this chapter, we briefly discuss some of the more prominent results to date. This is not in any way meant to provide a comprehensive review, but is rather only a sampling of the kinds of science that has been done. As interferometry becomes more mature and the new interferometric arrays become operational, the scientific output is expected to increase dramatically, and one hopes that this review will rapidly become out of date!

11.2 Stellar measurements and imaging

11.2.1 Stellar diameters and limb darkening

Fundamental understanding of stellar structure and stellar atmospheres depends amongst other things on the accurate determination of stellar diameters. Combined with the measured total flux from the star, the emergent flux density and effective temperature can be directly determined. Consistent models for stellar atmospheres have been created by combining this information with spectroscopic and spectrophotometric measurements. Since the angular diameters of even the supergiant stars are only tens of milliarcseconds and those of the closest main-sequence stars are a few milliarcseconds or less, long-baseline interferometry is the only way of measuring stellar diameters over the full range of spectral types. Measurement of a star's diameter is complicated by the effects of limb darkening. Limb darkening (section 3.2.2) is the apparent change of intensity from the center of the star to its limb. This is due to the star's being a sphere, not a flat disk. Because of the different angles of incidence, the observer sees different layers of a partially absorbing (gray) stellar atmosphere when looking at the star's center or limb. Near the limb, the higher atmospheric layers that one observes are cooler and so not as bright as

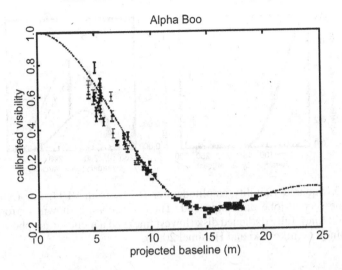

Fig. 11.1. Visibility amplitude for α-Bootis at 905 nm measured at COAST. Notice the negative values determined by phase closure. The fit is to a limb-darkening model (Haniff et al. 2003).

the center. Limb darkening must be accurately modeled if the true stellar diameter is to be measured and, in turn, the emergent flux estimated. Now, a stellar diameter is determined by fitting a model of the stellar disk to the measured visibilities, but when the diameter is unknown, the shape of the central lobe of the visibility function is very insensitive to limb darkening and differs from the visibility for a uniform disk of arbitrary diameter by only about 0.1%, far less than the typical errors in interferometric measurements. Therefore, fitting a uniformly bright disk to visibilities from a limb darkened star can result in as much as 15% error in the estimated diameter. There are two approaches to solving this problem. The first technique is to use models of limb darkening that are adjusted for each star based on high-resolution spectroscopy and spectrophotometry. Tango and Davis (2002) provide a detailed discussion of such corrections. The second approach is to measure the second lobe of the visibility function. Limb darkening has a much greater effect on this lobe (see figure 3.19). However, measuring the second lobe with any precision is difficult since the baselines must be long and the visibilities are small, necessarily less than the value for a uniform disk which has greatest value 0.13. Figure 11.1 shows measurements at COAST of the visibility extending into the second lobe, and Figure 11.2 shows a comparison of second lobe data with limb-darkening models made at VLTI.

The earliest measurements of stellar diameters were those of Michelson and Pease (1921) and Pease (1931), who reported the diameters of eight stars measured using the 20-ft beam interferometer. Their measurements were quite difficult since

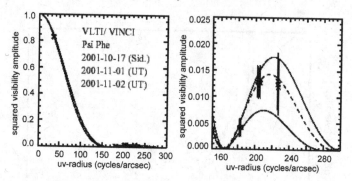

Fig. 11.2. Squared visibility amplitudes for ψ-Phe from VLTI. The right panel shows the second-lobe data expanded. Three models are shown: uniform disk (upper full line); fully darkened disk (lowest line) and a model atmosphere (center dashed line); (Wittkowski and Hummel 2003).

they were based on a visual estimate of when the fringes disappear as a function of the separation of the interferometer's two subaperture mirrors. In fact, none of Michelson's publications shows any visibility data, despite the fact that he explains how to measure visibility semiquantitatively, and his results are based only on this visual estimate.

The 20-ft beam was designed to be long enough to test Eddington's theoretical prediction of 51 mas for the diameter of the red giant Betelgeuse (α-Ori). The measured value was 47 mas.

Later on, Hanbury Brown et al. (1974) measured the diameters of 32 stars with the intensity interferometry. In spite of the much lower sensitivity of this indirect interferometry method, by using large collectors and relatively long baselines they were able to achieve the first accurate diameter measurements of stars over a wide range of spectral types. Hutter et al. (1989) measured diameters of 24 giant and supergiant stars using the Mark III astrometric interferometer and Dyke et al. (1996) reported another 37 stars measured at CERGA (GI2T). Nordgren et al. (2001) compared the stellar angular diameter measurements made with the Mark III to new measurements made with NPOI for 22 stars. They also compared the measurements to estimates of diameter using the infrared flux method and found in most cases that there was very good agreement between this analytical estimate and the measurements. Wittkowski et al. (2001) made direct measurements of limb-darkened intensity profiles of three late-type giant stars using NPOI. They were able to show that the estimates of limb darkening derived from model atmospheres were consistent with the direct measurements. Aufdenberg et al. (2003) tested their stellar atmosphere models by simultaneously matching the spectral energy distribution for two stars with diameter measurements of those stars made with the NPOI. They showed that their detailed models predicted quite accurately the measured uniform disk stellar diameters. The achievement of 0.5% accuracy in stellar diameter

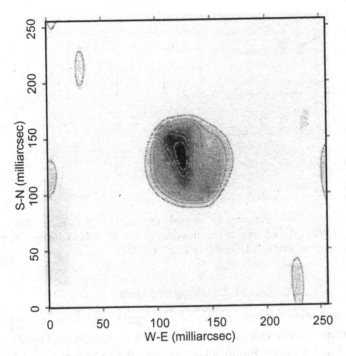

Fig. 11.3. Image reconstruction of Betelgeuse in early 2004 in the TiO band at 782 nm, showing a hot spot (Haniff et al. 2004).

measurements using limb-darkening models for τ-Cet at VLTI (Pijpers et al. 2003) suggests that astroseismic behavior might be studied using interferometry.

11.2.2 Star-spots, hot spots

Supergiant stars have radii that are hundreds of times larger than the Sun's radius. Supergiants are less common than main-sequence stars so there are only a few that are fairly close to us. Because these stars are so large, interferometry can be used to produce images of their photospheres with at least a few pixels across the disk. Larger telescope arrays should soon provide higher resolution images, but the results to date are sufficient to show structure on the stellar surface. Several observers have imaged surface structure on the supergiant star Betelgeuse (α-Orionis). The COAST interferometer group (Young et al. 2000) imaged Betelgeuse in multiple wavelengths in the red and near-infrared. The images show strong evidence for bright spots on the surface that emit 10–20% of the flux. To obtain acceptable images of extended objects, good coverage of the (u, v) plane is required, which is obtained by combining data from multiple apertures with Earth rotation. Figure 11.3 shows an image taken in the TiO absorption band at 782 nm. In general, the hot spots have shown up best at the shorter wavelengths; their scientific explanation is still under discussion.

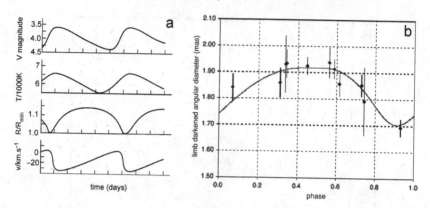

Fig. 11.4. (a) Periodic variation of several parameters of δ-Cephei, after Carroll and Ostlie (1996). (b) Change of angular diameter of a Cepheid during one period of pulsation, measured at VLTI (Kervella et al. 2004).

11.2.3 Pulsating stars

Cepheids are pulsating supergiant stars whose regular variations in brightness correspond to expansion and compression of the stellar atmosphere. Observed periods are between 1 and 50 days. Figure 11.4(a) shows the light curve for δ-Cephei, the prototype for this type of star. The radii of Cepheids can be calculated by observing spectroscopically the Doppler effect as the surface of the pulsating star moves toward and away from the observer. Integrating the radial velocity curve during a half-period gives the total displacement, which is the variation of stellar radius, and additional photometric data give the size, according to the Baade–Wesselink method that relates the pulsation period of the star to its diameter. The mean linear radius and the mean luminosity are then used to estimate the distance to the Cepheid. If the calibration of luminosity is accurate, then Cepheids found in distant galaxies can be used as calibrators of the distance to those galaxies, and these distances are important steps in the extragalactic distance scale. The accuracy of distance measurements with Cepheids is dependent on the reliability of the Baade–Wesselink method. But this, in turn, depends on the accuracy of the relationship between luminosity, radius and temperature of the star. Uncertainties in these parameters lead to what is called the "zero-point" error, since the luminosity of any particular Cepheid depends on such parameters as the metallicity of the star. The model-dependent errors can be reduced by measurements of the angular diameter change of nearby Cepheids using long-baseline optical interferometry (Sasselov and Karovska 1994), since measuring the angular diameter variation and combining it with the linear radius variation, as measured absolutely by Doppler shifts, allows a direct estimate of the distance to the star. Observations of Cepheids by interferometry require long baselines and very accurate visibility measurements, since galactic Cepheids are rare and distant, and the variations in their radii are submilliarcsecond in scale. Mourard et al. (1997) measured the diameter of δ-Cep, the prototype Cepheid and Armstrong et al. (2001) measured diameters of four additional Cepheids, all with accuracies of better than 1%. Measurements of the variations in diameter throughout the pulsation period have been made at VLTI on seven Cepheids (Kervella et al. 2004) (figure 11.4b). Interferometric data has been used by Nordgren et al. (2002) to calculate the absolute distance and size of δ-Cep, with results comparing well with parallax measurements.

RR Lyrae stars are also giant stars with regular pulsations with periods in the range 1.5–24 hours, similar to Cepheids. RR Lyraes have simpler light curves and are more common than Cepheids. They are found in globular clusters and so allow calibration of distances to those clusters. However, the precision of the period – luminosity relationship is less accurate than for Cepheids; so measurement of their angular diameters, combined with radial velocity measurements, is important if precise distances are to be found. RR Lyraes have smaller radii than Cepheids, so their measurement requires even longer baselines than the Cepheids. Limb darkening of pulsating stars adds further complication to using pulsating stars as distance indicators. Corrections for limb darkening must be of the same accuracy as the measurements and this requires accurate modeling.

11.2.4 Miras

Miras are M Giant stars in the latter stages of their evolution that pulsate with the largest amplitude of any pulsating star. The pulsations are due to expansion and collapse of the star's atmosphere. Pulsation periods are in the region of hundreds of days, and the star's brightness can change by many stellar magnitudes during a period. Interferometric observations have shown that the apparent diameter of Mira-type variables changes dramatically with the wavelength of observation, as observed by speckle interferometry (Bonneau and Labeyrie 1973), and the stars are asymmetric in shape (Karovska et al. 1991). Because these stars are cool and their atmospheres are very extended, they can be easily studied with infrared interferometry. Measurements of angular size and its variation with the phase of the pulsations have been carried out on a large number of Mira variables. Haniff et al. (1995) used nonredundant mask techniques (section 6.2) to image 10 Miras. Van Belle et al. (1996) measured diameters of 18 Mira variables using IOTA at 2.2 μm and Hofmann et al (2002) used the FLUOR fiber-optic beam combiner at IOTA to measure very precise diameters of five more Mira variables. Dyke and Nordgren (2002) measured 18 Miras with the NPOI and showed the correlation of diameter with TiO absorption in the stellar atmospheres.

11.2.5 Young stellar object disks and jets

Understanding the formation of stars requires observation of young stellar objects (YSOs) throughout various stages of evolution. Dramatic and complex activity occurs due to the presence of accretion disks, jets and highly collimated stellar winds. Understanding of these phenomena mostly comes from modeling and indirect measurements of infrared excess flux and spectra. The Hubble Space Telescope, with its 0.1 arcsecond resolution, has produced images of disks and other circumstellar structure around many young stars; however, improving on the models for these objects requires higher resolution measurements and imaging. Interferometers operating in the infrared, such as the PTI, IOTA and NPOI, are ideal for observing YSOs since their flux peaks in the infrared. A series of observations with two different interferometers (IOTA and PTI) were combined in the study of a YSO, FU Orionis. FU Ori has irregular outbursts in which its brightness increases by 4–6 magnitudes. A suggested model for this system is that it is a binary with an accretion disk around one of its components. In addition, they found that the fit to the accretion disk did not match the theoretical model. High-resolution imaging and measurement are indispensable for achieving a better understanding of such complex phenomena. Monnier and Millan-Gabet (2002) summarize interferometric measurements (mostly made with IOTA) of Herbig Ae/Be and T Tauri disk sizes and discuss their statistical properties. They show that a simple ring model for the disk works well and that the sizes are closely related to the radius at which stellar radiation sublimates the dust. Tuthill et al. (2002) combined data from Keck I and ISI to image Herbig Ae/Be star LkHα-101. They show that the star has a circular disk with a central hole or cavity. They also found that the change in apparent size of the disk with wavelength (going from 1.6 to 11.15μm) was inconsistent with standard power-law temperature profiles. Eisner et al. (2004) used PTI to resolve the inner disk structure at 2.2μm. Berio et al. (1999) observed γ Cassiopeiae, a Be star with an expanding, rotating envelope. They used the GI2T for their observations over many epochs. The results reveal asymmetric variations and an apparent one-armed oscillation precessing in the equatorial disk.

11.2.6 Dust shells, Wolf–Rayets

Infrared interferometry has been particularly useful for examining dust shells emitted as outbursts from stars that have episodic mass ejection such as α-Ori and R-Aqr. Danchi et al. (1994) have observed 15 stars with dust shells using the ISI. From these observations, they have estimated the inner radii of the shells and their temperature at the inner radii. Monnier et al. (1999, 2002) have studied two Wolf–Rayet binaries, WR 98a and WR 140, using nonredundant masking of the Keck I

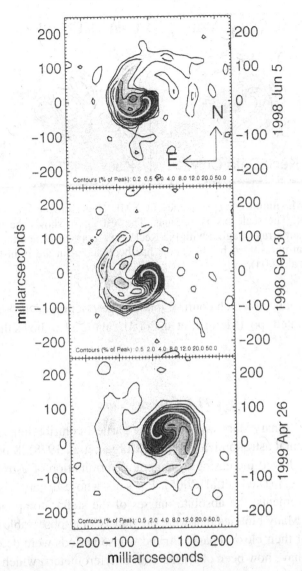

Fig. 11.5. Three epochs of the 2.2 μm emission from WR 98a showing a rotating spiral structure. The white line indicates the best fit to a plume morphology (Monnier et al. 1999).

telescope. Wolf–Rayet stars are very massive and highly evolved. The observations show that the dust shells surrounding the stars have a "pinwheel-like" shape to them. Models for the observations suggest that the pinwheels are due to colliding winds from the binary components. Figure 11.5 from Monnier et al. (1999) shows the evolution of the pinwheel over slightly more than a 1 year period. The addition of long-baseline information from IOTA to the Keck data allows maximum-entropy

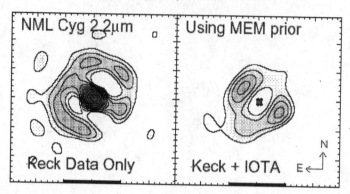

Fig. 11.6. Maximum entropy images of NML Cyg showing the circumstellar environment. The scale bar is 100 mas. The left figure shows the image from non-redundant array imaging with the Keck I telescope, to which the IOTA data indicating an unresolved bright source has been added in the right-hand figure (Monnier et al. 2004).

images to be produced which confirm that a large fraction of the emission comes from a concentrated spot (Monnier et al. 2004). Figure 11.6 shows the effect of this combination.

11.2.7 *Binary stars*

Measurement of binary stars was one of the earliest contributions of interferometry to fundamental astronomical science (Morgan et al. 1978). Since then, interferometry has allowed increasingly accurate determination of astrometric binary orbits (figure 11.7) (Davis et al. 2005). For stars whose distance is known from parallax measurements, the absolute masses of the stellar components can then be determined. Many binaries that were previously only observable spectroscopically because of their close separation, and whose periods were determined from Doppler shifts, have now been observed with interferometers, which have dramatically improved the accuracy of the stellar mass function. A recent, particularly interesting result is the detection of a very young quadruple system, HD 98800, using speckle interferometry and adaptive optics with the Keck telescopes (Prato et al. 2001). It was a known visual binary that has now been resolved into four components – each star was actually itself a binary – and that the age of the stars has been determined to be about 10^7 years. The authors show that most or all of the mid-infrared excess flux is coming from HD98800 B and that a dust disk around that binary pair is the source of the flux. They also conclude that the disk has an inner gap of about 2 AU and a height of < 1 AU.

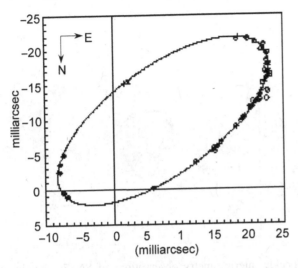

Fig. 11.7. Binary orbit of β-Centauri determined by SUSI (Davis et al. 2005).

11.3 Galactic and extragalactic sources

11.3.1 SN1987a

SN1987A, the explosion of a blue supergiant star in the Large Magellanic Cloud (LMC), a satellite galaxy of the Milky Way, which was first seen on 23 February 1987, was the brightest such event in almost 400 years. The last supernova that was visible to the naked eye exploded in 1604 and was observed by the great astronomer, Johannes Kepler. Because SN1987A was so bright (it reached a visual magnitude of 3 at its peak) and relatively close (the LMC is about 50 kiloparsec away), it was quite probably the most studied astronomical object in the modern era. After a supernova explodes, debris from the explosion is ejected with very high velocities of several thousand km s^{-1}. This debris forms a "nebula" masking the region close to the center of the explosion. Eventually, this ejected material spreads over a wide region, forming an extended supernova remnant similar to those seen in our own and other galaxies. The LMC is at $-69°$ declination and so it was only observable from the southern hemisphere. Several groups observed the supernova using speckle interferometry at early epochs in order to try to document the early formation of the nebula. The result provided some major surprises, showing the importance of high angular resolution observations. Using speckle interferometry, Karovska et al. (1989, 1991) measured the expanding size of the nebula. The apparent velocities of the debris (about 3500 km s^{-1}) agreed very well with the velocities measured by spectroscopy from Doppler shifts. More recent measurements of debris velocity, made using Hubble Telescope images, also agree with the early speckle images

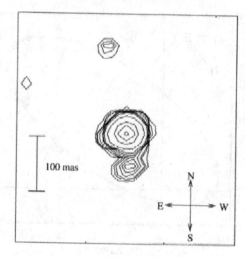

Fig. 11.8. Speckle interferometry observations of SN87a showing two ejected blobs (Nisenson and Papaliolios 1999).

(Wang et al. 2002). The speckle images also showed that the debris shape was elongated and asymmetric (Papaliolios et al. 1989). Again this was confirmed by HST imaging.

The mystery spot

Speckle interferometry observations, made just 30 and 38 days after the explosion of supernova SN1987A, showed evidence for a bright source, separated from the supernova by only 60 mas (Nisenson et al. 1988). This source was also detected using speckle 50 days after the explosion by Meikle et al. (1987). More recently, new reconstructions from the early speckle data (Nisenson and Papaliolios 1999) showed that there appeared to be a second bright source on the opposite side of the SN from the originally detected source (figure 11.8). By assuming that both of these sources were due to some sort of ejection from the supernova at the time of the explosion, they were able to model the geometry of the system.

11.3.2 R136a

R136a was an extremely luminous object in the 30 Doradus region of the Large Magellenic Cloud galaxy. Until it was observed using speckle masking techniques (section 6.4.2), it was believed to be the most massive star known – too massive for any reasonable theory. Speckle images by Weigelt and Baier (1985) showed that R136a was, in fact, a cluster of stars which indeed included some very massive stars, but none over the predicted limit for a massive star. Figure 6.11 is the reconstructed image. The subsequent image taken with the Hubble Telescope revealed many additional fainter stars, but also verified the accuracy of the speckle reconstruction.

11.3.3 The galactic center

Infrared speckle interferometry has been used to study the central region of our own galaxy, the Milky Way, by observing the stars close to the galactic center. Ghez et al. (1998) observed a 6×6 arcsecond region using the 10-m Keck I telescope and applied speckle shift-and-add reconstruction techniques (section 6.4) to monitor the proper motions of 90 stars. Fitting orbits to the stellar motions showed that the stellar surface density and the velocity dispersion were consistent with the position of Sagitarius A*, the unusual radio source that is suspected to be a black hole. The observations are also consistent with Sagitarius A* being located close to or at the center of the galaxy. Their measurements allow an independent estimate of the central density which is at least $10^{12} M_{sun}$ per cubic pc. Their conclusion is that these results are independent evidence that the center of the Milky Way harbors a massive black hole. Tanner et al. (2002) have also used speckle interferometry on the Keck I telescope to image a cool source near the Galactic center, IRS 21, an enigmatic object that until these observations had eluded classification. Mid-infrared imaging revealed that IRS 21 is a self-luminous source rather than a heated clump of gas. The best model for this source (and, the authors argue, for other similar sources) is that they are massive stars experiencing bow shocks as they move through the northern arm of the galaxy.

11.3.4 Astrometry

Interferometry when applied to astrometry (measuring the positions and proper motion of stars and other astronomical objects) is a very different paradigm from the high-resolution imaging and measurement obtained with the more conventional interferometric techniques. Instead of measuring visibilities or reconstructing images, astrometric interferometry uses the interferometer to determine the precise center of an object and then to measure the angle between that object and another object. Wide-field astrometry using interferometry has proven to be very difficult from the ground due to rapidly varying atmospheric dispersion and the difficulty of precisely and continuously monitoring the baseline length. The operation of the Mark III and the NPOI astrometric interferometer are discussed in detail in chapter 8. Both interferometers have made significant measurements of turbulence statistics, including Fried's inner and outer scales. The Mark III was used to measure the positions of 11 stars from the FK5 astrometric catalog (Hummel et al. 1994). Systematic errors limited the astrometric precision to about 13 mas. Still this was sufficient to determine that some of the stars had measurable differences in their astrometric positions from the FK5 positions. Whether these differences were due to errors in the catalog or to proper motion of the stars was undetermined because the time period over which observations were made (four years) was insufficient. Differential interferometric astrometry (section 8.4.8) is a technique for detection of very low-mass companions from the astrometric motion of the primary stars. Colavita (1994) applied differential astrometry to the visual binary α Gem using the Mark III interferometer and found that the repeatability of the measurements were of order $20\,\mu$as. This sensitivity would easily be sufficient to detect Jupiter-sized planets orbiting nearby stars. Implementation of differential interferometry for planet detection are planned for Keck, PTI and the VLT interferometer. Benedict et al. (2002) used the fine guidance sensors of HST in interferometric mode to make astrometric measurements of a very low-mass binary star GL 791.2. The interferometric precision allowed them to measure the masses of each component of the binary to an accuracy of better than 2%.

11.4 Solar system

11.4.1 The Galilean satellites

The very first interferometric observations were of solar system objects and were carried out by Michelson using the 12-inch refractor at Lick Observatory (Michelson 1891). Michelson observed the four Galilean satellites by placing a two-slit mechanism over the aperture of the telescope (section 1.1). Adjusting the separation of the slits while viewing the fringes from the slits until the fringe contrast went to zero allowed him to measure, in turn, the diameter of the satellites. Nisenson et al. (1981) imaged Saturn's satellite, Titan, using speckle interferometry and Knox–Thompson reconstruction techniques (section 6.4.1) on data from the University of Hawaii's 2.24-m telescope on Mauna Kea, Hawaii. Titan was found to be relatively featureless, but a precise radius and a best-fit limb darkening of the disk were determined. More recently, Gibbard et al. (1999) used infrared speckle interferometry with the Keck telescopes to image Titan. Using multiple colors, they were able to both determine the optical depth of Titan's haze layer and to construct surface albedo maps.

11.4.2 Asteroid imaging

Speckle interferometry, applied to imaging large asteroids, allowed observation of asteroid shapes and variations in albedo before the recent NASA fly-bys. Baier and Weigelt (1983) estimated the diameters and shapes of the asteroids Juno and

Amphitrite. Drummond et al. (1988) imaged the asteroid Vesta and matched the varying images with the photometric light curve. They found that the light curve was mostly determined by albedo structure rather than the shape of the asteroid. The same group showed images of the asteroids Davida, Herculina and Eros in previous papers. More recently, Ragazzoni et al. (2000) imaged the asteroids Hygiea and Eunomia using speckle. They estimated the ellipticity of both asteroids, and they were able to fit the image of Eunomia to either an egg-shape or a close double asteroid model.

11.4.3 Pluto–Charon

Pluto's large moon, Charon, was originally discovered from conventional photographic exposures where the planet's image appeared slightly elongated. Bonneau and Foy (1980) used speckle interferometry to measure the astrometric orbits and the diameters of each component. Hege et al. (1982) used visible speckle interferometry to measure separations and positions of Charon from Pluto and to revise the estimated orbit. Improved measurements of the diameter of each body were obtained by Baier and Weigelt (1987), using a photon-counting version of speckle interferometry, and resulted in a further revised orbit that included the speckle observations of several other groups.

11.5 Brown dwarfs

Detection of planets and brown dwarfs[†] orbiting other stars has become a major goal of many astronomers (chapter 10). Adaptive optics combined with a coronagraph was used to find the first known brown dwarf, GL229b (Oppenheimer et al. 1995). Since then, there have been many observations carried out using infrared speckle interferometry (the ratio of star to planet is more favorable in the infrared) and several candidate brown dwarfs have been found (along with many previously undetected M and L dwarfs), and recently two companions with size of order 1–10 Jupiters have been imaged directly in the infrared, using adaptive optics (but not interferometry) at VLT (Chauvin et al. 2004; Neuhäuser et al. 2005).

11.6 Solar feature imaging and dynamics measurements

High-resolution imaging of solar surface features has proven very useful for understanding many of the very complex phenomena observed in the photosphere and the chromosphere. Solar imaging is difficult because the Sun is so extended that

---

[†] The conventional definition of a brown dwarf is a substellar companion with mass between about 10 and 100 Jupiter masses

only small regions (isoplanatic patches) can be reconstructed. Each small region must be then interleaved to produce a high-resolution image. In addition, many of the features of interest are low contrast, so very good seeing is required before the speckle algorithms converge. von der Lühe (1994) successfully reconstructed a number of small regions that contained continuum bright points and Krieg et al. (2000) measured the velocity of granules in the photosphere using speckle. Sütterlin et al. (2001) used speckle in the Ba II 4554 Å line with data taken on the new Dutch Optical Telescope to construct Dopplergrams that show concentrated downflows of 1.2–2 km s^{-1}.

References

Armstrong, J. T., T. E., Nordgren, M. E., Germain *et al.* (2001). *Astron. J*, **121**, 476.

Aufdenberg, J. P., Hauschildt, P. H. and E. Baron, (2003), in *Stellar Atmosphere Modeling*, ed. I. Hubeny, I. Milahas and K. Werner, San Francisco: Astron. Soc. Pac.

Baier G. and G. Weigelt (1983). *Astron. Astrophys.*, **121**, 137.

Baier G. and G. Weigelt (1987). *Astron. Astrophys.*, **174**, 295.

Van Belle, G. T., H. M., Dyck, J. A., Benson *et al.* (1996). *Astron. J*, **112**, 2147.

Benedict, G. F. *et al.* (2002). *Astron. J*, **123**, 473.

Berio, P., P., Stee, F. Vakili *et al.* (1999). *Astron. Astrophys.*, **345**, 203.

Bonneau D. and R. Foy (1980). *Astron. Astrophys.*, **92**, L1.

Bonneau, D. and A. Labeyrie, (1973). *Astrophys. J*, **181**, L1.

Carroll B. W. and D. A. Ostlie (1996). *An Introduction to Modern Astrophysics*, Reading MA: Addison-Wesley

Chauvin, G., A.-M. Lagrange, C. Dumas *et al.* (2004). *Astron. Astrophys.*, **425**, L29.

Colavita, M. M. (1994). *Astron. Astrophys.*, **283**, 1027.

Davis, J., A. Mendez, E. B. Seneta *et al.* (2005). *Mon. Not. R. Astron. Soc.* **356**, 1362.

Danchi, W. C., M. Bester, L. J. Greenhill *et al.* (1994). *Amplitude and Intensity Spatial Interferometry,* ed. J. B. Breckinridge, Proc. SPIE **2200**, 286.

Drummond, J., A. Eckart and E. K. Hege (1998). *Icarus*, **73**, 1.

Dyke, H. M., G. T. Van Belle and S. T. Ridgeway (1996). *Astron. J.*, **111**, 1705.

Dyck, H. M. and T. E. Nordgren (2002). *Astron. J.*, **124**, 541.

Eisner J. A., B. F. Lane, L. A. Hillenbrand *et al.* (2004). *Astrophys. J.*, **613**.

Ghez, A. M., B. L. Klein, M. Morris *et al.* (1998). *Astrophys. J.* **509**, 678.

Gibbard, S. G., B. Macintosh, C. E. Max *et al.* (1999). *American Astronomical Society*, DPS meeting no. 31, no. 41.04, 31.

Hanbury Brown R., J. Davis and L. R. Allen (1974). *Mon. Not. R. Astron. Soc*, **167**, 121.

Haniff, C. A., M. Scholz, and P. G. Tuthill (1995). *Mon. Not. R. Astron. Soc*, **276**, 640.

Haniff, C. A. *et al.* (2004). *New Frontiers in Stellar Interferometry,* ed. W. A. Traub, Proc SPIE **5491**, 511.

Haniff, C. A., *et al.* (2003). *Interferometry for Optical Astronomy II,* ed. W. A. Traub, Proc SPIE, **4838**, 19.

Hege, E. K., E. N. Hubbard, J. D. Drummond *et al.* (1982). *Icarus*, **50**, 72.

Hofmann, K. H. *et al.* (2002). *New Astron.*, **7**, 9.

Hummel, C. A. and J. T. Armstrong (1994). *IAU Symposium Very High Angular Resolution Imaging*, **158**, 410.

Hutter, D. J. *et al.* (1989). *Astrophys. J.* **340**, 1103.

Karovska, M., L. Koechlin, P. Nisenson *et al.* (1989). *Astrophys. J.* **340**, 435.

Karovska, M., P. Nisenson, C. Papaliolios *et al.* (1991). *Astrophys. J.* **374**, L51.

Kervella P., N. Nardetto, D. Bersier *et al.* (2004). *Astron. Astrophys.*, **416**, 941–953.

Krieg, J., F. Kneer, M. Koschinsky *et al.* (2000). *Astron. Astrophys.*, **360**, 1157.

Von der Luehe, O. (1994). *Astron. Astrophys.*, **281**, 889.

Meikle, W. P. S., S. J. Matcher and B. L. Morgan (1987). *Nature*, **329**, 608.

Michelson A. A. (1891). *Nature*, **45**, 160.

Michelson A. A. and F. G. Pease (1921). *Astrophys. J.*, **53**, 249.

Monnier, J. D., P. G. Tuthill and W. C. Danchi (1999). *Astrophys. J.*, **525**, L97.

Monnier, J. D., P. G. Tuthill and W. C. Danchi (2002). *Astrophys. J.*, **567**, L137.

Monnier, J. D. and R. Millan-Gabet (2002). *Astrophys. J.*, **579**, 694.

Monnier J. D., R. Millan-Gabet, P. J. Tuthill *et al.* (2004). *Astrophys. J.*, **605**, 436.

Mourard, D., D. Bonneau, L. Koechlin *et al.* (1997). *Astron. Astrophys.*, **317**, 789.

Morgan, B. L., D. R. Beddoes, R. J. Scaddan *et al.*, (1978). *Mon. Not. R. Astron. Soc.* **183**, 701.

Neuhäuser, R., E. W. Guenther, G. Wuchterl *et al.* (2005). *Astron. Astrophys*, **435**, L13.

Nisenson, P., J. Apt, P. Horowitz *et al.* (1981). *Astron. J.*, **86**, 1690.

Nisenson, P., C. Papaliolios, M. Karovska *et al.* (1988). *Astrophys. J.*, **324**, L35.

Nisenson, P. and C. Papaliolios (1999). *Astrophys. J.* **518**, L29.

Nordgren, T. E., J. J. Sudol, and D. Mozurkewich (2001). *Astron. J.*, **122**, 2707.

Nordgren, T. E., B. F. Lane, R. B. Hindsley *et al.* (2002). *Astron. J.*, **123**, 3380.

Oppenheimer, B. R., S. R. Kulkarni, K. Matthews *et al.* (1995). *Science*, **270**, 1478.

Papaliolios, C., M. Krasovska, L. Koechlin *et al.* (1989). *Nature*, **338**, 565.

Pease, F. G. (1931). *Ergebnisse der Exakten Naturwissenschaften,* **10**, 84.

Pijpers F. P., T. C. Teixeira, P. J. Garcia *et al.* (2003). *Astron. Astrophys.*, **406**, L15.

Prato, L. *et al.* (2001). *Astrophys. J.*, **549**, 590.

Ragazzoni, R., A. Baruffolo, E. Marchetti *et al.* (2000), *Astron. Astrophys.*, **354**, 315.

Sasselov, D. and M. Karovska (1994). *Astrophys. J.* **432**, 367.

Sütterlin, P., R. J. Rutten and V. I. Skomorovsky (2001). *Astron. Astrophys.*, **378**, 251.

Tango, W. J. and J. Davis (2002). *Mon. Not. R. Astron. Soc.*, **333**, 642.

Tanner, A., A. M. Ghez, M. Morris *et al.* (2002). *Astrophys. J.*, **575**, 860.

Tuthill, P. G., A. B. Men'shchikov, D. Schertl *et al.* (2002). *Astron. Astrophys.*, **389**, 889.

Wang, L. *et al.* (2002). *Astrophys. J.*, **579**, 671.

Weigelt, G. and G. Baier (1985). *Astron. Astrophys.*, **150**, L18.

Wittkowski M., C. A. Hummel, K. J. Johnston *et al.* (2001). *Astron. Astrophys.*, **377**, 981.

Wittkowski, M. and C. A. Hummel (2003). *Proc SPIE* **4838**, 219.

Young J. S., J. E. Baldwin, R. C. Boysen *et al.* (2000). *Mon. Not. R. Astron. Soc.*, **315**, 635.

12

Future ground and space projects

12.1 Future ground-based projects

A few decades ago, there appeared to be two distinct evolutionary paths for astro-nomical imaging instruments: toward larger telescopes and longer-baseline inter-ferometers. A telescope would be based on a single massive monolithic mirror, for which the limiting size appeared to be on the order of 8 m. An interferometer could involve two or more telescopes spaced tens or hundreds of meters apart, possibly up to ten kilometers. More recently, the emphasis in interferometers has turned to larger numbers of subapertures rather than larger sizes or longer baselines, in order to improve (u, v) coverage (e.g. the Magdalena Ridge Observatory Interferome-ter), and considerable effort is being devoted to planning space interferometers, which will be discussed later in this chapter (section 12.2). On the other hand, following the success of the mosaic mirror Keck telescopes of 10-m diameter, the elements of which are carried by a common pointing mount, projects for larger versions up to 100 m, are now being studied under the generic name "Extremely Large Telescopes" (ELT).

Alternative paths have also emerged. Mosaics of many smaller apertures are also being considered, at scales much larger than ELTs, these being diluted mosaics in which the problems of path-length compensation are overcome by designing the diluted optics as if it were a single giant telescope. In terms of their optical scheme, these mosaics may be considered as "exploded" versions of ELT's and they operate according to the hypertelescope principle (chapter 9). However, a big difference between ELTs and hypertelescopes is in their mechanical structure. The former have a single large pointing mount, as large as those built for the largest steerable radio telescopes, reaching 100 m in diameter. The latter are too large for a pointing mount and achieve the pointing by other means, such as a mobile beam combiner like the Arecibo radio telescope, as was discussed in chapter 9. Hybrid instruments can also considered for a valuable synergy in observing performance, whereby an ELT is coupled with a large dilute array (figure 12.1).

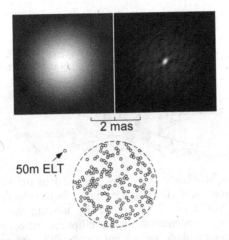

Fig. 12.1. Simulated image of a point source, formed by a coupled ELT and hypertelescope, having respectively a single 50-m mirror and 200 mirrors of 1-m diameter. The pupil densification is unequal, providing subpupils of equal size in the densified exit pupil. Left: the PSF of the 50-m telescope; right: that of the coupled system. The high-resolution interference peak thus obtained is seen to concentrate most energy, thus combining the advantages of both instruments. The sketch below shows the nonuniformly densified exit pupil where the 50-m and 1-m apertures appear with identical sizes.

12.1.1 New ground-based long-baseline interferometers

Some of the pioneering interferometers discussed in chapter 8, particularly COAST and Mark-III and its predecessors, were expressly designed as prototypes, the lessons from which would be incorporated in later instruments. Indeed several of the interferometers described in that chapter have been based on these lessons, and some of them are not yet operating at the time of writing. One interferometer in this class which is still on the drawing-board is the the Magdalena Ridge Observatory Interferometer (MROI) in New Mexico which is to be a multi-element imaging interferometer operating at wavelengths between 0.6 and 2.4 μm with baselines of up to 400 m. It will consist of eight to ten 1.4-m telescopes and a single conventional 2.4-m telescope, all with adaptive optics. The use of movable telescopes in a "Y" configuration will allow its reconfiguration from a very compact array with baselines of tens of meters to a long-baseline configuration with baselines up to 400 meters, giving 0.25 mas resolution. Another interferometer, the Antarctic Planet Interferometer, is planned to take advantage of the excellent seeing conditions (section 5.1) and unrestricted space available at the Concordia station in Antarctica. This will be in many ways similar to the Palomar Testbed Interferometer (section 8.4.8), with a baseline initially 100-m long, eventually to be extended to 400 m. It is aimed to work in the near to mid-infrared, taking advantage of the expected increase in contrast between a star and its planet in this spectral region (section 10.2)

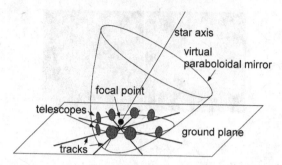

Fig. 12.2. The OVLA scheme originally proposed in the late 1970s involves tens of mobile telescopes, all feeding a common focus. The optical path lengths are kept balanced while the Earth rotates by moving the telescopes during the observation. They must remain on the elliptical locus, which is the intersection with the ground plane of a giant paraboloid kept pointed toward the star.

12.1.2 *The optical very large array (OVLA)*

As sketched in figure 12.2, the OVLA scheme has a number of mobile telescopes along a deformable elliptical ring. In this way, similar to the use of a circular track in the Narrabri intensity interferometer, the problem of variable delay-lines is avoided. The geometry can be described as follows. A given point star can be imaged to a geometrical image at the focus of a parabolic reflector whose axis passes through the source, and any smaller off-axis paraboloidal mirror which is cut out of this reflector would create the same geometrical image. The paraboloid intersects the ground plane in an ellipse, and a series of such cut-out mirrors could be moved along radial tracks on the ground to the points where they intersect this ellipse, thereby creating a stigmatic geometrical image. The ellipse is continuously changing as the Earth rotates, and the exact form of each mirror would have to be continuously modified to the required off-axis paraboloid by an active optical system. In practice, no shape variation would be needed if the mirrors were replaced by telescopes. Of course, the need for variable delay lines has been replaced by movable telescopes, which have to be positioned just as accurately, but a great advantage of a system of this sort is that it creates a direct broad-band image of a complex source, without algorithmic intervention. Variants with fixed telescopes, arrayed in two dimensions, can also be considered, but these will require delay lines as well in order to maintain the balance of the optical paths. As the notion of densified pupil, or "hypertelescope", imaging emerged, this mode of beam combination has become part of the recent designs for OVLA-type arrays. A similar array concept has been proposed for the "Kiloparsec Explorer for Optical Planet Search" (KEOPS) array at the Dome C site in the Antarctic.

12.1.3 Toward large Carlina hypertelescopes

As we have seen in chapter 9, a different way of accommodating many apertures can be contemplated if a fixed array of concave mirror segments is arranged in the form of separated dilute elements of a single fixed giant mirror, providing a welcome simplification of the optics, the mechanical structure and the control. This giant mirror might have a paraboloidal shape, providing perfect phasing for a star located on axis, but thereby introducing unacceptable aberrations for stars in other positions. It can have a different shape if the aberrations of the focused wavefront are corrected closer to the focus. Then, a spherical shape is obviously of interest because it does not have a unique axis and provides a broad primary field, spanning tens of degrees, which is needed since the diluted mirror is not pointable. Such fixed mirrors will allow effective aperture diameters much larger than the 100 m limit of a steerable mirror.

The idea of a fixed spherical mirror was implemented many years ago in the form of the Arecibo radio telescope in Puerto Rico, which has a continuous spherical aluminum sheet mirror spanning a sink-hole 330 m in diameter. Its effective aperture is smaller, owing to the limited angular acceptance of the aberration corrector located near the focal surface. This corrector, with its attached detector, moves along a curved track suspended from cables supported from three high towers on its periphery (figure 9.10).

In an optical version of such a telescope, the cost of mirror elements per unit area is much higher, and a diluted primary mirror must therefore be adopted. The mirror elements can be fixed and rigidly attached to the stable bedrock as parts of a sphere, providing a spherical paraxial focal surface, according to the Carlina concept (section 9.4). The focal corrector, with the attached pupil densifier and camera, are then suspended from a balloon (see figure 9.11). Depending on the crater or valley sites to be exploited, effective aperture sizes as large as one or two kilometers may become feasible. The number of primary mirror elements can reach hundreds or thousands, thus providing a rich (u, v) coverage and providing snapshot images containing much information. Adaptive optics is highly desirable to approach diffraction-limited performance in such images; since the subaperture mirror diameters are not much larger than r_0, they can be corrected, essentially piston and tip–tilt, by established techniques (section 5.8.3).

12.1.4 Comparison of OVLA and Carlina concepts

On Earth, suitable flat sites of size reaching perhaps 10 km could be equipped with OVLAs. The largest existing crater sites, with size 5 km, could support Carlina arrays, but the effective aperture would in practice amount to one or two km at most. This effective aperture arises, as in the Aricebo radio telescope, from the size of the focal corrector needed to collect the spherically aberrated beams.

The absence of the need for active components in Carlina designs favors a large number of subapertures, hundreds or thousands, whereas OVLA designs are likely to be limited to tens of elements by the complex drives required for each mirror. It is therefore of interest to compare a 10-km OVLA, with relatively few elements of size 10 or 20 m, a 2-km Carlina with thousands of fixed mirrors, and an ELT consisting of a single mirror. The same collecting area, 2000 m², for example will be assumed in all three cases, so that all will provide the same limiting magnitudes when adaptive phasing is achieved on a brighter reference star. Using an array of pupil densifiers in the OVLA or the Carlina, multifield exposures can be recorded, and full-field mosaics can be reconstructed from them, if the field crowding does not exceed the limitation imposed by the diluted aperture (section 9.3).[†] The sky field covered in a single exposure is extremely narrow; in a periodic array it is of the order of λ/s, where s is the center-to-center spacing of the mirrors. An array of pupil densifiers within the beam combiner (figure 9.8) provides multiple fields, each within a cell of size λ/d determined by the subaperture diameter d. The number of such cells is limited by the geometric aberrations, which can be separately corrected in each cell. One may compare the field aberrations of an ELT primary mirror with those of an "exploded" Carlina version having the same optical design, from the primary to the Fizeau focus. The Carlina version has larger physical dimensions and focal length, and so its diffraction limited field is proportionally smaller since, in the focal plane, the transverse aberrations of optical systems scale with size, thus providing an invariant angular value on the sky. But the advantage of "exploding" an ELT is to improve the diffraction-limited resolution. The diffraction-limited field, where aberrations remain smaller than diffraction, thus decreases in principle when exploding the ELT while keeping the same optical design. However, the field cells of a multiple densifier can be separately corrected for field aberrations, and this may reduce the field loss.

12.1.5 Comparing compact and exploded ELTs

The large mosaic mirror of an ELT can in principle be exploded to obtain a diluted mirror with the same total area but larger bounding diameter, thus having increased

---

[†] A relevant question is whether the crowding limitation thus arising when "exploding" the ELT into a 2-km or a 10-km diluted aperture is likely to affect the observation of the deep universe, where many faint remote galaxies may be contained within a square arcsecond. Extrapolations of the Hubble Deep Field and other exposures made with large telescopes indicate that there will be no crowding problem with an exploded ELT at the scales considered.

resolution. Let us compare the science achievable with the two variants. Assuming suitable adaptive optics, the limiting magnitude is invariant, since this depends on the collecting area. A price to pay for the increased angular resolution, however, is the appearance of the field-crowding limitation. This could be objectionable for deep-sky imaging and its cosmological applications. The question is therefore whether there are too many faint galaxies per square arcsecond for hypertelescope imaging.

A rule of thumb is that the point sources, called "active resels" (i.e. resolution elements which contain signals), on the sky must be more diluted than the aperture itself. Equivalently, it can be stated that the limiting number of active resels per steradian is invariant with respect to aperture dilution, at constant collecting area. The limit for a periodic aperture with N elements of diameter d (section 4.3.2) can be expressed as $2.35 \cdot 10^{-11} N d^2 \lambda^{-2}$ point sources per square arc-second (Labeyrie et al. 2001). For example, with 10 000 apertures of 1-m diameter, arranged as a periodic lattice and observing in the visible, extended objects having on the average fewer than three active resels per square milli-arcsecond can be imaged. If the global aperture $D = 1$ km, providing 100 μas resolution, remote galaxies with 100 active resels should be spaced 6 mas apart on the average to avoid crowding (the averaging scale is about 100 mas if $d = 1$ m). At the limiting magnitude considered, which is about 36 for this collecting area, this promises fruitful observing programs in cosmology.

12.1.6 Coupling telescopes through fibers: the OHANA project at Mauna Kea

On the peak of Mauna Kea, there are 10 large telescopes of various sizes, dedicated to different techniques and wavelength regions, in an area of about 900×400 m in lateral extent. The project OHANA (Optical Hawaiian Array for Nanoradian Astronomy, also meaning "family" in the Hawaiian language) considers interferometric coupling of seven of these telescopes using optical fibers. This idea was originally proposed by the late Jean Marie Mariotti et al. (1996). Together, these telescopes would create a maximum E-W baseline of 800 m, leading to a resolution of about 0.5 mas at $\lambda = 2 \mu$m (Perrin et al. 2004). Unfortunately, there does not seem to be a way of using fibers to make path-length equalizers. The two Keck telescopes were designed to be connected interferometrically, whereas the others will need modification for this project. The first connection has recently been accomplished. Figure 12.3 shows a map of the telescope positions, and examples of calculated point spread functions for two cohesive groups and the complete array.

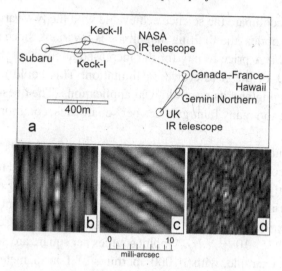

Fig. 12.3. (a) Schematic layout of the telescopes on Mauna Kea which may even-tually be linked in project OHANA. (b – d) Calculated instantaneous point spread functions at 2.0 μm for an source at zenith for (b) the northern four telescopes, (c) the eastern three, (d) all the telescopes combined interferometrically. The first two baselines to operate will be Keck-I to Keck-II and Gemini to CFHT.

12.2 Future space projects

The alternative to larger projects on Earth is to put interferometers into orbit. The advantages are obvious. First, there is no atmosphere to distort or attenuate the wavefront, so that one source of instrumental noise is absent, and the choice of wavelength is not restricted by atmospheric transmission. Second, space is three dimensional (at least!) and so the problems of path-length correction can be replaced by problems of positioning. Which problem is easier to solve has not yet been determined. Third, the maximum baseline is, for practical purposes, unrestricted. Fourth, extraordinary stability is to be expected, disturbed only by collisions with occasional dust particles; a central laser beacon can provide nanometric accuracy by fringe-counting techniques. And fifth, observing time can be continuous if the interferometer is positioned appropriately with the Sun behind it or at an angle to its pointing direction. Some of these advantages would also apply to an interferometer based on the Moon. Several projects for the design of space interferometers are being actively pursued. Their purposes are mainly to detect extrasolar planets and if possible to determine whether there are signs of life on them. The approaches include very accurate astrometry, in order to determine planetary masses and orbits, and imaging sensitive enough to see a dim planet in the presence of its star. Some of the basic problems were discussed in section 10.2.

It should be appreciated that all the space projects are still on the drawing-board and the concepts are in a state of flux. The descriptions given here should therefore be considered as illustrations of designs which satisfy the various optical and mechanical requirements, and certainly not as definitive pictures of systems which are actually going to be launched. Much of the information given here has been gleaned from reports available through the web-sites of the various projects.

12.2.1 Flotillas of mirrors

The idea of using multiple spacecrafts to carry elements of a large interferometer in space was first considered in 1982, and was initially discussed in some detail at a workshop organized in Cargése (Corsica) by the European Space Agency (ESA) on "Kilometric Optical Baselines in Space". A few years later, the National Aeronautics and Space Administration (NASA) supported studies on permanent Moon-based optical arrays and ESA conducted a comparative study of free-flier flotillas and Moon-based arrays. The conclusion was in favor of free-fliers, owing to their lower cost and higher design flexibility, especially for long baselines. The outcomes were proposals for Darwin, a six-element version of Bracewell's infrared nulling interferometer (section 10.5) using several free-flying spacecraft reaching baselines up to 300 m, and for the Terrestrial Planet Finder (TPF), built on a single truss supporting several mirrors. Both concepts have been the subject of industrial studies and negotiations between ESA and NASA toward a joint project. NASA encouraged a broad exploration of alternative concepts, which included an Exo-Earth Discoverer hypertelescope, and also visible coronagraphy with a 3-m space telescope, as described in section 10.3.

The prospect of formation flying with the extreme accuracy typical of interferometry and coronagraphy led ESA and NASA to prepare tests of flotilla control. The control techniques needed to drive such flotillas are indeed of crucial importance for future interferometric missions. If efficient techniques can be developed, at an affordable cost, they will provide a fast and elegant path toward optical arrays having kilometric and even much larger sizes. Other uses are also foreseen for such flotillas, such as the detection of gravitational waves with laser interferometers spanning millions of kilometers (LISA). Unlike earth-based interferometers, those involving flotillas in space have a cost which is little influenced by the spacing of the elements.

12.2.2 Darwin

Darwin is a version of Bracewell's interferometer, enlarged by using three or six free-flying Cassegrain telescopes symmetrically arranged around a central

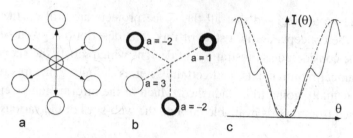

Fig. 12.4. Darwin concept, using six telescopes which can move radially in a nulling configuration, and a central beam-combining spacecraft. (a) Spacecraft configuration. (b) Form of one individual nulling interferometer; the "area" a represents the relative wave amplitude (including its sign) from that aperture which is used in the interferometer. When three such interferometers are superimposed at 0 and $\pm120°$, the sum of the values of a^2 at each mirror is 9. (c) Fringe profile of one interferometer; the dashed line shows, for comparison, the form of $\sin^2\theta$ fringes with the same fundamental period.

beam-combining hub satellite. Its goal is to detect exo-Earths at mid-infrared wavelengths (7–17 μm). The telescopes can move radially from the combiner hub between distances of 50 and 100 m. The interferometer will be situated at the second Lagrange point L2 of the Sun–Earth system, distant about 1.5×10^6 km from Earth on the side remote from the Sun[†].

Two concepts for nulling interferometry with Darwin have been studied, using respectively six and three telescopes. In the six-telescope version, there are seven spacecraft in which six each support a 1.5-m Cassegrain telescope. In these telescopes, a folding mirror with a central hole located on the axis in the focal plane diverts light from the peripheral field to a camera for field stabilization using remote stars. A similar concept is used at Palomar (section 8.4.8). The central hole transmits light from the investigated star to be collimated and beamed toward the seventh spacecraft which contains the beam-combiner. In the beam-combiner, the light is divided by calibrated beam-splitters to form three nulling interferometers, each involving four subapertures (figure 12.4). This type of nulling interferometer is based on a concept due to Angel et al. (1986) and is called a "generalized Angel's cross." As shown in the figure, the dark fringe has θ^4 variation with angle along the axis normal to the fringes, which reduces starlight-leakage (section 10.4.4). The outputs from the three nulling interferometers, which create fringe patterns at orientations at 120° to one another, can then be combined.

In the three-telescope version, there are three spacecraft arranged as an equilateral triangle; on all three of them are 3-m telescopes and one is used additionally for

---

[†] Lagrangian point L2 is on the night side, away from the Sun. A body placed there is more distant from the Sun and therefore should orbit it more slowly than the Earth; but the extra pull of the Earth adds to the Sun's pull, and this allows the body to move faster and keep up with the Earth, so that it is forever in the Earth's shadow.

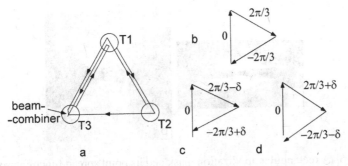

Fig. 12.5. Darwin concept using three spacecraft in an equilateral triangle: (a) the optical paths from the three telescopes to the beam-combiner, each path being twice the length of one side of the triangle; (b) phasor combination for the three interfering waves for a source on axis; (c) phasor combination for an off-axis source which creates phase shifts $\pm\delta$ at the second and third inputs; (d) as (c), when the phases of the second and third beams are interchanged.

beam-combination and communications. The paths between the spacecraft are used for path-length equalization as shown in figure 12.5. The outputs from the three telescopes are combined with periodically switchable phases. In one half of the period, the phases are $(0, \pi/3, 2\pi/3)$, and this combination is switched during the other half to $(0, 2\pi/3, \pi/3)$. For an on-axis source, in both cases there is complete destructive interference. But an off-axis source, which adds small phase changes to the inputs, is not nulled, and the intensity is modulated at twice the switching frequency, assuming that all light is obscured during the period that the phases are changing. Beam combination will use an optical fiber on the end of which all three inputs are focused, after the appropriate phase changes have been made. A question, which will be left to the reader, is: where does the light energy go when the three beams are combined destructively this way? In another three-telescope concept, a similar mode of periodically switched beam combination is carried out on a fourth spacecraft, maintained at a position equidistant from the other three, but not necessarily in the same plane.

12.2.3 Terrestrial planet finder (TPF)

NASA is studying the TPF in two different versions. One version, called TPF-C, is based on the realization that exo-Earths might be best detectable at visible wavelengths with refined forms of coronagraphy and apodization using a $3 \times 8\,\mathrm{m}$ telescope in space. As was described in section 10.3.1, an apodization mask of a form shown in figure 12.6 can achieve the necessary contrast over segments of the field of view when supplemented by active optics to correct mechanical wavefront distortions, although the resultant transmitting area is only 25% of the total, which

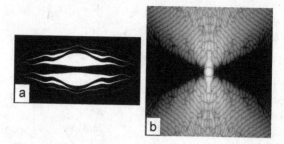

Fig. 12.6. A rectangular apodization mask and its point spread function providing quadrant regions of intensity below 10^{-10}. Courtesy of R. J. Vanderbei.

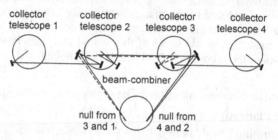

Fig. 12.7. Terrestrial Planet Finder-I. The optical scheme to create two independent nulling interferometers, which can be coherently combined. Four telescopes and a beam-combiner are situated on five satellites.

is disappointing. A possible solution exists in the form of phase-induced amplitude apodization (section 10.3.2), but this has not yet proved itself. Along these lines, a TPF-C instrument is expected to fly around 2014.

The second version, called TPF-I, is a form of nulling interferometer (section 10.5) planned to work in the thermal wavelength range 6.5–13 μm, using optics cooled to 40 K, where a nulling ratio of 10^{-6} should be sufficient to discover a terrestrial planet. This instrument consists of four or five spacecraft carrying 3- to 4-m infrared telescopes flying in precise formation. It is to be launched before 2020. The concept, illustrated by figure 12.7, contains two independent nulling interferometers, which can be combined coherently with a variable phase shift to give a broader null which varies as θ^4 and not θ^2 as in sinusoidal interference fringes (section 10.4.4).

12.2.4 Space interferometry mission (SIM)

SIM PlanetQuest, scheduled for launch by NASA in 2009, is an astrometric space interferometer based on one spacecraft with four fixed 9–10 m baselines operating in the visible waveband (0.4–0.9 μm). Its field of view for one orientation will be 15°, and its position will be stabilized by using two interferometers to lock onto members of a grid of reference stars situated at distances $>10^3$ parsec, where stellar wobble

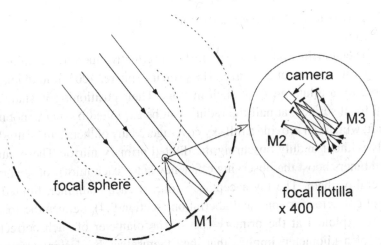

Fig. 12.8. The Exo-Earth Imager concept in bubble form. This space version of a Carlina hypertelescope has a primary spherical locus M1 which is entirely, but sparsely, paved with mirrors. These can be fixed in space. Focal beam-combiners, each incorporating a clam-shell corrector which itself is a flotilla of small mirrors on loci M2 and M3 (inset), are movable on the half-radius focal sphere to acquire various stars. With many combiners, independently movable on the focal sphere, each primary mirror segment can feed several combiners simultaneously, thus increasing the observing efficiency. For a system of this size, all the elements can be plane mirrors.

due to companions should be indiscernible. The stabilization accuracy expected in a single measurement is $12\,\mu$as. The other two interferometers will then be used for astrometry. Each one has two main telescopes with 0.35-m apertures which compress the beams to about 30-mm diameter before they pass through delay lines to the beam combiner. They have a goal of $4\,\mu$as astrometric accuracy for measurements of stellar separations less than $1°$. The limiting magnitude for measurement is expected to be 12.

SIM will be launched into an Earth-trailing orbit to circle the Sun while lagging increasingly behind the Earth as the mission proceeds. The search for exo-Earths is planned to include about 250 candidate stars and to search for wobble at the level of $1\,\mu$as by observing each one about 80 times during the first five years of the mission. During the following five-year period, it is intended to make more detailed observations of the most likely candidates for exo-Earths.

12.2.5 The exo-Earth imager (EEI)

In the longer term, the detailed observation of exo-Earths detected by these instruments will in principle be achievable with hypertelescopes larger than 100 km. Imaging simulations with a 150-aperture instrument of this kind, called the Exo-Earth Imager (EEI) (figure 12.8), indeed show that continents and vegetated areas are potentially detectable in the images thus obtained at visible wavelengths with μas resolution.

The EEI version illustrated in the figure has an effective aperture spanning 150 km containing 150 mirrors of 3-m size. The primary sphere, 400 km in diameter, can be covered by a dilute array of such mirrors. Their positioning is static but the spherical focal surface, at half the radius, can be exploited by many mobile focal spacecraft, which can provide a full sky coverage. Many of them can be used simultaneously without unduly obscuring the diluted primary mirror. These numerous focal combiners boost the observing efficiency since each mirror of the array can feed several of them. The ray acceptance cone of each combiner, limited by the clam-shell corrector for spherical aberration (section 9.4), defines the focal ratio effectively exploited at the primary focus. The diameter of such correctors, of the order of a kilometer, implies that they themselves be designed as a flotilla of small mirrors! The focal combiners themselves have to be agile for efficient repointing.

For maximal luminosity in the planet's image, pupil densification is adjusted to shrink as much as possible the diffractive envelope defining the hypertelescope's direct imaging field (section 9.2). The array being preferably nonredundant, for a minimal number of free-fliers, the filling of the densified pupil is limited by the shortest baselines, which have to provide adjacent but non overlapping subpupils.

Coronagraphic components (section 10.5) are essential for removing most contaminating light from the parent star of the observed exoplanet. The subapertures, being several meters in size, typically resolve an exo-Earth from its parent star, in the visible, at distances up to several tens of parsecs. Each subaperture should therefore be apodized or equipped with a highly efficient coronagraph, serving to remove most starlight (by a factor of about 10^{-9} in terms of residual intensity per speckle) from the corresponding low-resolution subimage, before combining them into the high-resolution image.

12.3 Simulated images of exo-Earths obtainable with the Exo-Earth Imager

It is interesting to speculate on how a vegetated planet might look from such enormous distances. The simulated image of an exo-Earth shown in figure 9.1(c) has 50×50 resolution elements (resels). It was calculated from a satellite picture of the Earth, convolved with the spread function of a nonredundant 150-aperture array shown in figure 9.1(c). It was then multiplied by the image envelope, as shrunk by the pupil densification. Some contrast enhancement was finally applied to correct the loss caused by the convolved feet of the spread function.

To preserve a usable level of contrast in the raw image, the number of apertures arrayed nonredundantly must exceed the planet's diameter, expressed in units of the angular resolution (section 9.3). If 150 mirrors are used to observe an Earth at 10 light-years, their size has to be at least 3 m in order to collect enough photons per resel in a 30-minute exposure, short enough to avoid the blurring caused by the planet's rotation. Somewhat smaller mirrors could probably be used if image processing techniques are used to generate "de-rotated" images from multiple short exposures.

12.3.1 Some speculations on identifying life from colored patches

Since a recognized objective of exo-planet imaging is to discover places where life exists elsewhere in the Universe, it is interesting to speculate on how we might recognize life from the necessarily low-resolution image of a planet. One can argue *ad infinitum* about "what is life" and what are the conditions which would support it, so it has been generally agreed that the search would be limited to regions where liquid water could exist, this being the common factor of all life as we know it. This limits the region around a star to a fairly narrow shell, where temperatures between 0 and 100 °C could be expected to be found. One way is to look for spectral biosignatures, such as spectroscopic absorption lines of O_2, O_3, CO_2, H_2O or other molecules believed to be associated with life. However, resolved exoplanet images, if achieved with a spectro-imaging attachment providing low-resolution spectra of every resel, can in principle evidence colored photosynthetic features similar to the green spots of the terrestrial Amazon, Congo, Siberia, etc. This is a rather direct way of searching for extrasolar life similar to our own, with more sensitivity than allowed by the nonresolving instruments. The direct detection of photosynthetic or "exo-chlorophyll" spectral absorption bands, with information on their latitudinal and seasonal changes, can help discriminating against colored mineral features.

It is the seasonal variations which would be important in discriminating living colors. Mineral colorations, such as produced by iron oxides, are insensitive to the temperature or humidity, although other abiotic materials such as opals and liquid crystals could conceivably have such sensitivity and provide confusing signals. Opals are self-ordered three-dimensional arrays of silica nanospheres, which can display bright colors through wavelength-selective resonant reflection, according to the Lippmann–Bragg effect. This is the interference of light waves reflected by successive ordered layers of spheres of transparent material with diameters of the order of half a wavelength, now known as a "photonic crystal." Resonant reflection occurs when Bragg's condition $2d \sin \theta = m\lambda$ is fulfilled. The color of the reflected light is therefore angle-sensitive, which tends to broaden the spectral band if the opal grains have random orientations although, if the refractive index difference

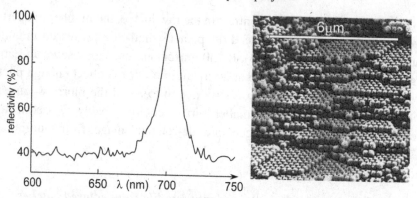

Fig. 12.9. Reflectance spectrum at normal incidence and an electron microgram of a synthetic opal. Courtesy of Z. V. Vardeny, University of Utah.

between the spheres and the matrix is strong, a contrasted band can persist in the reflection spectrum (figure 12.9). Moreover, because of its sensitivity to the exact values of the refractive indices of the constituents, the spectrum might conceivably respond to seasonal temperature and humidity variations. Temperature-sensitive colors are also displayed by some synthetic liquid-crystal materials.

A further discrimination between temperature-sensitive abiotic materials and photosynthetic life forms is possible through the more complex response of "light harvesting" by living organisms and micro-organisms. The occurrence of green and red-colored snow on some mountains, when weather conditions favor the growth of certain micro-organisms, shows that photosynthesis is more responsive to the availability of visible sunlight than to temperature. On Earth, the yearly temperature cycle lags in time with respect to the illumination cycle, mainly due to the thermal inertia of oceanic masses, and much of the spring-time greening at mid-latitudes occurs before the temperature rise. The fall's flamboyant Indian summer colors appear before the temperature drop. Observing this phase difference in the temperature and spectral cycles, should also help discriminating the somewhat unlikely possibility of variable colors from abiotic materials. The angular resolution and spectro-imaging capability of an EEI allow in principle such detailed observations.

12.4 Extreme baselines for a Neutron Star Imager; maximal size of interferometers and hypertelescopes in space

Among the few neutron stars known, the Crab Pulsar has a visual magnitude of the order of 18 in spite of its very small size, believed to be of the order of 20 km. It will take huge interferometric baselines, approaching 10^6 km, to resolve such objects into a usable number of resels. A problem with such long baselines is the diffractive spreading of the light beams propagating from the primary mirror elements to the

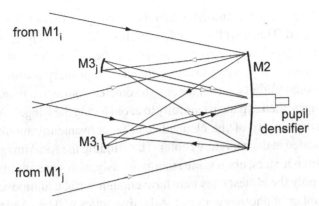

Fig. 12.10. Beam-combination scheme for extreme baselines. A single large M2 concave mirror receives the Fizeau image at the common focus of the primary elements. It must capture most light from the star observed, and therefore must be larger than the central lobe of the Fizeau envelope. This defines the minimal sizes of mirrors at both ends. Several M3 mirrors receive the relayed subapertures and form a combined image at the entrance of the pupil densifier.

beam combiner. Rather large mirrors are required at both ends if sufficient light is to be collected in the combiner. Indeed, to capture most of the light diffracted from a subaperture of size d into a beam-combiner located at distance L requires a collecting aperture of size d' larger than the Airy peak. The condition can be written $\lambda L/d < d'$. If a uniform size is preferred for all mirrors, then $d = d' = \sqrt{\lambda L}$. If, instead, reasons of cost-efficiency justify a larger collecting mirror with area d'^2 matching the total area Nd^2 of the N primary segments, the primary segments can be smaller and then $d' = d\sqrt{N}$ and the condition imposes $d = (\lambda L)^{\frac{1}{2}} N^{-\frac{1}{4}}$ and $d' = (\lambda L)^{\frac{1}{2}} N^{-\frac{3}{4}}$.

At the 121.6-nm ultraviolet wavelength of hydrogen's Lyman alpha emission, mirrors of 10 m size at both ends allow focal distances approaching a million kilometers, providing an angular resolution 25 pico-arcsec or $1.2 \cdot 10^{-16}$ radian. This is enough to resolve 53×53 resels on a 20 km-size neutron star located 100 parsecs away.

A spherical, rather than paraboloidal, primary array has spherical aberration requiring large perifocal optics for correction. This collecting optics can however itself be diluted, with segments carried by a flotilla of free fliers, but these will have to carry rather large mirrors, according to the above calculation. A proposed layout is shown in figure 12.10.

For primary mirror elements of 10-m size, having a focal length of 2×10^5 km, it makes no difference whether they are off-axis paraboloidal segments, as required for perfect stigmatism in the sense of geometric optics, or spherical or even flat. The quadratic term in a series expansion of the paraboloid's equation indicates the shape difference with a flat mirror, and this difference amounts to about 30 nanometers in the case considered. This is well within Rayleigh's tolerance at visible wavelengths.

Finding the fringes may at first glance seem to be a daunting task, since it will be difficult to achieve a good accuracy of the diluted primary mirror. In principle, however, one could initially shrink the interferometer flotilla to a manageable size, acquire the fringes, then expand it carefully to avoid losing the fringes. Alternatively, the accurate cosphericity of the elements can be conveniently monitored with a laser source located at the curvature center. Then finding the focal image is the main difficulty. Reference stars, observable simultaneously with separate combiners, can help although only the hottest ones may have enough surface luminosity to provide a sufficient number of photons per resel. Adaptive optics will be required to correct "gravitational seeing", the effect of moving masses on the optical wavefront.[†]

The prospect of attaining such high resolution is fascinating. The resolution gain and the added knowledge eventually becoming accessible to the terrestrial civilization is also of interest for other types of objects. An exo-Earth located at 3 parsecs is resolvable with 50 m detail, but impractically long exposures would be needed to collect a usable number of photons in such resels, unless a very large number of mirrors is used. Stellar and galactic physics will be much easier, considering the higher surface luminosities of stars.

References

Angel, J. R., A. Y. Cheng and N. J. Woolf (1986). *Nature*, **322**, 341.

Labeyrie, A. *et al.* (2001). in *Semaine de l'Astrophysique Francaise*, EDP-Sciences Conf. Series, 505.

Mariotti, J. M. *et al.* (1996). *Astron. and Astrophys. Suppl. Series*, **116**, 381.

Perrin, G. *et al.* (2004). *New Frontiers in Stellar Interferometry*, ed. W. A. Traub, Proc SPIE **5491**, 391.

[†] The radiative component, i.e. the gravitational waves believed to be generated by fast binary stars, especially during their final merging, produce no shadow pattern according to general relativity but can provide such a pattern according to scalar-tensor theories of gravitation.

Appendix A

In this appendix, we shall develop some theoretical tools necessary for a detailed understanding of astronomical interferometry, and also establish the notation and nomenclature used in the book. Many equivalent discussions can be found in other books (e.g. Lipson et al. 1995; Goodman 1996; Born and Wolf 2000); there are some differences between the various books in notation and approach, but the physics is the same!

A.1 Electromagnetic waves: a summary

We characterize an isotropic, insulating and charge-free medium by two scalar constants, a dielectric constant ϵ and a magnetic permittivity μ, which have the values ϵ_0 and μ_0 in free space. The dynamics of the electromagnetic fields are then described by Maxwell's equations:

$$\nabla \cdot \mathbf{D} = \epsilon \nabla \cdot \mathbf{E} = 0, \tag{A.1}$$

$$\nabla \cdot \mathbf{B} = \mu \nabla \cdot \mathbf{H} = 0, \tag{A.2}$$

$$\nabla \times \mathbf{H} = \frac{\partial \mathbf{D}}{\partial t} = \epsilon \frac{\partial \mathbf{E}}{\partial t}, \tag{A.3}$$

$$\nabla \times \mathbf{E} = -\frac{\partial \mathbf{B}}{\partial t} = -\mu \frac{\partial \mathbf{H}}{\partial t}. \tag{A.4}$$

These lead to two associated wave equations

$$\nabla^2 \mathbf{E} = \epsilon \mu \frac{\partial^2 \mathbf{E}}{\partial t^2}, \tag{A.5}$$

$$\nabla^2 \mathbf{H} = \epsilon \mu \frac{\partial^2 \mathbf{H}}{\partial t^2}. \tag{A.6}$$

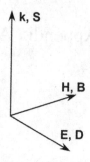

Fig. A.1. The wavevector **k**, Poynting vector **S** and electric and magnetic field vectors in a plane electromagnetic wave.

A.1.1 Plane and spherical electromagnetic waves

These equations have a first basic and useful solution in the form of a *transverse plane wave* involving both the electric field **E** and the magnetic field **H**, which have orthogonal directions (figure A.1):

$$\mathbf{E}(\mathbf{r}, t) = \mathbf{E}_0 \cos[\omega t - \mathbf{k} \cdot \mathbf{r} + \phi], \tag{A.7}$$

$$\mathbf{H}(\mathbf{r}, t) = \mathbf{H}_0 \cos[\omega t - \mathbf{k} \cdot \mathbf{r} + \phi], \tag{A.8}$$

$$\mathbf{E} \cdot \mathbf{H} = 0, \tag{A.9}$$

$$\mathbf{E} \cdot \mathbf{k} = 0, \tag{A.10}$$

$$\mathbf{H} \cdot \mathbf{k} = 0, \tag{A.11}$$

$$\frac{E}{H} = \left(\frac{\mu}{\epsilon}\right)^{\frac{1}{2}} \equiv Z, \tag{A.12}$$

where Z is the *wave impedance*. The wave velocity is $v = (\epsilon\mu)^{-\frac{1}{2}}$ and in free space this is the velocity of light, the fundamental constant $c \equiv (\epsilon_0\mu_0)^{-\frac{1}{2}}$ which is now defined to have the value $2.997\,924\,58 \cdot 10^8$ m. s^{-1} exactly. In an optical medium, the *refractive index* is the ratio

$$\frac{c}{v} = \left(\frac{\epsilon\mu}{\epsilon_0\mu_0}\right)^{-\frac{1}{2}} \equiv n. \tag{A.13}$$

Since most optical materials have $\mu \approx \mu_0$, this is generally shortened to $n = \epsilon_r^{-\frac{1}{2}}$, in which $\epsilon_r \equiv \epsilon/\epsilon_0$, and $Z = Z_0/n$.

A second useful solution is a *spherical wave*, emanating from a point source. This is not an easy solution, and we'll only describe it at large distances from the source, i.e. at radius $r \gg$ both λ and the size of the source. This is called the *far field*. The wave is then expressed by

$$E(\mathbf{r}, t) = \frac{1}{r} E_0 \cos[\omega t - kr + \phi], \tag{A.14}$$

$$H(\mathbf{r}, t) = \frac{1}{r} H_0 \cos[\omega t - kr + \phi], \tag{A.15}$$

together with (A.10)–(A.12). Maybe the most important feature to emphasize here is that the amplitude of the wave decays like r^{-1}, and so the intensity of the wave decays like r^{-2}, which is called the *inverse square law*. Notice that this law doesn't apply to the plane wave, which is essentially a spherical wave whose source is at $-\infty$.

Combinations of plane waves, particularly in the form of circularly or elliptically polarized waves, also commonly occur. A *right-handed circularly polarized* wave is generally defined as having field vectors $\mathbf{E}$ and $\mathbf{H}$ which rotate clockwise when observed by an observer receiving the wave, i.e. looking in the $-\mathbf{k}$ direction. For a wave traveling along $+z$, the field is then the linear combination $E_x + iE_y$, as defined in (A.8) and (A.9). Likewise, a *left-handed circularly polarized* wave is given by $E_x - iE_y$. Elliptically polarized waves are defined by other linear combinations $aE_x + bE_y$, where a and b are complex numbers.

A.1.2 Energy and momentum in waves

The rate at which energy is transported by an electromagnetic wave is given by Poynting's vector

$$\mathbf{S} = \langle \mathbf{E} \times \mathbf{H} \rangle_t = \tfrac{1}{2} E_0 H_0 = \tfrac{1}{2} E_0^2 / Z, \tag{A.16}$$

where $\langle ... \rangle_t$ indicates averaging over time. The dimensions of this vector are power per unit area. Now when the transported energy is absorbed by a receiver, the energy is converted to heat or some other form of response, and the quantity we associate with this is *irradiance*, usually loosely referred to as intensity; thus we have the useful relationship that the intensity is proportional to the square of the amplitude of the wave.

At the same time, the wave transfers momentum to the receiver. It is easiest to see this by considering a specific means of absorption, such as $\mathbf{E}$ generating a current density $\mathbf{j}$ in the absorber (figure A.2). But this current density, since it is parallel to $\mathbf{E}$, is normal to the field $\mathbf{B} = \mu_0 \mathbf{H}$ and so there is a Lorenz force $\mathbf{j} \times \mathbf{B}$ exerted on it. Since from (A.8)–(A.9) $\mathbf{j}$ and $\mathbf{H}$ are in phase with one another, this force has a nonzero time-average and corresponds to momentum absorbed from the wave; it is called *light pressure* and has the magnitude $\mathbf{S}/c$.

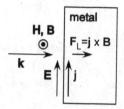

Fig. A.2. The Lorenz force between the wave's magnetic field and the current induced by its electric field results in light pressure on a conducting reflector.

A.2 Propagation of electromagnetic waves in coiled structures: geometrical phase

When an electromagnetic wave propagates along a three-dimensional route, such as in an optical fiber or as defined by the mirrors in a three-dimensional interferometer, an additional phase called the "geometrical" or "topological phase"[†] is added to the normal kinetic phase ($\omega t - \mathbf{k} \cdot \mathbf{r}$) of the propagating wave. This phase is related to the helicity of the route along which the light travels, and is important when an achromatic phase shift is required, such as in nulling interferometry (chapter 10). The geometrical phase was first described as a general phenomenon in quantum mechanics by Berry (1984), and is often known as "Berry's phase," but applies to quantum and classical situations (such as Foucault's pendulum) alike. Its value can be easily determined when the route of the light wave is drawn on a sphere representing the directions of $\mathbf{k}$. It is assumed that the changes in direction of $\mathbf{k}$ are continuous (*adiabatic* case), so that their locus on the sphere of directions is uniquely defined; this is not true for systems with mirrors, but the results are apparently still correct even in this nonadiabatic case. A simple derivation (Lipson 1990; Lipson et al. 1995), based on that for the Foucault pendulum, is as follows.

We postulate observers of the electromagnetic wave who travel slowly along each of the routes, measuring the vector field in their own frame of reference, which changes continuously so that the local z-axis always coincides with $\mathbf{k}$. In order to maintain this situation, the observers have to rotate their frame, and we define $\alpha(t)$ as their angular velocity with respect to the laboratory frame.

In the frame rotating at α, we relate the time-derivative of a general vector $\mathbf{V}$ to that in the inertial frame by

$$\left(\frac{\partial \mathbf{V}}{\partial t}\right)_{\text{rotating frame}} = \left(\frac{\partial \mathbf{V}}{\partial t}\right)_{\text{inertial frame}} + \alpha \times \mathbf{V}. \qquad (A.17)$$

[†] According to J. Hannay and M. Berry (private communication) the commonly used term "topological phase" is incorrect, since the phase is a continuous function of the geometrical parameters, whereas topological functions are discrete.

We can apply this to Maxwell's equations (A.3)–(A.4) which yield, in free space,

$$\nabla \times \mathbf{H} = \epsilon_0 \left(\frac{\partial \mathbf{E}}{\partial t} + \alpha \times \mathbf{E} \right) \tag{A.18}$$

$$\nabla \times \mathbf{E} = -\mu_0 \left(\frac{\partial \mathbf{H}}{\partial t} + \alpha \times \mathbf{H} \right). \tag{A.19}$$

The other two Maxwell equations (A.1)–(A.2) are unchanged. The wave equation which replaces (A.5) then follows as

$$\frac{\partial^2 \mathbf{E}}{\partial t^2} - \alpha \times \frac{\partial \mathbf{E}}{\partial t} = c^2 \nabla^2 \mathbf{E}. \tag{A.20}$$

This equation is analogous to that obtained in classical mechanics for Foucault's pendulum swinging on a rotating Earth, in which the second term arises from the Coriolis force.

In contrast to the usual wave equation (A.5), a linear polarized plane wave (A.8) is not a solution of (A.20); however, left- and right-handed circularly polarized waves are solutions. Substituting waves

$$\mathbf{E}_\pm = E_0(1, \pm i, 0) \exp[i(\omega t - ks)], \tag{A.21}$$

traveling in direction s parallel to the wavevector $\mathbf{k}$, we immediately find the dispersion relation

$$c^2 k^2 = \omega^2 \pm \alpha_s \omega \quad \Rightarrow \quad \omega_\pm \approx ck \pm \alpha_s/2 \tag{A.22}$$

when $\alpha \ll \omega$. The origin of the phase difference between the waves along the two routes of an interferometer such as shown in figure 10.9 or 10.10 is in the slight difference between the velocities which arises if α_s is positive for one and negative for the other, as a result of their opposite helicities.

The phase difference between beginning and end of a route is

$$\int_0^{s,t} (k \, ds - \omega \, dt) \pm \tfrac{1}{2} \int_0^t \alpha_s \, dt = \Delta\Phi_0 \pm \tfrac{1}{2} \int_0^t \alpha_s \, dt \equiv \Delta\Phi_0 + \gamma_B, \tag{A.23}$$

$\Delta\Phi_0$ indicating the usual (kinetic) phase difference expected from the optical path length of the route. The extra term γ_B can be easily interpreted by the construction on the sphere whose radius vectors represent the wave propagation directions. The direction of $\mathbf{k}$ has angular position (θ, ϕ), figure A.3. Then the locus is the curve $\theta(\phi)$ and α is the sum of the two orthogonal components

$$\alpha(t) = \frac{d\phi}{dt}\hat{z} + \frac{d\theta}{dt}\hat{j}, \tag{A.24}$$

where $\hat{z}$ is the unit vector to the pole and $\hat{j}$ is the common normal to $\hat{z}$ and $\mathbf{k}$. Thus the projection of α on the $\hat{s}$-axis in the rotating coordinates, defined as being

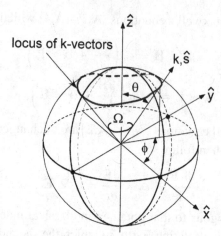

Fig. A.3. Construction of a general **k**-route on the surface of the sphere of wave-propagation directions.

parallel to **k**, is

$$\alpha_s = \frac{\mathrm{d}\phi}{\mathrm{d}t}(\hat{s} \cdot \hat{z}) = \frac{\mathrm{d}\phi}{\mathrm{d}t}\cos\theta. \tag{A.25}$$

Integrating this along the route gives

$$\gamma_B = \tfrac{1}{2}\int_0^t \alpha_s \,\mathrm{d}t = \tfrac{1}{2}\int_0^t \frac{\mathrm{d}\phi}{\mathrm{d}t}\cos\theta \,\mathrm{d}t = \tfrac{1}{2}\int \cos\theta(\phi)\mathrm{d}\phi. \tag{A.26}$$

For a closed loop, for which the last integral is from 0 to 2π, the solid angle subtended at the center of the sphere is $\Omega = \oint[1 - \cos\theta(\phi)]\,\mathrm{d}\phi = 2(\pi - \gamma_B)$. Thus, when we have drawn on the sphere the **k**-routes corresponding to the two arms of the interferometer, $\gamma_B = \pi - \Omega/2$, where Ω is the solid angle subtended by the enclosed segment. We recall that this γ_B corresponds to one of the circularly polarized waves. The wave with opposite sense gives $-\gamma_B$, and the *difference* between the two, namely $2\gamma_B$, is directly measurable since $\Delta\Phi$ in (A.23) cancels out. Thus the observable geometric phase difference is $2\gamma_B = 2\pi - \Omega$, which is the solid angle contained between the diametric plane and the locus of the **k**-vector on the sphere. Experiments on nonplanar interferometers by Chiao et al. (1988) and on propagation in helically coiled fibres (Tomita and Chiao 1986 confirm this result. It can be directly applied to the achromatic nulling experiments in section 10.4.3.

A.3 Fourier theory

J. B. J. Fourier (1763–1830) was one of the French scientists of the time of Napoleon who raised French science to extraordinary heights. He was an applied mathematician, and the work by which his name is now known was his contribution to the theory of heat transmission. He was faced with the problem of solving the linear heat-diffusion equation for the development of the temperature distribution $T(x, t)$ in a body, under conditions where boundary conditions are known: e.g. the value of $T(t)$ at some x and the behavior of $T(x)$ for all times on the boundaries of the medium. He developed a theory for temporal and spatial variations which were periodic in space and time, and used superposition of harmonics with frequencies equal to integer multiples of the basic frequency (the fundamental) to express arbitrary periodic functions. These were called *Fourier series*. Today, we use Fourier theory also for nonperiodic functions, by an extension of Fourier's ideas to what is called the "Fourier transform," and this extension is relevant to almost every field of wave propagation. Moreover, powerful computer programs, based on the fast Fourier transform (FFT) algorithm invented by Cooley and Tukey in 1965, allow calculations expressed in terms of Fourier transforms to be carried out numerically very efficiently. Again, without going into too much detail of the proofs, we shall summarize the background necessary to understanding astronomical interferometry. The development will at first be in one spatial dimension; extension to two dimensions (images) is relatively straightforward and will be made at a later stage when needed.

A.3.1 The Fourier transform

The *Fourier transform* of a function describes mathematically how a given function can be built up by superposition of elementary sinusoidal waves with different frequencies and phases. To make the description easier, we use the complex numbers, in which $\exp(i\theta) = \cos\theta + i\sin\theta$. A nonperiodic function is synthesized from sinusoidal waves of all frequencies, not just the harmonics of a fundamental. Following Fourier's idea, we therefore *define* the Fourier transform of a function $f(x)$ of the spatial variable x as:

$$F(k) = \int_{-\infty}^{\infty} f(x)\exp(-ikx)\mathrm{d}x \tag{A.27}$$

where the variable k, the *spatial frequency*, has already been introduced in section 2.2; it has dimensions of inverse distance. The function $f(x)$ can be complex, and also generally is the transform $F(k)$. Notice also the convention that lower-case and capital letters are transforms of one another, which we'll use as far as possible. The square-modulus of the Fourier transform, $|F(k)|^2$, is an important observable, and is called the *power spectrum* of $f(x)$. In optical diffraction experiments it is the intensity of the diffraction pattern, and is usually the only thing that can be measured directly. The phase itself is more elusive.

Do all functions have Fourier transforms? The answer is "no." In order to have a transform, a function must have a finite amount of energy, i.e. $\int_{-\infty}^{\infty} |f(x)|^2 dx$ must be finite. In addition, there must be only a finite number of discontinuities in $f(x)$. Of course, all practical functions, which refer to real physical situations, fulfill these conditions; however, we often use "representative" functions which don't. The problem is rarely of importance in physics, so we'll not overemphasize it here.

There are several useful properties of $F(k)$ which we shall list here, and can be proved with little difficulty directly from the definition.

(i) If $f(x)$ is real, then $F(-k) = F^*(k)$, in which * represents the complex conjugate. The power spectrum of a real function is therefore symmetrical about the origin, since $|F(-k)|^2 = |F^*(k)|^2 = |F(k)|^2$.

(ii) If $f(x)$ is a real symmetrical (even) function, i.e. $f(x) = f(-x)$, then its transform is real and symmetrical (i.e. it can be built up from cosines only); likewise, a real anti-symmetrical function $f(x) = -f(-x)$ has an imaginary antisymmetrical transform (i.e. is built up from sines only).

(iii) By writing a complex $f(x) = f_R(x) + i f_I(x)$, where f_R and f_I are real, it follows that the transform of the complex conjugate $f^*(x)$ is $F^*(-k)$.

(iv) The transform of $f(ax)$, where a is a real number, is $|a|^{-1} F(k/a)$. This means that on increasing the scale of a function by a certain factor, the scale of the transform decreases by the same factor.

(v) The transform of $f(x - x_0)$ is $\exp(-ikx_0)F(k)$. Shifting the origin of the function changes only the phase of the transform, and the power spectrum is unaffected.

(vi) Fourier inversion: carrying out the Fourier transform twice reverts to the original function with inverted direction. If $F(k)$ is the transform of $f(x)$, then the transform of $F(x)$ is $(1/2\pi)f(-k)$.[†]

(vii) The energy in the function is equal to that in its transform, up to a factor 2π:

$$\int_{-\infty}^{\infty} |F(k)|^2 dk = 2\pi \int_{-\infty}^{\infty} |f(x)|^2 dx . \tag{A.28}$$

This relationship is known as *Parseval's theorem*.

A.3.2 Some simple examples

Some simple functions have Fourier transforms which will be used over and over again. The integrals involved in proving them are fairly straightforward, and in any case can be found in any of the standard texts. We remark on a certain degree of variation between texts concerning prefactors and exponent factors of 2π which

---

[†] The transforms from x to k and from k to x are defined *exactly the same* in this statement. Some works define the inverse transform without the $-$ in the exponent, and with an extra factor of 2π, which tidy up this point and the next one, but produce complications elsewhere. In physical optics, the production of a real image is described mathematically by two successive Fourier transforms, and the inversion of the direction after the second transform explains neatly why the real image is inverted.

depend on the definition of the transform (A.27), but have of course no physical significance once the transform is interpreted in a particular experimental situation. The examples below are first in one dimension, but later in two; the extension from one to two dimensions is quite straightforward.

(i) **A square pulse, or window function.** This is defined as $f(x) = 1$ when $-1 < x < 1$, otherwise 0, and has the transform $F(k) = \sin(k)/k$. These functions are conventionally known as $f(x) \equiv \mathrm{rect}(x)$ and $F(k) \equiv \mathrm{sinc}(k)$. If the window has width $2a$ and height h, so that $f(x) = h\,\mathrm{rect}(x/a)$, then the rules above lead directly to the transform $F(k) = 2ah\,\mathrm{sinc}(ka)$ (figure A.4a).

(ii) **A Dirac δ-function.** This function is the limit of $\frac{1}{2}a\,\mathrm{rect}(x/a)$ as $a \to 0$, and has enclosed area $= 1$. It has transform unity: $F(k) = 1$ (figure A.4b). The Dirac δ-function is a spike of infinite height at $x = 0$ and is zero elsewhere, and has numerous very useful applications. One of the most important is "sampling:" by multiplying an arbitrary function $g(x)$ by a δ-function shifted to the point $x = b$, and integrating from $-\infty$ to ∞, we sample the function at b, i.e.

$$\int_{-\infty}^{\infty} g(x)\delta(x-b)\mathrm{d}x = g(b). \tag{A.29}$$

(iii) **The cosine function** $f(x) = \cos(px)$ has a transform consisting of two Dirac δ-functions of strength π at positions $\pm p$, i.e. $F(k) = \pi[\delta(k+p) + \delta(k-p)]$. Similarly, since $\sin(px)$ is the same function shifted along the x-axis by $\pi/2p$, its transform is $i\pi[\delta(k+p) - \delta(k-p)]$ (figure A.4c).

(iv) **The Gaussian function** $f(x) = [1/(2\pi\sigma^2)]^{\frac{1}{2}} \exp(-x^2/2\sigma^2)$ transforms to the Gaussian $F(k) = \exp(-k^2/2)\sigma^2$. This is one example of a "self-Fourier function" which transforms into itself for the particular scale factor $\sigma = 1$ (figure A.4d). There are many such functions; see, for example Caola (1991) and Lipson (1993).

(v) **The "comb" function** is an infinite series of equally-spaced δ-functions. Formally, this does not have a Fourier transform because it does not satisfy the two conditions mentioned above. However, by considering it as the limit of the transform of a finite number N of equally spaced δ-functions as $N \to \infty$, it can also be shown to transform into itself for the scale factor $a = \sqrt{2}/\pi$ (figure A.4e):

$$f(x) = \sum_{n=-\infty}^{\infty} \delta(x - na) \Rightarrow F(x) = (2\pi/a)\sum_{m=-\infty}^{\infty} \delta(k - 2\pi m/a). \tag{A.30}$$

(vi) **The circular disk** of radius a is an important two-dimensional Fourier transform. It is described by $f(r) = \mathrm{circ}(r/a) = 1$ if $r \le a$, or 0 if $r > a$. The two-dimensional transform of a function $f(x, y)$ is defined by

$$F(k_x, k_y) = \iint_{-\infty}^{\infty} f(x, y)\exp[-i(k_x x + k_y y)]\mathrm{d}x\mathrm{d}y. \tag{A.31}$$

Very often, this double integral can be factored into the product of two single integrals, but not in this case! Here we convert the Cartesian integral into a polar integral, in

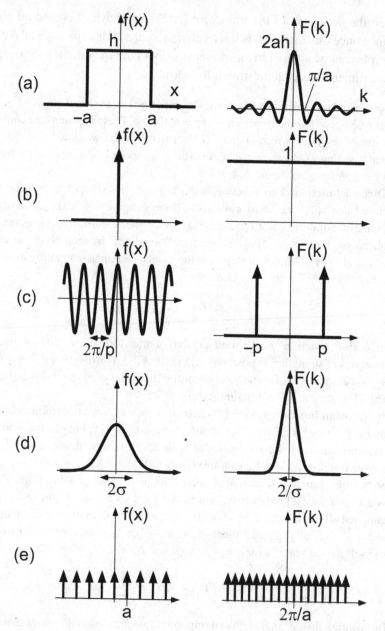

Fig. A.4. Sketches of five simple one-dimensional functions $f(x)$ and their Fourier transforms $F(k)$. A Dirac δ-function is represented by a vertical arrow, and is assumed to have zero width and unit area.

which $(k_x, k_y) \equiv k_r(\cos\phi, \sin\phi)$:

$$F(k_r, \phi) = \int_0^{2\pi} \int_0^{\infty} f(r, \theta) \exp[-i(k_r r \cos(\theta - \phi))] \, r \, dr \, d\theta. \qquad (A.32)$$

For the function $f(r) = \text{circ}(r/a)$, which is axially symmetric, we can put $\phi = 0$ and the integral becomes:

$$F(k_r) = \int_0^{2\pi} \int_0^{a} \exp[-i(k_r r \cos\theta)] \, r \, dr \, d\theta \qquad (A.33)$$

$$= 2\pi a^2 J_1(k_r a)/(k_r a) \qquad (A.34)$$

where J_1 is the first-order Bessel function (figure A.5a). The proof of this is given in many books (e.g. Lipson et al. 1995), so we have skipped the details.

(vii) **A circular ring** of radius R and having a δ-function cross-section, $f(r) = \delta(r - R)$ is also an important function in aperture synthesis. In the same way as described above, its Fourier transform can be shown to be $2\pi R \, J_0(k_r R)$ (figure A.5b).

It is most convenient to see these functions visually, and they are shown in figure A.4. Simple variations on the functions can be obtained by using the properties enumerated in section A.3.1. Functions can also be combined; addition and subtraction are straightforward because Fourier transformation is a linear operation, but multiplication and division require the convolution operation, which will be discussed next.

A.3.3 Convolution

Convolution is a mathematical description of blurring. It is important in Fourier theory because convolution in Fourier space (k) results from multiplication in real space, and vice versa. Suppose we consider a camera which is out of focus. Photographing a point source results in a blurred image which looks like the aperture of the lens (on a scale depending on how far the camera is out of focus). In general (in one dimension) this can be represented by some function $g(x)$, when the point source image should have been at $x = 0$. Then a point source at $x = x'$ will produce a blurred image $g(x - x')$. Considering each point on an object whose sharp image would have been $f(x')$, we see immediately that the blurred image is described by

$$c(x) = \int_{-\infty}^{\infty} f(x')g(x - x')dx'. \qquad (A.35)$$

This is called the *convolution* of the functions $f(x)$ and $g(x)$, and will be represented by $f(x) \star g(x)$, although several other conventions also exist. It is easy to show that $f(x) \star g(x) = g(x) \star f(x)$. Convolution between two one-dimensional functions is illustrated by figure A.6 and in figure A.7 for two-dimensional

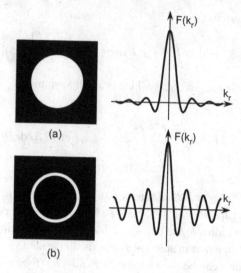

Fig. A.5. Two-dimensional Fourier transforms: (a) a circular aperture; (b) an annular aperture.

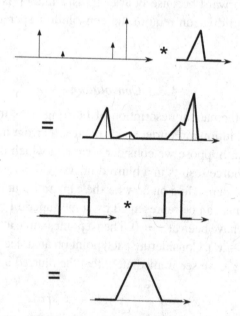

Fig. A.6. Convolutions between one-dimensional functions: (a) one function is a set of δ-functions; (b) two rect functions with different widths.

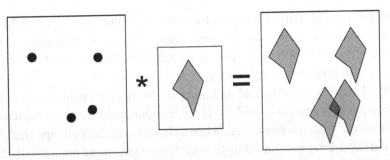

Fig. A.7. Convolution between a two-dimensional array of δ-functions and a polygon.

functions. The importance of the convolution operation in Fourier theory is that the Fourier transform of $f(x) \star g(x)$ is the product $F(k)G(k)$, and of course by Fourier inversion, the transform of $f(x)g(x)$ is $F(k) \star G(k)$. It is interesting to note that multiplication and convolution do not associate, i.e.

$$[f(x) \star g(x)] \cdot h(x) \neq f(x) \star [g(x) \cdot h(x)]; \qquad (A.36)$$

this has several interesting physical consequences.

A particular form of convolution is the *autocorrelation function* which is the convolution of a function with the inverse of its complex conjugate. This function essentially describes the range of self-similarity of the function. It is defined by:

$$h_{AC}(x) = \int_{-\infty}^{\infty} f(x')f^*(x' - x)\,dx' \equiv f(x) \star f^*(-x), \qquad (A.37)$$

and its Fourier transform is therefore $F(k) \cdot F^*(k) = |F(k)|^2$, the power spectrum. The result that the power spectrum is the Fourier transform of the autocorrelation function is known (in time–frequency space) as the *Wiener–Khinchin theorem*. One should notice that since the power spectrum is real, $h_{AC}(x)$ must be symmetrical about $x = 0$.

A.3.4 Sampling and aliasing

Any signal which is measured undergoes sampling of some sort, because all measuring systems have a resolution limit. For example, an image is recorded by a CCD camera, which has pixels of a certain size. A pixel can only output one value, which is determined by the total light falling on it during its exposure time. This value is a sample of the signal: a CCD with 10×10 μm pixels samples the image on a grid which has spacing 10 μm in each dimension. The same is true for measuring a signal as a function of time. The measuring system has a certain response time, during which it integrates its response to the signal, and the output represents the

Appendix A

value of this average. No extra information would be gained by reading the output at intervals shorter than than the response time, since adjacent readings will then approximately represent the same integrated value. We therefore, optimally, sample the output at intervals about equal to the response time.

We shall discuss the effect of sampling in the time domain (t, ω); it is one-dimensional and in general considerably more important in the temporal than in the spatial domain. What happens, then, when a signal really does change significantly in a period shorter than the sampling time t_0? How is the output related to the the real signal? Can anything be done to restore the real signal? We represent the process of sampling by multiplying the signal $s(t)$ by a periodic series of δ-functions, one for each sampling point, spaced by t_0, the sampling interval[†]. Then each δ-function has a value (its area) equal to the value of $s(t)$ at that time, generating an output $g(t)$:

$$g(t) = \sum_{n=-\infty}^{\infty} s(t) \cdot \delta(t - nt_0). \qquad (A.38)$$

Given this sampled function, we often reproduce $s(t)$ as a continuous function as well as possible by interpolating smoothly between the values of the δ-functions. But now let's see how the Fourier transform of the sampled function $G(\omega)$ is related to the transform $S(\omega)$ of the true function; using the transform of the "comb" function (A.3.2)

$$G(\omega) = S(\omega) \star \sum_{m=-\infty}^{\infty} \delta(\omega - m2\pi/t_0). \qquad (A.39)$$

Describing this in words, the transform $S(\omega)$ is repeated indefinitely in Fourier space at intervals or $2\pi/t_0$, as in figure A.8(a). Now if $S(\omega)$ is limited in extent to a region of length less than $2\pi/t_0$, as in figure A.8(a), nothing is lost or corrupted in this spectrum, because we can simply cut out a cell of length $2\pi/t_0$ containing one of the repeated spectra and we have the complete $S(\omega)$ back again. The signal $s(t)$ is then called "band-limited" with bandwidth $< 2\pi/t_0$. But if its bandwidth is greater than this value, when we cut out a cell we get a corrupted function because the tails of the repeats overlap (figure A.8b). There is nothing to be gained by cutting out a longer cell because the function repeats exactly, so it seems that the sampling process causes information to be lost if the bandwidth is greater than $2\pi/t_0$. The bandwidth limitation can be expressed in a different way. Supposing that the signal is sinusoidal with a slowly varying envelope (modulation), of the form $s_1(t) = a(t)\cos(\omega_0 t)$. The transform of this consists of two narrow peaks

---

[†] This is not quite the same as integrating during a period of t_0 around each sampling point, but the difference is marginal, and affects the noise rather than the signal itself, which we are discussing here. We also assume here that the samples are equally spaced in time, but of course they might not be.

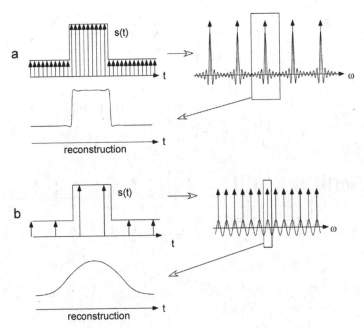

Fig. A.8. (a) A function $s(t)$ with bandwidth much smaller than $2\pi/t_0$ sampled at intervals t_0, its Fourier transform and the reconstruction from the cell of size $2\pi/t_0$. (b) The same when the bandwidth is close to $2\pi/t_0$ giving a poor reconstruction of $s(t)$. Note that the vertical arrows represent δ-functions, and the ordinate axis has been omitted to avoid confusion.

(because $a(t)$ is only varying slowly) around frequencies $\omega = \pm\omega_0$. In order that this signal be sampled properly, these two peaks must both be within the cell of length $2\pi/t_0$. This means that $2\pi/t_0 > 2\omega_0$, or $t_0 < \pi/\omega_0$ for the spectrum to be sampled correctly. Now the period of the sinusoidal function is $2\pi/\omega_0$, so that this means that for reliable sampling of a periodic function it must be sampled at least twice in every period. This is called *Nyquist's theorem*.

When a periodic function is incorrectly sampled, a process called *aliasing* occurs. Suppose that the sinusoidal function $s_1(t)$ has $\omega_0 > 2\pi/t_0$. Then, when it is sampled at intervals t_0, the repeated spectrum is shown in figure A.9(a). The spectrum is identical to what would be obtained for the same envelope $a(t)$ modulating a sinusoidal wave with frequency $\omega_0 - 2\pi/t_0$, which is now within the cell. In general, for any ω_0 we should write "the value of $\omega_0 - 2m\pi/t_0$ which lies within the cell". This is called the *alias frequency*, and is what we will get when we reconstruct the wave from its sampled Fourier transform. Clearly, if ω_0 is known approximately, the aliased signal can be converted exactly into the real signal (see section 6.2); otherwise there is no way of distinguishing between them. When the

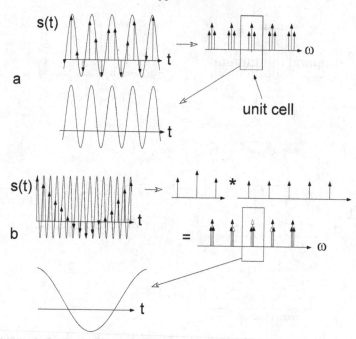

Fig. A.9. (a) A periodic function correctly sampled, its spectrum and reconstruction from the spectrum in the unit cell $2\pi/t_0$. (b) The same when the periodic signal is undersampled, showing the aliased signal reconstructed from the unit cell. Note that the vertical arrows represent δ-functions, and the ordinate axis has been omitted to avoid confusion.

Fig. A.10. Moiré fringes between overlaid grids with similar spatial frequencies.

signal is has a broad band, larger than $2\pi/t_0$, the sampled function is corrupted by aliasing, and can not be reconstructed faithfully.

In the spatial domain, aliasing often corrupts images which are sampled by imaging devices, and very obvious artifacts arise. A well-known example, which you can probably find in this book, is the result of dot-matrix sampling which is used for printing gray-scale pictures. If the picture contains periodic detail near the dot-matrix frequency, then often one observes spurious low-frequency fringes crossing the image. Sometimes these are very prominent, as when a television news-reader wears a tweed jacket and the period of the weave is close to that of the camera CCD in the image; then sometimes you see very brightly colored aliasing fringes. The fringes are called "Moiré fringes" and simple examples are shown in figure A.10. These fringes can be avoided if the sampling function is not periodic. This does not solve the problem when the function is under-sampled, but the artifacts are not necessarily periodic and are less eye-catching!

A.4 Fraunhofer diffraction

Suppose that a uniform spatially coherent monochromatic plane wave illuminates normally a two-dimensional mask which modifies the plane wave in some fashion, by attenuating it and changing its phase in a way which varies from point to point. If we now look at the light distribution in some plane at a distance L from it we see its *diffraction pattern*. If L is quite short, the diffraction pattern will resemble a shadow of the mask, but as L increases, it becomes less and less similar to the mask, and also weaker. As L becomes large, the distribution approaches a fixed pattern, which just gets larger in scale as $L \to \infty$. The patterns at small L are called *near-field* or *Fresnel diffraction patterns*, and the limiting pattern as $L \to \infty$ is called the *far-field* or *Fraunhofer diffraction pattern*. When described in angular coordinates, the Frauhofer pattern is seen to be independent of L, and can be described in terms of the two-dimensional Fourier transform of the function describing the mask.

It is usual to observe Fraunhofer patterns not by going to infinity, but by inserting a converging lens after the mask. Then, the focal plane of the converging lens is conjugate to infinity, and Fraunhofer diffraction patterns are observed in this plane. A typical set-up for seeing these patterns is shown in figure 3.4(a). The distance between the mask and the lens is not important in general. In the figure, the mask is in the back focal plane of the lens; this ensures that the phases of the diffraction patterns are correct too, but since we cannot usually observe phase, this is not always necessary. Casual observation of Fraunhofer diffraction patterns can be done by using the eye lens as the converging lens. Then, we look at a fairly distant point source (as was described in section 3.1.3) with the mask in front of the pupil and we see the patterns directly (figure 3.4b). The relationship between the Fraunhofer diffraction pattern and the Fourier transform is as follows. Consider the geometry of figure 3.3(a), which is shown in three dimensions in figure A.11. The illuminating plane wave, propagating parallel to the z-axis, illuminates the mask M which has a transmission function $f(x, y)$ and is situated in the plane $z = 0$, which is the back focal plane of the converging lens. The transmission function describes the ratio

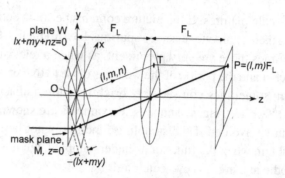

Fig. A.11. Geometry for Fraunhofer diffraction by a two-dimensional mask in the plane $z = 0$.

between the amplitude leaving the mask at point (x, y) to that incident on it at the same point; it will generally be a complex number $f = |f| \exp[-i\phi(x, y)]$ with $|f| \leq 1$. The lens is assumed to be ideal and paraxial and has focal length F_L. A wavefront W through the origin O (the focal point) having normal with direction cosines (ℓ, m, n), is focused by the lens to position $P \equiv (x, y) = F_L(\ell, m)$ in the focal plane $z = 2F_L$. But the amplitude received at W is the same as that leaving the mask, except for a phase change resulting from the distance $-\ell x - my$ that the light has traveled from the mask at (x, y) to W; this phase change is $-ik_0(\ell x + my)$. Now since the light amplitude A_P at P is the integrated amplitude over the whole of W, we immediately see that this is:

$$A_P = \exp(ik_0\overline{OTP}) \iint f(x, y) \exp[-ik_0(\ell x + my)]\mathrm{d}x\,\mathrm{d}y, \qquad (A.40)$$

where the integrals are over the whole wavefront, i.e. mathematically between $\pm\infty$. Note that $\overline{OTP}$ is constant as a result of Fermat's principle, since it is the optical path from O to a point on the plane normal to z after the lens, and $OC = F_L$. If we write the variables in the exponent as $u \equiv k_0\ell$ and $v \equiv k_0m$ we see immediately that (A.40) is the two-dimensional Fourier transform:

$$A_P = \exp(ik_0\overline{OTP}) \iint_{-\infty}^{\infty} f(x, y) \exp[-i(ux + vy)]\mathrm{d}x\,\mathrm{d}y$$
$$= \exp(ik_0\overline{OTP})F(u, v). \qquad (A.41)$$

When we observe the diffraction pattern, we see the intensity $I(u, v) = |A_P|^2 = |F(u, v)|^2$, and the actual coordinate scale of the pattern in the plane F_2 is given by $(x, y) = F_L(\ell, m) = F_L(u, v)/k_0$.

Fraunhofer diffraction patterns are an excellent way of visualizing Fourier transforms, and as such have played an enormous role in the development of subjects such as X-ray diffraction analysis of crystals and electronic filter theory. Many books have also been based on this relationship, e.g. Goodman (1996), Lipson et al. (1995). Several examples of Fraunhofer diffraction patterns are shown in chapter 3.

A.4.1 Random objects and their diffraction patterns: speckle images

A particular class of masks which have importance in astronomical interferometry are random masks, which represent the effect of atmospheric turbulence on the seeing of a telescope. This will also serve as an example, if not the simplest one, of a Fraunhofer diffraction calculation. When a telescope is pointed at a distant star, the image is the Fraunhofer pattern of the telescope aperture (figure 3.4c). The physical telescope aperture is the function $|f| = \text{circ}(r/R)$ but because of atmospheric turbulence, the phase is modified by random fluctuations in refractive index of the air in its vicinity, so that the appropriate mask function is $f(x, y) = \text{circ}(r/R) \exp[i\phi(x, y)]$. Atmospheric statistics determine the properties of $\phi(x, y)$ and are discussed in chapter 5.

We'll first consider a simplified related problem, to illustrate the idea. In section 3.1.3 we introduced the diffraction pattern of an array of identical apertures arranged on a *periodic* two-dimensional lattice. Now, suppose we have a one-dimensional mask with an array of N such apertures centered at *random* positions in the field $-R < x < R$. The individual aperture, relative to its origin, is described by $g(x)$ and the position of the origin of the jth aperture is $x = x_j$. Then the aperture function $f(x)$ is described by the convolution between $g(x)$ and N δ-functions at positions x_j, the whole multiplied by a window function $\text{rect}(x/R)$ (figure A.12a):

$$f(x) = \left[g(x) \star \sum_{j=1}^{N} \delta(x - x_j) \right] \cdot \text{rect}(x/R). \qquad (A.42)$$

The Fourier transform of this is the amplitude, and its square modulus is the intensity of the diffraction pattern:

$$F(u) = \left\{ G(u) \cdot \sum_{j=1}^{N} \exp[iux_j] \right\} \star \text{sinc}(uR), \qquad (A.43)$$

$$|F(u)|^2 = \left\{ |G(u)|^2 \cdot \sum_{j=1}^{N} \exp[iux_j] \sum_{j=1}^{N} \exp[-iux_j] \right\} \star \text{sinc}^2(uR)$$

$$= \left\{ |G(u)|^2 \cdot \sum_{j=1}^{N} \sum_{k=1}^{N} \exp[iu(x_j - x_k)] \right\} \star \text{sinc}^2(uR), \qquad (A.44)$$

in which we have introduced a new counting index k in order to change the product of two sums into a double sum. The first thing to notice about the double sum is

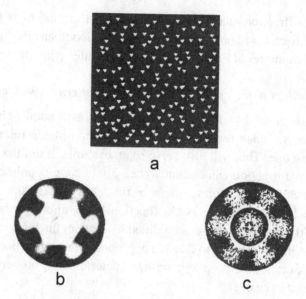

Fig. A.12. (a) An aperture is repeated at random positions within a square region. (b) Experimental diffraction pattern $|G(u)|^2$ of one element of the array. (c) Diffraction pattern of the complete array in (a). The circular central region of the pattern was photographically underexposed in order to make the bright spot at the origin visible. From Lipson et al. (1995).

that when $u = 0$ it has the value $N^2|G(0)|^2$, since every element in the sum has unit value. Moreover, when $u \neq 0$ the complex exponentials, which are sine waves with various spatial frequencies, more-or-less cancel one another out except for the cases where $j = k$, which all have unit value and thus the sum is N, plus noise or speckle, because the cancellation is only "more-or-less." The result is that (A.44) can be described as

$$|F(u)|^2 = [(N \text{ plus speckle})|G(u)|^2 + N^2|G(0)|^2\delta(u)] \star \text{sinc}^2(uR) \quad \text{(A.45)}$$

which is N times a noisy version of the diffraction pattern intensity $|G(u)|^2$ from one aperture, plus a bright spot of intensity N^2 at the origin, all convolved with the diffraction pattern of the window function. This is a well-known result, and is important to understanding the hypertelescope (chapter 9); it is illustrated in figure A.12(c), where the bright spot is emphasized by underexposing the central region. The noisy version of $|G(u)|^2$ is called a "speckle pattern."

One way of explaining the formation of a speckle pattern involves a random walk in phasor space. Indeed, any point in the image zone where speckles appear receives randomly phased vibrations from the N apertures. Even if the vibrations are phased at the apertures, their random location in its plane indeed causes different and random propagation path length toward the image point considered. Adding the vibration contributions at this point, using "phasor" vectors, amounts to doing a random walk in the phasor plane. Classical results of random walk theory show that the resulting sum phasor can have any modulus value, between zero and N times the elementary modulus. The statistical distribution is Gaussian. At a neighboring image point, located more than λ/D away, the contributing phasors are decorrelated, and therefore generate a different sum; hence the contrasted features of the speckle pattern. Now in fact, for the problem of atmospheric turbulence, the problem is similar but a bit more complicated than this. We'll continue to develop it in one dimension. The wavefront entering the telescope is uniform in amplitude but has random phase fluctuations from place to place. The phases are correlated within a region of size r_0. We'll describe this analytically by a model in which there are N "apertures" j as before, but each one has its own phase ϕ_j, and the function $g(x)$ extends out to the average separation between the apertures, which is put equal to r_0, i.e. $g(x) = \text{rect}(2x/r_0)$, so that the telescope aperture is more-or-less filled with patches having random phases. The only difference from the above treatment is that we must add in the effect of the phases ϕ_j. Then (A.44) becomes

$$|F(u)|^2 = \left\{ |G(u)|^2 \cdot \sum_{j=1}^{N} \sum_{k=1}^{N} \exp\{i[u(x_j - x_k) + \phi_j - \phi_k]\} \right\} \star \text{sinc}^2(uR).$$

$$(A.46)$$

But now, when $u = 0$, the sum is no different from other values of u, because the elements in the sum have value $\exp[i(\phi_j - \phi_k)]$ which sums to N (plus noise). So the central spot has disappeared. The result depends on the fact that the phases ϕ_j are distributed throughout $(0, 2\pi)$ with no bias (see figures 5.11 and 5.12a). The important features of this function are as follows:

- There is an envelope or "diffraction function" (section 3.1.3) $|G(u)|^2$, the Fourier transform of the small "aperture" $g(x)$ which multiplies the whole intensity pattern and determines its lateral extent to be about $2\pi/r_0$.
- There is a uniform noise, called a "speckle pattern" with no central peak, resulting from the interference between the random aperture positions.
- The noise pattern is convolved with the function $\text{sinc}^2(uR)$, which is the transform of the external window function, outside which the amplitude $f(x)$ is zero. As a result, features of the speckle pattern are blurred to that extent, which is equivalent to saying that the external aperture $2R$ determines the size of the smallest features in the pattern.

Appendix B

Table B.1. *Standard spectral bands for photometry; e_0 is the irradiance from a zero-magnitude star in the band. After Léna et al. (1998)*

Name	$\lambda_0/[\mu\text{m}]$	$\Delta\lambda/[\mu\text{m}]$	$e_0/[\text{w m}^{-2}\,\mu\text{m}^{-1}]$	$e_0/[\text{photons}$ $\text{m}^{-2}\text{s}^{-1}\,\mu\text{m}^{-1}]$	Color
U	0.36	0.068	4.35×10^{-8}	7.9×10^{10}	UV
B	0.44	0.098	7.20×10^{-8}	1.6×10^{11}	Blue
V	0.55	0.089	3.92×10^{-8}	1.1×10^{11}	Green
R	0.70	0.22	1.76×10^{-8}	6.2×10^{10}	Red
I	0.90	0.24	8.3×10^{-9}	3.7×10^{10}	NIR
J	1.25	0.30	3.4×10^{-9}	2.1×10^{10}	NIR
H	1.65	0.35	7.0×10^{-10}	5.8×10^{9}	NIR
K	2.20	0.40	3.9×10^{-10}	4.3×10^{9}	NIR
L	3.40	0.55	8.1×10^{-11}	1.3×10^{9}	NIR
M	5.00	0.30	2.2×10^{-11}	5.5×10^{8}	NIR
N	10.2	5.0	1.23×10^{-12}	6.4×10^{7}	MIR
Q	21.0	8.0	6.8×10^{-14}	7.2×10^{6}	FIR

References

Berry, M. V. (1984). *Proc. Roy. Soc.* London, A**392**, 45.

Born, M. and E. Wolf (2000). *Principles of Optics*, 7th edn., Pergamon, Oxford.

Caola, M. (1991). *J. Phys.* A **24**, 1143.

Chiao, R. Y., A. Antaramian, K. M. Ganga *et al.* (1988). *Phys. Rev. Lett.*, **60**, 1214.

Goodman, J. W. (1996). *An Introduction to Fourier Optics*, 2nd edn, New York, McGrow Hill.

Léna, P., F. Lebrun and F. Mignard (1998). *Observational Astrophysics*, Berlin: Springer.

Lipson, S. G., H. Lipson and D. S. Tannhauser (1995). *Optical Physics*, 3rd edn., Cambridge; Cambridge University Press.

Lipson, S. G. (1990). *Opt. Lett.*, **15**, 154.

Lipson, S. G. (1993). *J. Opt. Soc. Am.*, A**10**, 2088.

Tomita, A. and R. Y. Chiao (1986). *Phys. Rev. Lett.* **57**, 937.

Index

Stars and stellar objects

Printed in the United States
By Bookmasters